Official Guide to Certified SOLIDWORKS Associate Exams:
CSWA, CSWA-SD, CSWSA-S, CSWA-AM

SOLIDWORKS 2021 - SOLIDWORKS 2023

An authorized CSWA preparation exam guide with additional information on the CSWA-SD, CSWSA-FEA and CSWA-AM exams

David C. Planchard
CSWP & SOLIDWORKS Accredited Educator

SDC
PUBLICATIONS

SDC Publications
P.O. Box 1334
Mission, KS 66222
913-262-2664
www.SDCpublications.com
Publisher: Stephen Schroff

Examination Copies
Books received as examination copies are for review purposes only and may not be made available for student use. Resale of examination copies is prohibited.

Electronic Files
Any electronic files associated with this book are licensed to the original user only. These files may not be transferred to any other party.

Trademarks
SOLIDWORKS®, eDrawings®, SOLIDWORKS Simulation®, SOLIDWORKS Flow Simulation, and SOLIDWORKS Sustainability are a registered trademark of Dassault Systèmes SOLIDWORKS Corporation in the United States and other countries; certain images of the models in this publication courtesy of Dassault Systèmes SOLIDWORKS Corporation.

Microsoft Windows®, Microsoft Office® and its family of products are registered trademarks of the Microsoft Corporation. Other software applications and parts described in this book are trademarks or registered trademarks of their respective owners.

The publisher and the author make no representations or warranties with respect to the accuracy or completeness of the contents of this work and specifically disclaim all warranties, including without limitation warranties of fitness for a particular purpose. No warranty may be created or extended by sales or promotional materials. Dimensions of parts are modified for illustration purposes. Every effort is made to provide an accurate text. The author and the manufacturers shall not be held liable for any parts, components, assemblies or drawings developed or designed with this book or any responsibility for inaccuracies that appear in the book. Web and company information was valid at the time of this printing.

The Y14 ASME Engineering Drawing and Related Documentation Publications utilized in this text are as follows: ASME Y14.1 1995, ASME Y14.2M-1992 (R1998), ASME Y14.3M-1994 (R1999), ASME Y14.41-2003, ASME Y14.5-1982, ASME Y14.5-1999, and ASME B4.2. Note: By permission of The American Society of Mechanical Engineers, Codes and Standards, New York, NY, USA. All rights reserved.

Download all needed model files from the SDC Publication website www.SDCpublications.com/downloads/978-1-63057-567-0.

| SOLIDWORKS 2020 |
| SOLIDWORKS 2021 |
| SOLIDWORKS 2022 |
| SOLIDWORKS 2023 |

ISBN-13: 978-1-63057-567-0
ISBN-10: 1-63057-567-4

Printed and bound in the United States of America.

INTRODUCTION

The **Official Guide to Certified SOLIDWORKS Associate Exams: CSWA, CSWA-SD, CSWA-S, CSWA-AM** is written to assist the SOLIDWORKS user to pass the associate level exams.

Information is provided to aid a person to pass the Certified SOLIDWORKS Associate (CSWA), Certified SOLIDWORKS Associate Sustainable Design (CSWA-SD), Certified SOLIDWORKS Associate Simulation (CSWA-S) and the Certified SOLIDWORKS Associate Additive Manufacturing (CSWA-AM) exam.

The **Certified SOLIDWORKS Associate (CSWA)** certification indicates a foundation in and apprentice knowledge of 3D CAD design and engineering practices and principles.

The CSWA Academic exam is provided either in a single 3-hour segment, or 2 - 90-minute segments.

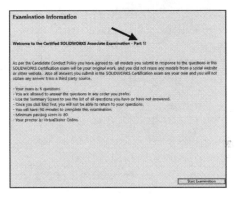

Segment 1 of the CSWA Academic exam is 90 minutes, minimum passing score is 80, with 6 questions. There are two questions in the Basic Part Creation and Modification category, two questions in the Intermediate Part Creation and Modification category and two questions in the Assembly Creation and Modification category.

Segment 2 of the CSWA Academic exam is 90 minutes, minimum passing score is 80 with 8 questions. There are three questions on the CSWA Academic exam in the Drafting Competencies category, three questions in the Advanced Part Creation and Modification category and two questions in the Assembly Creation and Modification category.

There are over 50 practice questions and examples covering Segment 1 and Segment 2 of the CSWA certification in the book, along with a practice exam.

Introduction

You need to pass both segments to obtain your CSWA Certification. Once you pass a segment, you will not have to take it again. You can take the segments in any order.

If you fail a segment, there is a 14-day waiting period before retaking the failed segment.

The CSWA 3-hour Academic exam consists of 14 questions worth a total of 240 points. All exams cover the same material. You need to wait 14 days before you can retake the exam.

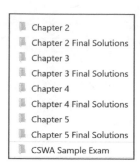
View the provided SOLIDWORKS CSWA Sample Exam folder included in the downloaded materials. The CSWA Sample Exam folder contains a pdf with information on the following: exam details, how to prepare for the exam, how to take the practice exam, taking the exam, sample test quesitons, test answers, and helpful sites.

SOLIDWORKS Certification Exam Guide & Practice Test

ASSOCIATE
Mechanical
Design

ĎS SOLIDWORKS

CSWA: Certified SOLIDWORKS Associate

Go to the 3DEXPERIENCE® Certification Center at https://3dexperience.virtualtester.com/#home to take the free online CSWA sample exam.

The 3DEXPERIENCE® Certification Center is where you are able to log in to manage your certificates, take certifications, and make changes to your account settings.

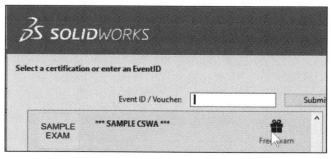

You need to wait 30 days before you can retake the free sample exam.

If your school is an academic certification provider, your instructor can allocate free exam credits for Segment 1 & 2. The instructor will require your .edu email address.

- SOLIDWORKS 2020
- SOLIDWORKS 2021
- SOLIDWORKS 2022
- SOLIDWORKS 2023

Download all needed model files (initial and final) and the SOLIDWORKS CSWA Sample Exam folder from the SDC Publications website www.SDCpublications.com/downloads/978-1-63057-567-0

The **Certified SOLIDWORKS Associate Sustainable Design (CSWA-SD)** certification indicates a foundation in and apprentice knowledge of demonstrating an understanding in the principles of environmental assessment and sustainable design.

The CSWA-SD is based off the SOLIDWORKS Sustainable Design Guide that incorporates concepts including sustainability, environmental assessment and life cycle impact assessment.

A copy of the Design Guide is included with this book. There are over 40 practice questions and examples in the chapter to help you pass the certification.

All questions are in a multiple choice/multi answer format. No SOLIDWORKS models need to be created for this exam.

If your school is an academic certification provider, your instructor can allocate a free CSWA-SD exam credit. The instructor will require your .edu email address.

If you fail the exam, you will need to wait 14-days before retaking it again.

The **Certified SOLIDWORKS Associate - Simulation (CSWA-S)** certification indicates a foundation in and apprentice knowledge of demonstrating an understanding in the principles of stress analysis and the Finite Element Method (FEM).

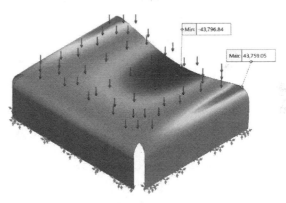

The main requirement for obtaining the CSWA-S certification is to take and pass the on-line 120-minute exam which consists of 20 questions. The questions consist of 3 hands-on problems, single answer, multiple choice, yes/no and multiple selection for a total of 100 points. Minimum passing score is 70.

The purpose of this section is not to educate a new or intermediate user on SOLIDWORKS Simulation or Finite Element Analysis theory, but to cover and to inform you on the required understanding types of questions, layout and what to expect when taking the CSWA-S exam.

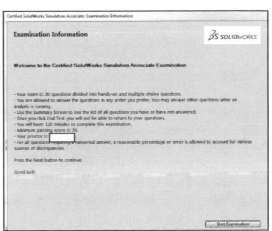

Introduction

There are over 100 practice questions and examples in the chapter to help you pass the certification. If your school is an academic certification provider, your instructor can allocate a free CSWA-S exam credit. The instructor will require your .edu email address.

If you fail the exam, you will need to wait 14-days before retaking it again.

The **Certified SOLIDWORKS Associate Additive Manufacturing (CSWA-AM)** certification indicates a foundation in and apprentice knowledge of today's 3D printing technology and market.

The CSWA-AM exam is meant to be taken after the completion of the 10-part learning path located on MySOLIDWORKS.com.

Note: The book is meant to provide all of the needed information from the online MySOLIDWORKS lessons.

After each lesson, there is a short online quiz covering the topic area. The lessons are focused on two types of 3D printer technology: Fused Filament Fabrication (FFF) and STereoLithography (SLA). There are a few questions on Selective Laser Sintering (SLS) technology and available software-based printing aids.

The CSWA-AM certification exam is 50 questions. Each question is worth 2 points. You are allowed to answer the questions in any order. Total exam time is 60 minutes. You need a passing score of 80 or higher. The exam is out of 100 points.

There are over 80 practice questions in the chapter to help you pass the certification.

If your school is an academic certification provider, your instructor can allocate a free CSWA-AM exam credit. The instructor will require your .edu email address.

If you fail the exam, you will need to wait 14-days before retaking it again.

Goals

The primary goal is not only to help you pass the CSWA, CSWA-SD, CSWA-S and CSWA-AM exams, but also to ensure that you understand and comprehend the concepts and implementation details of the four certification processes.

The second goal is to provide the most comprehensive coverage of CSWA, CSWA-SD, CSWA-S and CSWA-AM exam related topics available, without too much coverage of topics not on the exam.

CSWA Exam Audience

The intended audience for this book trying to take and pass the **Certified SOLIDWORKS Associate (CSWA)** exam is anyone with a minimum of 6 - 9 months of SOLIDWORKS experience and basic knowledge of engineering fundamentals and practices. SOLIDWORKS recommends that you review their SOLIDWORKS Tutorials on Parts, Assemblies and Drawings as a prerequisite and have at least 45 hours of classroom time learning SOLIDWORKS or using SOLIDWORKS with basic engineering design principles and practices. There are over 50 practice questions and examples covering Segment 1 and Segment 2 in the book, along with a practice exam.

CSWA-SD Exam Audience

The intended audience for this book trying to take and pass the **Certified SOLIDWORKS Associate Sustainable Design (CSWA-SD)** exam is anyone interested in Sustainable design and life cycle assessment. Although no hands-on usage of SOLIDWORKS is required for the CSWA-SD certification exam, it is a good idea to review the SOLIDWORKS SustainablityXpress and SOLIDWORKS Sustainability tutorials inside of SOLIDWORKS to better understand the actual workflow. A copy of the Design Guide is included with this book. There are over 40 practice questions and examples in the chapter to help you pass the certification.

CSWA-S Exam Audience

The intended audience for this book trying to take and pass the **Certified SOLIDWORKS Associate - Simulation (CSWA-S)** exam is anyone with a minimum of 6 - 9 months of SOLIDWORKS experience and knowledge in the following areas: Engineering Mechanics - Statics, Strength of Materials and Finite Element Method/Finite Element Analysis Theory. Also, applied concepts in SOLIDWORKS Simulation: namely Static Analysis, Solid, Shell, and Beam elements, Connections and Applying loads and boundary conditions and interpreting results. There are over 100 practice questions and examples in the chapter to help you pass the certification.

CSWA-AM Exam Audience

The intended audience for this book trying to take and pass the **Certified SOLIDWORKS Associate Additive Manufacturing (CSWA-AM)** exam is anyone interested in Additive Manufacturing.

The CSWA-AM exam fundamentally covers two 3D printing technologies: Fused Filament Fabrication (FFF) and STereoLithography (SLA). There are over 80 practice questions and examples in the chapter to help you pass the certification.

About the Author

David Planchard is the founder of D&M Education LLC. Before starting D&M Education, he spent over 35 years in industry and academia holding various engineering, marketing, and teaching positions. He holds five U.S. patents. He has published and authored numerous papers on Machine Design, Product Design, Mechanics of Materials, and Solid Modeling. He is an active member of the SOLIDWORKS Users Group and the American Society of Engineering Education (ASEE). David holds a BSME, MSM with the following professional certifications: CCAI, CCNP, CSWA-SD, CSWA-S, CSWA-AM, CSWP, CSWP-DRWT and SOLIDWORKS Accredited Educator. David is a SOLIDWORKS Solution Partner, a faculty member and the SAE advisor at Worcester Polytechnic Institute in the Mechanical Engineering department.

In 2012, David's senior Major Qualifying Project team (senior capstone) won first place in the Mechanical Engineering department at WPI.

In 2014, 2015 and 2016, David's senior Major Qualifying Project teams won the Provost award in the Mechanical Engineering department for design excellence.

In 2018, David's senior Major Qualifying Project team (Co-advisor) won the Provost award in the Electrical and Computer Engineering department. Subject area: Electrical System Implementation of Formula SAE Racing Platform.

In 2020, he was awarded Emeritus status at Worcester Polytechnic Institute (WPI). His ME design class achieved world recognition, featured in *Compass* published by Dassault Systèmes, in the technical article, "Open Innovation in a Pandemic."

In 2022, David's senior Major Qualifying Project team won second place in the Provost award in the Electrical and Computer Engineering department (ECE) at WPI. His FSAE Electric team won the IEEE Excellence in Electric Vehicle Award and achieved 4th place at the Formula Hybrid + Electric competition.

David Planchard is the author of the following books:

- **SOLIDWORKS® 2021 Reference Guide** 2020, 2019, 2018, and 2016

- **Engineering Design with SOLIDWORKS® 2023**, 2022, 2021, 2020, 2019, 2018, 2017, 2016, 2015, and 2014

- **Engineering Graphics with SOLIDWORKS® 2023,** 2022, 2021, 2020, 2019, 2018, 2016, and 2015

- **SOLIDWORKS® 2023 Quick Start**, 2022, 2021, 2020, 2019, and 2018

- **SOLIDWORKS® 2023 Tutorial**, 2021, 2020, 2019, 2018, 2017, and 2016

- **Drawing and Detailing with SOLIDWORKS® 2022**, 2014, 2012, and 2010

- **Official Certified SOLIDWORKS® Professional (CSWP) Certification Guide 2020 - 2023**, 2019 - 2020, 2015 - 2017

- **Official Guide to Certified SOLIDWORKS® Associate Exams: CSWA, CSWA-SD, CSWA-S, CSWA-AM 2020 - 2023**, 2019 - 2021, 2017 - 2019, and 2015 - 2017

Acknowledgements

Writing this book was a substantial effort that would not have been possible without the help and support of my loving family and of my professional colleagues. I would like to thank Professor John M. Sullivan Jr., Professor Jack Hall, Professor Mehul A. Bhatia and the community of scholars at Worcester Polytechnic Institute who have enhanced my life, my knowledge and helped to shape the approach and content to this text.

The author is greatly indebted to my colleagues from Dassault Systèmes SOLIDWORKS Corporation for their help and continuous support: Mike Puckett, Avelino Rochino, Yannick Chaigneau, Terry McCabe and the SOLIDWORKS Partner team.

Contact the Author

We realize that keeping software application books current is imperative to our customers. We value the hundreds of professors, students, designers, and engineers that have provided us input to enhance the book. Please contact me directly with any comments, questions or suggestions on this book or any of our other SOLIDWORKS books at dplanchard@msn.com or planchard@wpi.edu.

Note to Instructors

Please contact the publisher **www.sdcpublications.com** for classroom support materials (.ppt presentations, labs and more) and the Instructor's Guide with model solutions and tips that support the usage of this text in a classroom environment.

Trademarks, Disclaimer and Copyrighted Material

SOLIDWORKS®, eDrawings®, SOLIDWORKS Simulation®, SOLIDWORKS Flow Simulation, and SOLIDWORKS Sustainability are a registered trademark of Dassault Systèmes SOLIDWORKS Corporation in the United States and other countries; certain images of the models in this publication courtesy of Dassault Systèmes SOLIDWORKS Corporation.

Microsoft Windows®, Microsoft Office® and its family of products are registered trademarks of the Microsoft Corporation. Other software applications and parts described in this book are trademarks or registered trademarks of their respective owners.

Introduction

The publisher and the author make no representations or warranties with respect to the accuracy or completeness of the contents of this work and specifically disclaim all warranties, including without limitation warranties of fitness for a particular purpose. No warranty may be created or extended by sales or promotional materials. Dimensions of parts are modified for illustration purposes. Every effort is made to provide an accurate text. The authors and the manufacturers shall not be held liable for any parts, components, assemblies or drawings developed or designed with this book or any responsibility for inaccuracies that appear in the book. Web and company information was valid at the time of this printing.

The Y14 ASME Engineering Drawing and Related Documentation Publications utilized in this text are as follows: ASME Y14.1 1995, ASME Y14.2M-1992 (R1998), ASME Y14.3M-1994 (R1999), ASME Y14.41-2003, ASME Y14.5-1982, ASME Y14.5-1999, and ASME B4.2. Note: By permission of The American Society of Mechanical Engineers, Codes and Standards, New York, NY, USA. All rights reserved.

Additional information references the American Welding Society, AWS 2.4:1997 Standard Symbols for Welding, Braising, and Non-Destructive Examinations, Miami, Florida, USA.

References

- SOLIDWORKS Help Topics and What's New, SOLIDWORKS Corporation.

- Beers & Johnson, <u>Vector Mechanics for Engineers</u>, 6th ed. McGraw Hill, Boston, MA.

- Ticona Designing with Plastics - The Fundamentals, Summit, NJ, 2009.

FeatureManager, CommandManager, Graphics window, tabs and tree folders will vary depending on system setup, version and Add-ins.

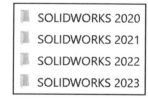

Download all needed model files from the SDC Publication website www.SDCpublications.com/downloads/ 978-1-63057-567-0.

If your school is an academic certification provider, your instructor can allocate free exam credits for the CSWA, CSWA-SD, CSWA-S, and CSWA-AM certifications. The instructor will require your .edu email address.

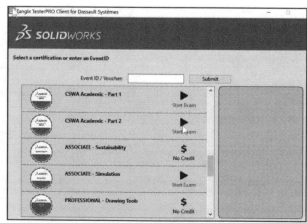

TABLE OF CONTENTS

Introduction

About the Book

You will find a wealth of information in this book. The book is written for new and intermediate users. The following conventions are used throughout this book:

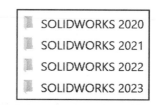

- The term document refers to a SOLIDWORKS part, drawing or assembly file.

- The list of items across the top of the SOLIDWORKS interface is the Menu bar menu or the Menu bar toolbar. Each item in the Menu bar has a pull-down menu. When you need to select a series of commands from these menus, the following format is used: Click **Insert**, **Reference Geometry**, **Plane** from the Menu bar. The Plane PropertyManager is displayed.

- The book is organized into chapters. Each chapter is focused on a specific certification category. Use the model files for the chapter exercises.

- The ANSI overall drafting standard and Third Angle projection is used as the default setting in this text. IPS (inch, pound, second) and MMGS (millimeter, gram, second) unit systems are used.

Download all needed model files from the SDC Publication website www.SDCpublications.com/downloads/978-1-63057-567-0

SOLIDWORKS 2020
SOLIDWORKS 2021
SOLIDWORKS 2022
SOLIDWORKS 2023

Take a sample online CSWA exam, visit https://3dexperience.virtualtester.com/#home.

If your school is an academic certification provider, your instructor can allocate free exam credits for the CSWA (Segment 1 & 2), CSWA-SD, CSWA-S, and CSWA-AM certifications. The instructor will require your .edu email address.

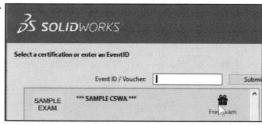

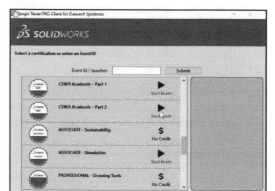

Chapter 1

Overview of SOLIDWORKS® and the User Interface

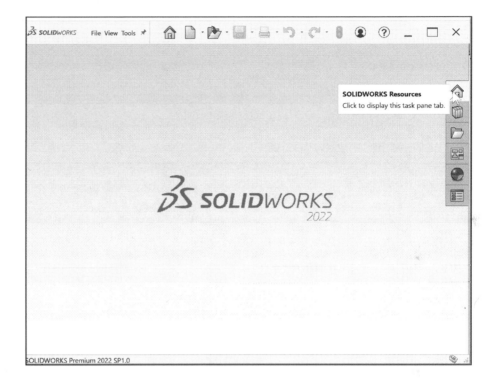

Below are the desired outcomes and usage competencies based on the completion of Chapter 1.

Desired Outcomes:	Usage Competencies:
• An understanding of the SOLIDWORKS® User Interface (UI) and CommandManager.	• Ability to establish a SOLIDWORKS session. • Aptitude to utilize the following items: Menu bar toolbar, Menu bar menu, Drop-down menus, Context toolbars, Consolidated drop-down toolbars, System feedback icons, Confirmation Corner, Heads-up View toolbar, Document Properties and more. • Open a new and existing SOLIDWORKS part. • Knowledge to zoom, rotate and maneuver a three-button mouse in the SOLIDWORKS Graphics window.

Notes:

CHAPTER 1: SOLIDWORKS USER INTERFACE

Objectives

SOLIDWORKS is a design software application used to model and create 2D and 3D sketches, 3D parts and assemblies and 2D drawings. Over the years, the SOLIDWORKS UI has changed, but will not affect an experienced user during the exam.

Chapter 1 covers the SOLIDWORKS User Interface and CommandManager. The following items are addressed: Menu bar toolbar, Menu bar menu, Context toolbars, Drop-down menus, Consolidated drop-down toolbars, Confirmation Corner, System feedback icons, Heads-up View toolbar and more.

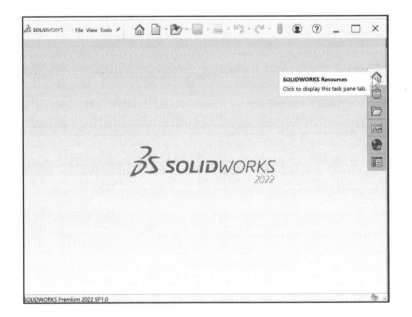

Download all needed model files from the SDC Publications website www.SDCpublications.com/downloads/978-1-63057-567-0.

The Graphics window, FeatureManager, CommandManager, tabs and tree folders will vary depending on system setup, version, and Add-ins.

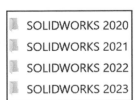

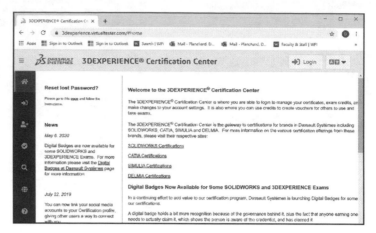 To obtain additional CSWA exam information and to take a sample online exam, visit https://3dexperience.virtualtester.com/#home

Chapter Objective

Provide a comprehensive understanding of the SOLIDWORKS® default User Interface and CommandManager: Menu bar toolbar, Menu bar menu, Drop-down menu, Right-click Pop-up menus, Context toolbars/menus, Fly-out tool button, System feedback icons, Confirmation Corner, Heads-up View toolbar and more.

On the completion of this chapter, you will be able to:

- Utilize the SOLIDWORKS Welcome dialog box.

- Establish a SOLIDWORKS session.

- Comprehend the SOLIDWORKS User Interface.

- Recognize the default Reference Planes in the FeatureManager.

- Open a new and existing SOLIDWORKS part.

- Utilize SOLIDWORKS Help and SOLIDWORKS Tutorials.

- Zoom, rotate and maneuver a three-button mouse in the SOLIDWORKS Graphics window.

What is SOLIDWORKS®?

- SOLIDWORKS® is a mechanical design automation software package used to build parts, assemblies and drawings that takes advantage of the familiar Microsoft® Windows graphical user interface.

- SOLIDWORKS is an easy to learn design and analysis tool (SOLIDWORKS Simulation, SOLIDWORKS Motion, SOLIDWORKS Flow Simulation, Sustainability, etc.), which makes it possible for designers to quickly sketch 2D and 3D concepts, create 3D parts and assemblies and detail 2D drawings.

- Model dimensions in SOLIDWORKS are associative between parts, assemblies and drawings. Reference dimensions are one-way associative from the part to the drawing or from the part to the assembly.

Start a SOLIDWORKS Session

Start a SOLIDWORKS session and familiarize yourself with the SOLIDWORKS User Interface. As you read and perform the tasks in this chapter, you will obtain a sense of how to use the book and the structure. Actual input commands or required actions in the chapter are displayed in bold.

The book does not cover starting a SOLIDWORKS session in detail for the first time. A default SOLIDWORKS installation presents you with several options. For additional information, visit http://www.SOLIDWORKS.com.

Activity: Start a SOLIDWORKS Session.

Start a SOLIDWORKS 20XX session.

1)   Type **SOLIDWORKS 20XX** in the Search window.

2)   Click the **SOLIDWORKS 20XX** application (or if available, **double-click** the SOLIDWORKS icon on the desktop). The SOLIDWORKS Welcome dialog box is displayed by default.

The Welcome dialog box provides a convenient way to open recent documents (Parts, Assemblies, and Drawings), view recent folders, access SOLIDWORKS resources, and stay updated on SOLIDWORKS news.

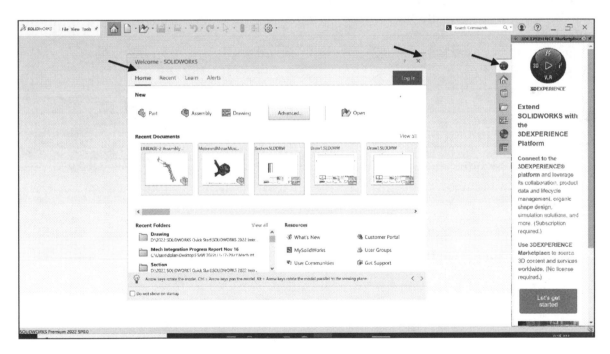

3)   **View** the options. Do not open a document at this time.

Home Tab

The Home tab lets you open new and existing documents, view recent documents and folders, and access SOLIDWORKS resources (*Part, Assembly, Drawing, Advanced mode, Open*).

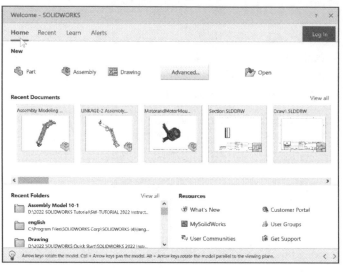

Recent Tab

The Recent tab lets you view a longer list of recent documents and folders. Sections in the Recent tab include Documents and Folders.

The Documents section includes thumbnails of documents that you have opened recently.

Click a thumbnail to open the document, or hover over a thumbnail to see the document location and access additional information about the document. When you hover over a thumbnail, the full path and last saved date of the document appears.

Learn Tab

The Learn tab lets you access instructional resources to help you learn more about the SOLIDWORKS software.

Sections in the Learn tab include:

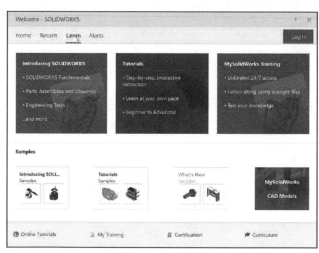

- **Introducing SOLIDWORKS**. Open the Introducing SOLIDWORKS book.

- **Tutorials**. Open the step-by-step tutorials in the SOLIDWORKS software.

- **MySolidWorks Training**. Open the Training section at MySolidWorks.com.

- **Introducing SOLIDWORKS (Samples)**. Open local folders containing sample models.

- **Tutorials (Samples)**. Open the SOLIDWORKS Tutorials (videos) section at solidworks.com.

- **What's New (Samples)**. List of new changes.

- **Online Tutorials**.

- **My Training**. Open the My Training section at MySolidWorks.com.

- **Certification**. Open the SOLIDWORKS Certification Program section at solidworks.com. You will need to create an account with a password.

- **Curriculum**. Open the Curriculum section at solidworks.com. You will need to create an account with a password.

- **MySOLIDWORKS - CAD Models**. Open models in the Community Library. You will need to create an account with a password.

🔅 When you install the software, if you do not install the Help Files or Example Files, the Tutorials and Samples links are unavailable.

Alerts Tab

The Alerts tab keeps you updated with SOLIDWORKS news.

Sections in the Alerts tab include Critical, Troubleshooting, and Technical.

🔅 The Critical section does not appear if there are no critical alerts to display.

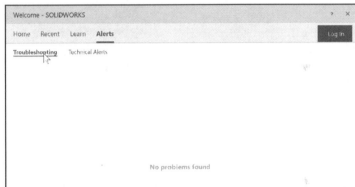

- Troubleshooting. Includes troubleshooting messages and recovered documents that used to be on the SOLIDWORKS Recovery tab in the Task Pane.

🔅 If the software has a technical problem and an associated troubleshooting message exists, the Welcome dialog box opens to the Troubleshooting section automatically on startup, even if you selected Do not show at startup in the dialog box.

- **Technical Alerts**. Open the contents of the SOLIDWORKS Support Bulletins RSS feed (Hotfixes, release news) at solidworks.com.

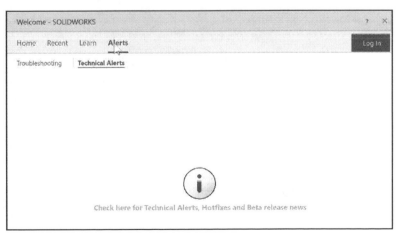

Close the Welcome - SOLIDWORKS dialog box.

4) Click **Close** ✕ from the Welcome - SOLIDWORKS dialog box. The SOLIDWORKS Graphics window is displayed. You can also click outside the Welcome - SOLIDWORKS dialog box, in the Graphics window.

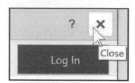

View the SOLIDWORKS Graphics window.

Menu Bar toolbar

The SOLIDWORKS (UI) is designed to make maximum use of the Graphics window. The Menu Bar toolbar contains a set of the most frequently used tool buttons from the Standard toolbar.

The following default tools are available:

- **Welcome to SOLIDWORKS** ⌂ - Open the Welcome dialog box, **New** ▯ - Create a new document; **Open** 🗁 - Open an existing document; **Save** 💾 - Save an active document; **Print** 🖶 - Print an active document; **Undo** ↩ - Reverse the last action; **Redo** ↪ - Redoes the last action that you reverse; **Select** ⬉ - Select Sketch entities, components and more; **Rebuild** 🚥 - Rebuild the active part, assembly or drawing;

File Properties  - Summary information on the active document;

Options ⚙ ▾ - Change system options and Add-Ins for SOLIDWORKS; **Login** ⊕ - Login to SOLIDWORKS. You will need to create an account with a password; **Help** ⑦ - access to help, tutorials, updates and more.

Menu Bar menu (No model open)

SOLIDWORKS provides a context-sensitive menu structure. The menu titles remain the same for all

three types of documents (Parts, Assemblies and Drawings), but the menu

items change depending on which type of document is active.

Menu Bar menu (Model open)

The Pin ⚲ option displays the Menu bar toolbar and the Menu bar menu as illustrated. Throughout the book, the Menu bar menu and the Menu bar toolbar are referred to as the Menu bar.

Drop-down menu (Open part document)

SOLIDWORKS takes advantage of the familiar Microsoft® Windows user interface. Communicate with SOLIDWORKS through drop-down menus, Context sensitive toolbars, Consolidated toolbars or the CommandManager tabs.

To close a SOLIDWORKS drop-down menu, press the Esc key. You can also click any other part of the SOLIDWORKS Graphics window or click another drop-down menu.

💡 The Graphics window, FeatureManager, CommandManager, tabs and tree folders will vary depending on system setup, version, and Add-ins.

Activity: Create a new Part Document.

A part is a 3D model, which consists of features. What are features? Features are geometry building blocks. Most features either add or remove material. Some features do not affect material (Cosmetic Thread).

Features are created either from 2D or 3D sketched profiles or from edges and faces of existing geometry.

Features are individual shapes that combined with other features make up a part or assembly. Some features, such as bosses and cuts, originate as sketches. Other features, such as shells and fillets, modify a feature's geometry.

Features are displayed in the FeatureManager.

The first sketch of a part is called the Base Sketch. The Base sketch is the foundation for the 3D model. The book focuses on 2D sketches and 3D features.

FeatureManager, CommandManager, and tree folders will vary depending on system setup, version, and Add-ins.

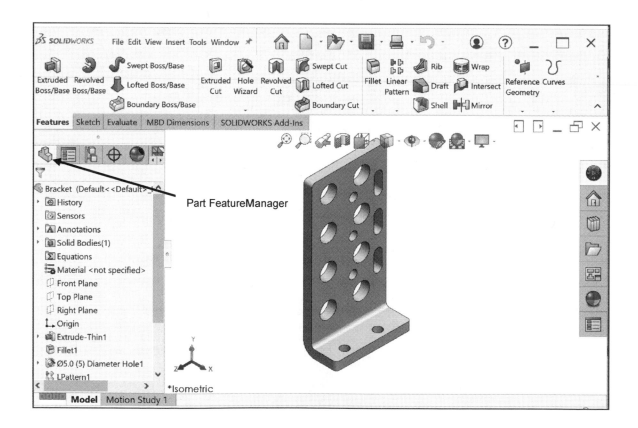

There are two modes in the New SOLIDWORKS Document dialog box: Novice and Advanced. The Novice option is the default option with three templates. The Advanced mode contains access to additional templates and tabs that you create in system options. Use the Advanced mode in this book.

Create a new part.

5) Click **New** ⬜ from the Main menu. The New SOLIDWORKS Document dialog box is displayed.

Select the Advanced mode.

6) If needed, click the **Advanced** tab. The below New SOLIDWORKS Document box is displayed.

7) Click the **Templates** tab.

8) Click **Part**. Part is the default template from the New SOLIDWORKS Document dialog box.

9) Click **OK** from the New SOLIDWORKS Document dialog box.

The Advanced mode remains selected for all new documents in the current SOLIDWORKS session. When you exit SOLIDWORKS, the Advanced mode setting is saved.

The default SOLIDWORKS installation contains three tabs in the New SOLIDWORKS Document dialog box: *Templates, MBD, and Tutorial*. The *Templates* tab corresponds to the default SOLIDWORKS templates. The *MBD* tab corresponds to the templates utilized in the SOLIDWORKS (Model Based Definition). The *Tutorial* tab corresponds to the templates utilized in the SOLIDWORKS Tutorials.

Part1 is displayed in the FeatureManager and is the name of the document. Part1 is the default part window name.

The Part Origin ⊥ is displayed in blue in the center of the Graphics window. The Origin represents the intersection of the three default reference planes: *Front Plane*, *Top Plane* and *Right Plane*. The positive X-axis is horizontal and points to the right of the Origin in the Front view. The positive Y-axis is vertical and points upward in the Front view. The FeatureManager contains a list of features, reference geometry, and settings utilized in the part.

Edit the document units directly from the Graphics window as illustrated.

CommandManager, FeatureManager, tabs, and Graphics window will vary depending on system setup, version, and Add-ins.

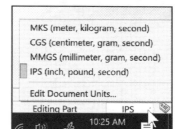

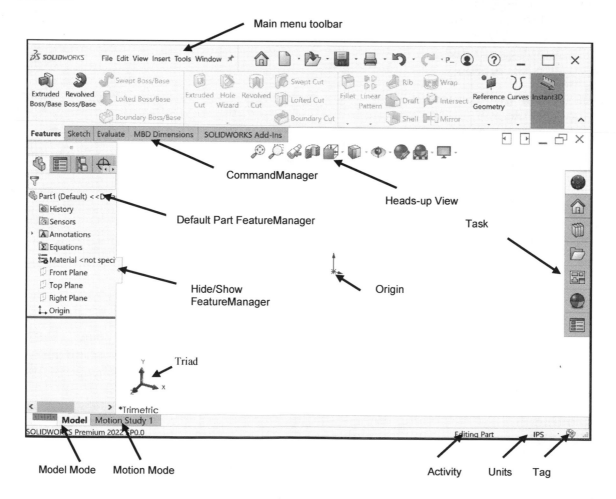

View the Default Sketch Planes.

10) Click the **Front Plane** from the FeatureManager.

11) Click the **Top Plane** from the FeatureManager.

12) Click the **Right Plane** from the FeatureManager.

13) Click the **Origin** from the FeatureManager. The Origin is the intersection of the Front, Top, and Right Planes. The Origin point is displayed.

14) Click **inside** the Graphics window.

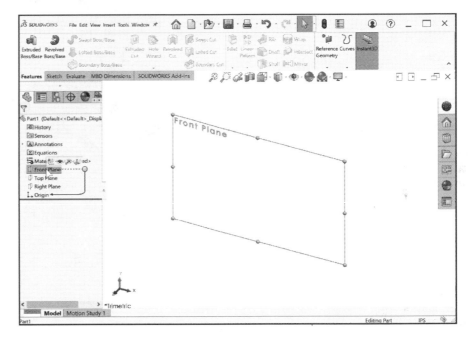

Download all model files from the SDC Publication website www.SDCpublications.com/downloads/978-1-63057-567-0. Open the Bracket part. Review the features and sketches in the Bracket FeatureManager. Work directly from a local hard drive.

| SOLIDWORKS 2020 |
| SOLIDWORKS 2021 |
| SOLIDWORKS 2022 |
| SOLIDWORKS 2023 |

Activity: Download the SOLIDWORKS folder. Open the Bracket Part.

Download the SOLIDWORKS folder. Open an existing SOLIDWORKS part.

15) **Download** the SOLIDWORKS 20XX folder.

16) **Unzip** the SOLIDWORKS 20XX folder. **Work** from the unzip folder.

17) Click **Open**  from the Main menu.

18) Browse to the **SOLIDWORKS 20XX\Bracket** folder.

19) Double-click the **Bracket** part. The Bracket part is displayed in the Graphics window.

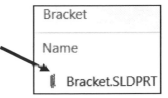

Bracket

Name

Bracket.SLDPRT

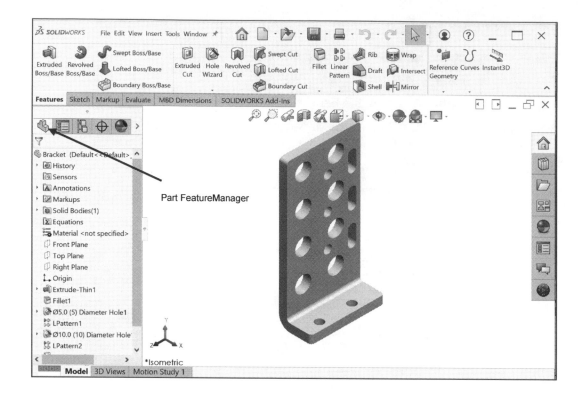

The FeatureManager design tree is located on the left side of the SOLIDWORKS Graphics window. The FeatureManager provides a summarized view of the active part, assembly, or drawing document. The tree displays the details on how the part, assembly or drawing document was created.

Use the FeatureManager rollback bar to temporarily roll back to an earlier state, to absorbed features, roll forward, roll to previous, or roll to the end of the FeatureManager design tree. You can add new features or edit existing features while the model is in the rolled-back state. You can save models with the rollback bar placed anywhere.

In the next section, review the features in the Bracket FeatureManager using the Rollback bar.

Activity: Use the FeatureManager Rollback Bar option.

Apply the FeatureManager Rollback Bar. Revert to an earlier state in the model.

20) Place the **mouse pointer** over the rollback bar in the FeatureManager design tree as illustrated. The pointer changes to a hand 🖑. Note the provided information on the feature. This is called Dynamic Reference Visualization.

21) Drag the **rollback bar** up the FeatureManager design tree until it is above the features you want rolled back, in this case 10.0 (10) Diameter Hole1.

22) **Release** the mouse button.

View the first feature in the Bracket Part.

23) Drag the **rollback bar** up the FeatureManager above Fillet1. View the results in the Graphics window.

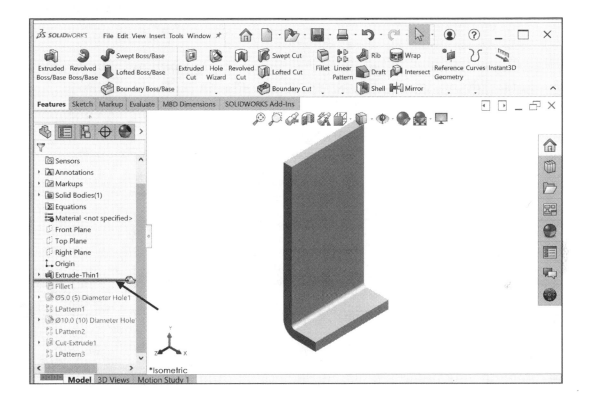

Return to the original Bracket Part FeatureManager.

24) Right-click **Extrude-Thin1** in the FeatureManager. The Pop-up Context toolbar is displayed.

25) Click **Roll to End**. View the results in the Graphics window.

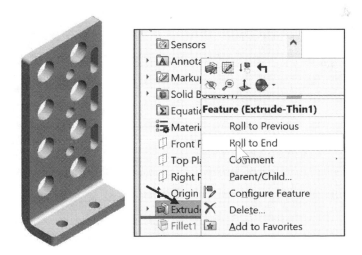

Heads-up View toolbar

SOLIDWORKS provides the user with numerous view options. One of the most useful tools is the Heads-up View toolbar displayed in the Graphics window when a document is active.

💡 *Dynamic Annotation Views* 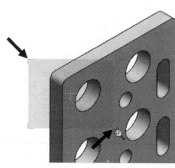: Only available with SOLIDWORKS MBD (Model Based Definition). Provides the ability to control how annotations are displayed when you rotate models.

In the next section, apply the following tools: Zoom to Fit, Zoom to Area, Zoom out, Rotate and select various view orientations from the Heads-up View toolbar.

Activity: Utilize the Heads-up View toolbar.

Zoom to Fit the model in the Graphics window.

26) Click the **Zoom to Fit** 🔍 icon. The tool fits the model to the Graphics window.

Zoom to Area on the model in the Graphics window.

27) Click the **Zoom to Area** 🔍 icon. The Zoom to Area 🔍 icon is displayed.

Zoom in on the top left hole.

28) Window-select the top left corner as illustrated. View the results.

De-select the Zoom to Area tool.

29) Click the **Zoom to Area** 🔍 icon.

Fit the model to the Graphics window.

30) Press the **f** key.

Rotate the model.

31) Hold the **middle mouse button** down. Drag **upward** ↺, **downward** ↺, to the **left** ↺ and to the **right** ↺ to rotate the model in the Graphics window.

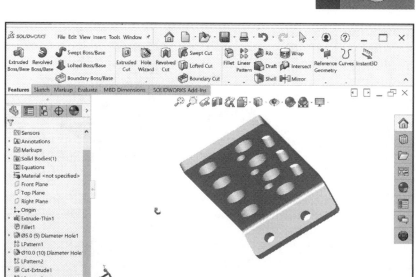

Display a few Standard Views.

32) Click **inside** the Graphics window.

33) Click **Front** ⬒ from the drop-down Heads-up view toolbar. The model is displayed in the Front view.

34) Click **Right** ⬓ from the drop-down Heads-up view toolbar. The model is displayed in the Right view.

35) Click **Top** ⬒ from the drop-down Heads-up view toolbar. The model is displayed in the Top view.

Display a Trimetric view of the Bracket model.

36) Click **Trimetric** ⬙ from the drop-down Heads-up view toolbar as illustrated. Note your options. View the results in the Graphics window.

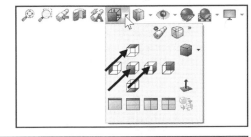

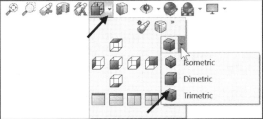

SOLIDWORKS Help

Help in SOLIDWORKS is context-sensitive and in HTML format. Help is accessed in many ways, including Help buttons in all dialog boxes and PropertyManager and

Help ⑦ on the Standard toolbar for SOLIDWORKS Help.

☼ CommandManager and FeatureManagers tabs will vary depending on system setup, version, and Add-ins.

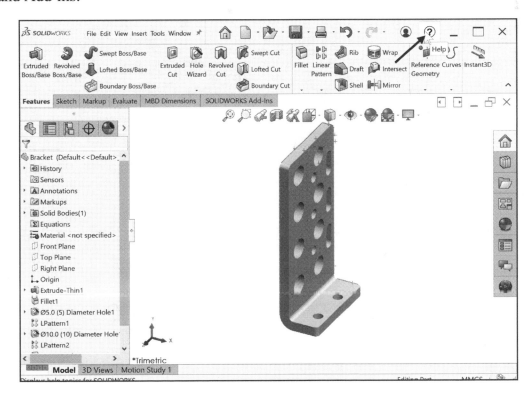

37) Click ⑦ from the Standard toolbar.

38) Click ⑦ **Help** from the drop-down menu. The SOLIDWORKS Home Page is displayed by default. View your options.

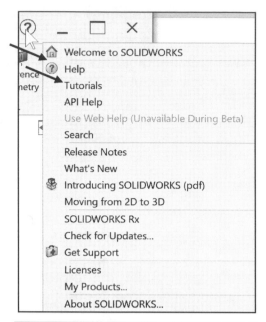

💡 SOLIDWORKS Web Help is active by default under Help in the Main menu.

Close Help. Return to the SOLIDWORKS Graphics window.

39) **Close** ❌ SOLIDWORKS Home.

SOLIDWORKS Tutorials

Display and explore the SOLIDWORKS tutorials.

40) Click ⑦ from the Standard toolbar.

41) Click **Tutorials**. The SOLIDWORKS Tutorials are displayed. The SOLIDWORKS Tutorials are presented by category.

42) Click the **Getting Started** category. The Getting Started category provides lessons on parts, assemblies, and drawings.

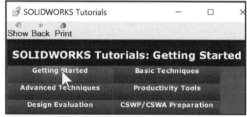

In the next section, close all models, tutorials and view the additional User Interface tools.

Activity: Close all Tutorials and Models.

Close SOLIDWORKS Tutorials and models.

43) **Close** ❌ SOLIDWORKS Tutorials.

44) Click **Window**, **Close All** from the Menu bar menu.

User Interface Tools

The book utilizes additional areas of the SOLIDWORKS User Interface. Explore an overview of these tools in the next section.

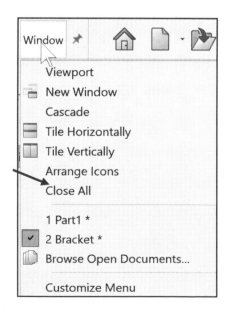

Right-click

Right-click in the Graphics window on a model, or in the FeatureManager on a feature or sketch to display the Context-sensitive toolbar. If you are in the middle of a command, this toolbar displays a list of options specifically related to that command.

Right-click an empty space in the Graphics window of a part or assembly, and a selection context toolbar above the shortcut menu is displayed. This provides easy access to the most commonly used selection tools.

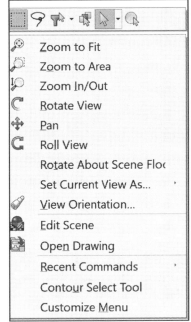

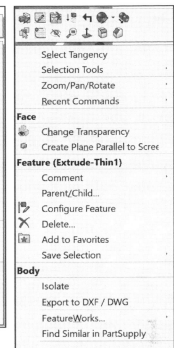

Consolidated toolbar

Similar commands are grouped together in the CommandManager. For example, variations of the Rectangle sketch tool are grouped in a single fly-out button as illustrated.

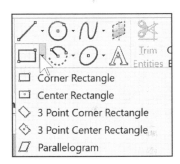

If you select the Consolidated toolbar button without expanding:

For some commands such as Sketch, the most commonly used command is performed. This command is the first listed and the command shown on the button.

For commands such as rectangle, where you may want to repeatedly create the same variant of the rectangle, the last used command is performed. This is the highlighted command when the Consolidated toolbar is expanded.

System feedback icon

SOLIDWORKS provides system feedback by attaching a symbol to the mouse pointer cursor.

The system feedback symbol indicates what you are selecting or what the system is expecting you to select.

As you move the mouse pointer across your model, system feedback is displayed in the form of a symbol, riding next to the cursor as illustrated. This is a valuable feature in SOLIDWORKS.

Confirmation Corner

When numerous SOLIDWORKS commands are active, a symbol or a set of symbols is displayed in the upper right hand corner of the Graphics window. This area is called the Confirmation Corner.

When a sketch is active, the confirmation corner box displays two symbols. The first symbol is the sketch tool icon. The second symbol is a large red X. These two symbols supply a visual reminder that you are in an active sketch. Click the sketch symbol icon to exit the sketch and to save any changes that you made.

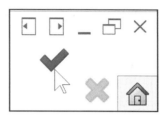

When other commands are active, the confirmation corner box provides a green check mark and a large red X. Use the green check mark to execute the current command. Use the large red X to cancel the command.

☀ Confirm changes you make in sketches and tools by using the D keyboard shortcut to move the OK and Cancel buttons to the pointer location in the Graphics window.

Heads-up View toolbar

SOLIDWORKS provides the user with numerous view options from the Standard Views, View and Heads-up View toolbar.

The Heads-up View toolbar is a transparent toolbar that is displayed in the Graphics window when a document is active.

You can hide, move or modify the Heads-up View toolbar. To modify the Heads-up View toolbar, right-click on a tool and select or deselect the tools that you want to display.

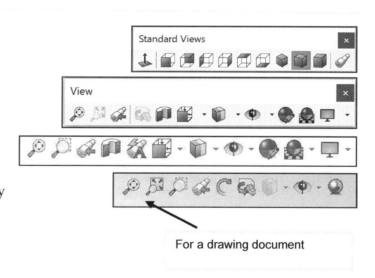

For a drawing document

The following views are available. Note: available views are document dependent.

- *Zoom to Fit* 🔍 : Fit the model to the Graphics window.

- *Zoom to Area* 🔍 : Zoom to the areas you select with a bounding box.

- *Previous View* 🔍 : Display the previous view.

- *Section View* 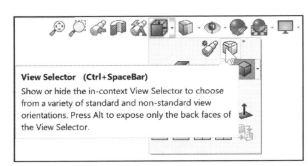: Display a cutaway of a part or assembly, using one or more cross section planes.

- *Dynamic Annotation Views* : Only available with SOLIDWORKS MBD. Control how annotations are displayed when you rotate a model.

The Orientation dialog has an option to display a view cube (in-context View Selector) with a live model preview. This helps the user to understand how each standard view orientates the model. With the view cube, you can access additional standard views. The views are easy to understand, and they can be accessed simply by selecting a face on the cube.

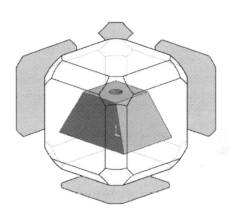

To activate the Orientation dialog box, press (Ctrl + spacebar) or click the View Orientation icon from the Heads-up View toolbar. The active model is displayed in the View Selector in an Isometric orientation (default view).

Click the View Selector icon in the Orientation dialog box to show or hide the in-context View Selector.

Press **Ctrl + spacebar** to activate the View Selector.

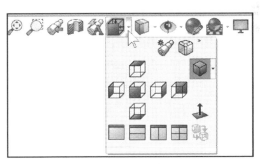

Press the **spacebar** to activate the Orientation dialog box.

- *View Orientation box* : Select a view orientation or the number of viewports. The options are: *Top, Left, Front, Right, Back, Bottom, Single view, Two view - Horizontal, Two view - Vertical, Four view*. Click the drop-down arrow to access Axonometric views: *Isometric, Dimetric* and *Trimetric*.

- *Display Style* : Display the style for the active view. The options are: *Wireframe, Hidden Lines Visible, Hidden Lines Removed, Shaded, Shaded With Edges*.

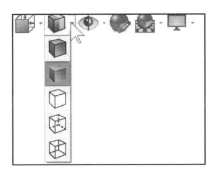

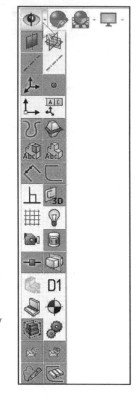

- *Hide/Show Items* ◉ ⁻ : Select items to hide or show in the Graphics window. The available items are document dependent. Note the View Center of Mass ✛ icon.

- *Edit Appearance* ● : Edit the appearance of entities of the model.

- *Apply Scene* ◕ ⁻ : Apply a scene to an active part or assembly document. View the available options.

- *View Setting* ⬚ ⁻ : Select the following settings: *RealView Graphics*, *Shadows In Shaded Mode*, *Ambient Occlusion*, *Perspective* and *Cartoon*.

- *Rotate view* ↻ : Rotate a drawing view. Input Drawing view angle and select the ability to update and rotate center marks with view.

- *3D Drawing View* ◈ : Dynamically manipulate the drawing view in 3D to make a selection.

To display a grid for a part, click Options ⚙ ⁻ , Document Properties tab. Click Grid/Snaps, check the Display grid box.

💡 Add a custom view to the Heads-up View toolbar. Press the space key. The Orientation dialog box is displayed. Click the New View tool. The Name View dialog box is displayed. Enter a new named view. Click OK.

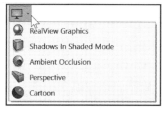

Use commands to display information about the triad or to change the position and orientation of the triad. Available commands depend on the triad's context.

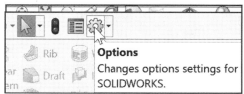

💡 Save space in the CommandManager, limit your CommandManager tabs. **Right-click** on a CommandManager tab. Click **Tabs**. View your options to display CommandManager tabs.

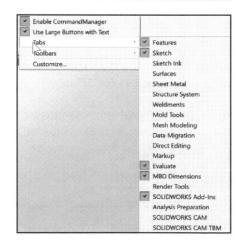

SOLIDWORKS CommandManager

The SOLIDWORKS CommandManager is a Context-sensitive toolbar. By default, it has toolbars embedded in it based on your active document type. When you click a tab below the CommandManager, it updates to display that toolbar. For example, if you click the Sketch tab, the Sketch toolbar is displayed.

For commercial users, SOLIDWORKS Model Based Definition (MBD) and SOLIDWORKS CAM is a separate application. For education users, SOLIDWORKS MBD and SOLIDWORKS CAM is included in the SOLIDWORKS Education Edition.

Below is an illustrated CommandManager for a **Part** document. Tabs will vary depending on system setup, system version and Add-ins.

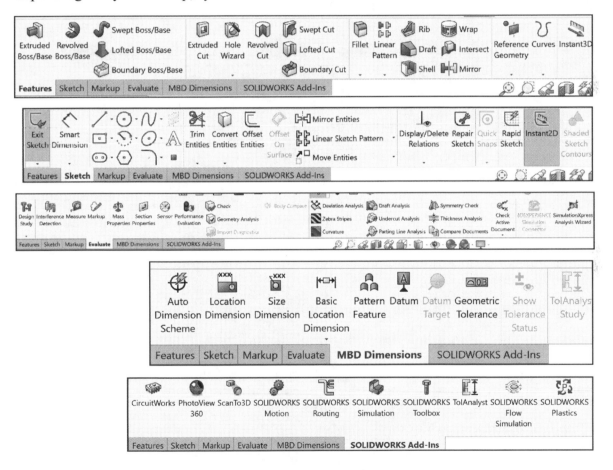

Set button size from the Toolbars tab of the Customize dialog box. To facilitate element selection on touch interfaces such as tablets, you can set up the larger Size buttons and text from the Options menu (Standard toolbar).

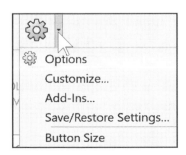

SOLIDWORKS User Interface

The SOLIDWORKS CommandManager is a Context-sensitive toolbar that automatically updates based on the toolbar you want to access. By default, it has toolbars embedded in it based on your active document type. The available tools are feature and document dependent.

Below is an illustrated CommandManager for a **Drawing** document. Tabs will vary depending on system setup, system version and Add-ins.

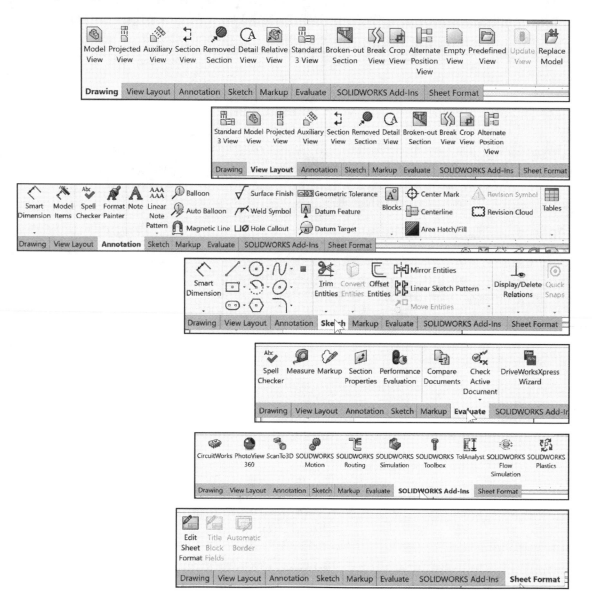

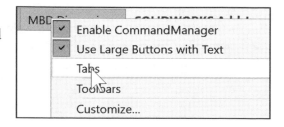 To add a custom tab, right-click on a tab and click Customize. You can also select to add a blank tab and populate it with custom tools from the Customize dialog box.

The SOLIDWORKS CommandManager is a Context-sensitive toolbar that automatically updates based on the toolbar you want to access. By default, it has toolbars embedded in it based on your active document type. The available tools are feature and document dependent.

Below is an illustrated CommandManager for an **Assembly** document. Tabs will vary depending on system setup, system version and Add-ins.

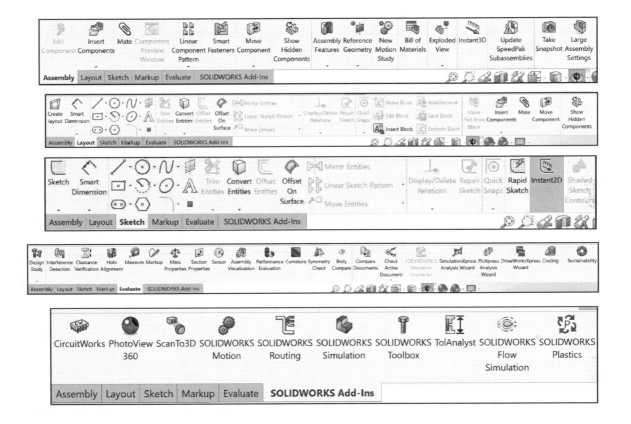

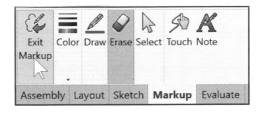 The Markup tab is displayed by default in the CommandManager on some systems. You can draw markups with a mouse on non-touch devices, display bounding boxes for markups, create markups in drawings, and use the context toolbar to access markup options.

The Markup toolbar displays different options

depending on the device. **Draw** Draw and **Touch**

Touch are **not** available for non-touch devices.

Float the CommandManager. Drag the Features, Sketch or any CommandManager tab. Drag the CommandManager anywhere on or outside the SOLIDWORKS window.

To dock the CommandManager, perform one of the following:

While dragging the CommandManager in the SOLIDWORKS window, move the pointer over a docking icon -

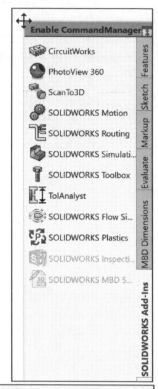

Dock above , Dock left , Dock right and click the needed command.

Double-click the floating CommandManager to revert the CommandManager to the last docking position.

Collapse the CommandManager

Collapse the CommandManager to only display the available tabs. No tools are shown.

Click the Collapsed CommandManager arrow to the right of the active CommandManager tab as illustrated. Only the tabs are displayed.

To display the CommandManager tools back, click on a CommandManager tab. Click the Pin CommandManager as illustrated.

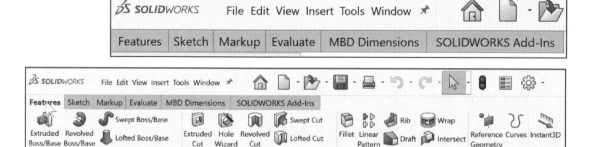

Part FeatureManager Design Tree

The Part FeatureManager consists of various tabs:

- *FeatureManager design tree* 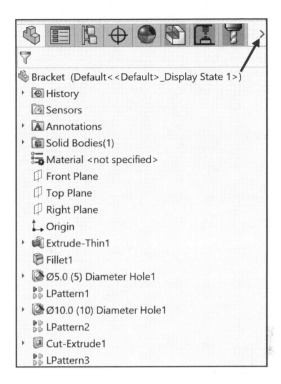 tab.

- *PropertyManager* 🗐 tab.

- *ConfigurationManager* 🗒 tab.

- *DimXpertManager* ⊕ tab.

- *DisplayManager* ◉ tab.

- *CAM FeatureManager tree* 🗒 tab.

- *CAM Operation tree* 🖳 tab.

- *CAM Tools tree* 🗡 tab.

Click the direction arrows to expand or collapse the FeatureManager design tree.

CommandManager and FeatureManager tabs and folder files will vary depending on your SOLIDWORKS applications and Add-ins.

Select the Hide/Show FeatureManager Area tab

as illustrated to enlarge the Graphics window for modeling.

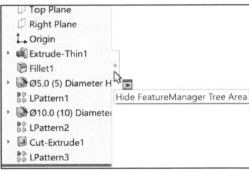

The Sensors tool 🗓 located in the FeatureManager monitors selected properties in a part or assembly and alerts you when values deviate from the specified limits. There are five sensor types: Simulation Data, Mass properties, Dimensions, Measurement and Costing Data.

Various commands provide the ability to control what is displayed in the FeatureManager design tree.

1. Show or Hide FeatureManager items.

🔆 Click **Options** ⚙️ ⁻ from the Menu bar. Click **FeatureManager** from the System Options tab. Customize your FeatureManager from the Hide/Show tree Items dialog box.

2. Filter the FeatureManager design tree. Enter information in the filter field. You can filter by *Type of features, Feature names, Sketches, Folders, Mates, User-defined tags* and *Custom properties*.

🔆 Tags are keywords you can add to a SOLIDWORKS document to make them easier to filter and to search. The Tags 🏷️ icon is located in the bottom right corner of the Graphics window.

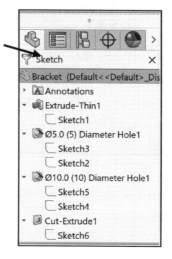

🔆 Collapse all items in the FeatureManager, **right-click** and select **Collapse items**, or press the **Shift + C** keys.

The FeatureManager design tree and the Graphics window are dynamically linked. Select sketches, features, drawing views, and construction geometry in either pane.

Split the FeatureManager design tree and either display two FeatureManager instances or combine the FeatureManager design tree with the ConfigurationManager or PropertyManager.

Move between the FeatureManager design tree, PropertyManager, ConfigurationManager, DimXpertManager, DisplayManager and others by selecting the tab at the top of the menu.

Split

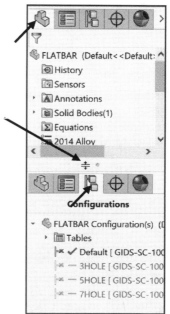

The ConfigurationManager 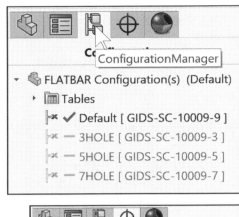 tab is located to the right of the PropertyManager tab. Use the ConfigurationManager to create, select and view multiple configurations of parts and assemblies.

The icons in the ConfigurationManager denote whether the configuration was created manually or with a design table.

The DimXpertManager ⊕ tab provides the ability to insert dimensions and tolerances manually or automatically. The options are: **Auto Dimension Scheme**, **Auto Pair Tolerance**, **Basic, Location Dimension**, **Basic Size Dimension**, **General Profile Tolerance**, **Show Tolerance Status**, **Copy Scheme**, **Import Scheme**, **TolAnalyst Study** and **Datum Target**.

☀ TolAnalyst is available in SOLIDWORKS Premium.

Fly-out FeatureManager

The fly-out FeatureManager design tree provides the ability to view and select items in the PropertyManager and the FeatureManager design tree at the same time.

Throughout the book, you will select commands and command options from the drop-down menu, fly-out FeatureManager, Context toolbar, or from a SOLIDWORKS toolbar.

☀ Another method for accessing a command is to use the accelerator key. Accelerator keys are special keystrokes, which activate the drop-down menu options. Some commands in the menu bar and items in the drop-down menus have an underlined character.

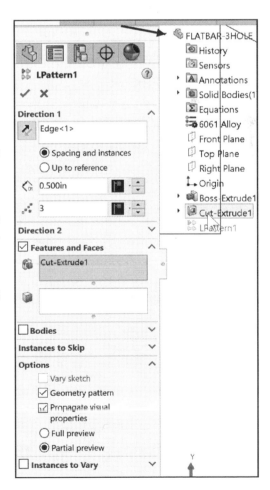

Task Pane

The Task Pane is displayed when a SOLIDWORKS session starts. You can show, hide, and reorder tabs in the Task Pane. You can also set a tab as the default so it appears when you open the Task Pane, pin or unpin to the default location.

The Task Pane contains the following default tabs:

- *3DEXPERIENCE Marketplace* .

- *SOLIDWORKS Resources* .

- *Design Library* .

- *File Explorer* .

- *View Palette* .

- *Appearances, Scenes and Decals* .

- *Custom Properties* .

Additional tabs are displayed with Add-Ins.

Use the **Back** and **Forward** buttons in the Design Library tab and the Appearances, Scenes, and Decals tab of the Task Pane to navigate in folders.

3DEXPERIENCE MARKETPLACE

Click the 3DEXPERIENCE MARKETPLACE icon to connect to the 3DEXPERIENCE platform and leverage its collaboration, product data and lifecycle management, organic shape design, simulation solutions, and more. (Subscription required).

Use 3DEXPERIENCE MARKETPLACE to source 3D content and services worldwide. (No license required.)

SOLIDWORKS Resources

The SOLIDWORKS Resources  icon displays the following default selections:

- *Welcome to SOLIDWORKS.*

- *SOLIDWORKS Tools.*

- *Online Resources.*

- *Subscription Services.*

Other user interfaces are available during the initial software installation selection: *Machine Design, Mold Design, Consumer Products Design, etc.*

Design Library

The Design Library  contains reusable parts, assemblies, and other elements including library features.

The Design Library tab contains default selections. Each default selection contains additional subcategories.

The default selections are:

- *Design Library.*

- *SOLIDWORKS Content (Internet access required).*

- *Toolbox.*

Activate the SOLIDWORKS Toolbox. Click Tools, Add-Ins.., from the Main menu. Check the SOLIDWORKS Toolbox Library and SOLIDWORKS Toolbox Utilities box from the Add-ins dialog box or click SOLIDWORKS Toolbox from the SOLIDWORKS Add-Ins tab.

To access the Design Library folders in a non-network environment, click Add File Location  and browse to the needed path. Paths may vary depending on your SOLIDWORKS version and window setup. In a network environment, contact your IT department for system details.

File Explorer

File Explorer ⌷ duplicates Windows Explorer from your local computer and displays:

- *Recent Documents.*
- *Samples.*
- *Open in SOLIDWORKS*
- *Desktop.*

View Palette

The View Palette 🔲 tool located in the Task Pane provides the ability to insert drawing views of an active document, or click the Browse button to locate the desired document.

Click and drag the view from the View Palette into an active drawing sheet to create a drawing view.

 The selected model is Bracket.

Appearances, Scenes, and Decals

Appearances, Scenes, and Decals 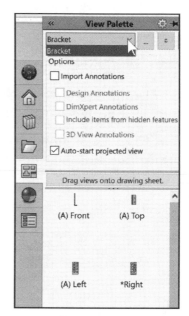 provide a simplified way to display models in a photo-realistic setting using a library of Appearances, Scenes, and Decals.

An appearance defines the visual properties of a model, including color and texture. Appearances do not affect physical properties, which are defined by materials.

Scenes provide a visual backdrop behind a model. In SOLIDWORKS they provide reflections on the model. PhotoView 360 is an Add-in. Drag and drop a selected appearance, scene or decal on a feature, surface, part or assembly.

Custom Properties

The Custom Properties  tool provides the ability to enter custom and configuration specific properties directly into SOLIDWORKS files.

Motion Study tab

Motion Studies are graphical simulations of motion for an assembly. Access the MotionManager from the Motion Study tab. The Motion Study tab is located in the bottom left corner of the Graphics window.

Incorporate visual properties such as lighting and camera perspective. Click the Motion Study tab to view the MotionManager. Click the Model tab to return to the FeatureManager design tree.

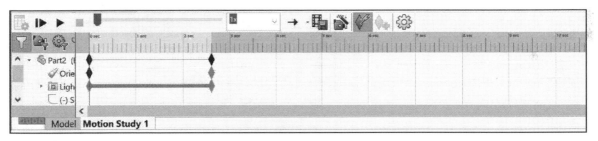

The MotionManager displays a timeline-based interface and provides the following selections from the drop-down menu as illustrated:

- *Animation:* Apply Animation to animate the motion of an assembly. Add a motor and insert positions of assembly components at various times using set key points. Use the Animation option to create animations for motion that do **not** require accounting for mass or gravity.

- *Basic Motion:* Apply Basic Motion for approximating the effects of motors, springs, collisions and gravity on assemblies. Basic Motion takes mass into account in calculating motion. Basic Motion computation is relatively fast, so you can use this for creating presentation animations using physics-based simulations. Use the Basic Motion option to create simulations of motion that account for mass, collisions or gravity.

If the Motion Study tab is not displayed in the Graphics window, click **View ➤ Toolbars ➤ MotionManager** from the Menu bar.

The Graphics window, FeatureManager, CommandManager, tabs and tree folders will vary depending on system setup, version, and Add-ins.

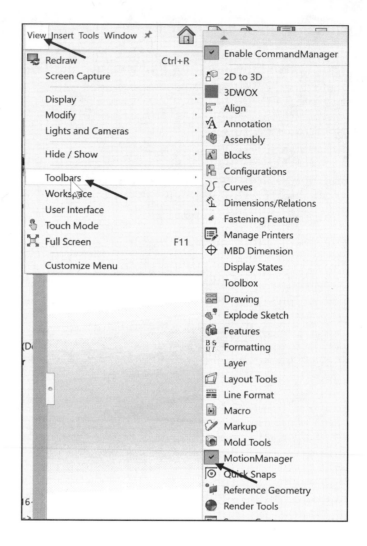

3D Views tab

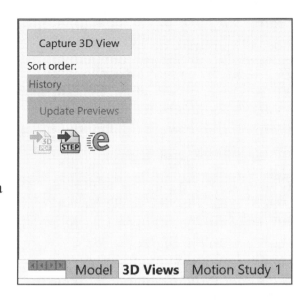

Only available in SOLIDWORKS MBD.
SOLIDWORKS MBD (Model Based
Definition) lets you create models without
the need for drawings giving you an
integrated manufacturing solution. MBD
helps companies define, organize, and
publish 3D product and manufacturing
information (PMI), including 3D model data
in industry standard file formats.

Create 3D drawing views of your parts and
assemblies that contain the model settings
needed for review and manufacturing. This
lets users navigate back to those settings as
they evaluate the design.

Use the tools in the MBD Dimensions CommandManager to set up your model with
selected configurations, including explodes and abbreviated views, annotations, display
states, zoom level, view orientation and section views. Capture those settings so that you
and other users can return to them at any time using the 3D view palette.

To access the 3D View palette, click the 3D Views tab at the bottom of the
SOLIDWORKS window or the SOLIDWORKS MBD tab in the CommandManager. The
Capture 3D View button opens the Capture 3D View PropertyManager, where you
specify the 3D view name, and the configuration, display state and annotation view to
capture. See help for additional information.

Dynamic Reference Visualization (Parent/Child)

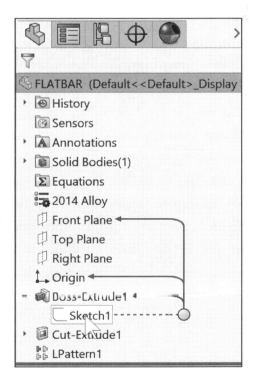

Dynamic Reference Visualization provides the
ability to view the parent/child relationships
between items in the FeatureManager design tree.
When you hover over a feature with references in
the FeatureManager design tree, arrows display
showing the relationships. If a reference cannot be
shown because a feature is not expanded, the arrow
points to the feature that contains the reference, and
the actual reference appears in a text box to the right
of the arrow.

Use Dynamic reference visualization for a part, assembly and mates.

To display the Dynamic Reference Visualization, click **View** ➢ **User Interface** ➢ **Dynamic Reference Visualization Parent/Child)** from the Main menu bar.

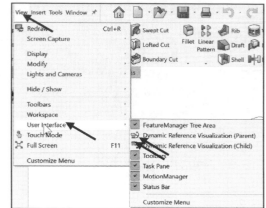

Mouse Movements

A mouse typically has two buttons: a primary button (usually the left button) and a secondary button (usually the right button). Most mice also include a scroll wheel between the buttons to help you scroll through documents and to Zoom in, Zoom out and rotate models in SOLIDWORKS. It is highly recommended that you use a mouse with at least a Primary, Scroll and Secondary button.

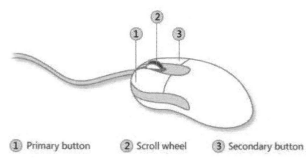

① Primary button   ② Scroll wheel   ③ Secondary button

Single-click

To click an item, point to the item on the screen, and then press and release the primary button (usually the left button). Clicking is most often used to select (mark) an item or open a menu. This is sometimes called single-clicking or left-clicking.

Double-click

To double-click an item, point to the item on the screen, and then click twice quickly. If the two clicks are spaced too far apart, they might be interpreted as two individual clicks rather than as one double-click. Double-clicking is most often used to open items on your desktop. For example, you can start a program or open a folder by double-clicking its icon on the desktop.

Right-click

To right-click an item, point to the item on the screen, and then press and release the secondary button (usually the right button). Right-clicking an item usually displays a list of things you can do with the item. Right-click in the open Graphics window or on a command in SOLIDWORKS, and additional pop-up context is displayed.

Scroll wheel

Use the scroll wheel to zoom-in or to zoom-out of the Graphics window in SOLIDWORKS. To zoom-in, roll the wheel backward (toward you). To zoom-out, roll the wheel forward (away from you).

Summary

The SOLIDWORKS (UI) is designed to make maximum use of the Graphics window for your model. Displayed toolbars and commands are kept to a minimum.

The SOLIDWORKS User Interface and CommandManager consist of the following main options: Menu bar toolbar, Menu bar menu, Drop-down menus, Context toolbars, Consolidated fly-out menus, System feedback icons, Confirmation Corner and Heads-up View toolbar.

The Part CommandManager controls the display of tabs: *Features*, *Sketch*, *Evaluate*, *MBD Dimensions* and various *SOLIDWORKS Add-Ins*.

The FeatureManager consists of various tabs:

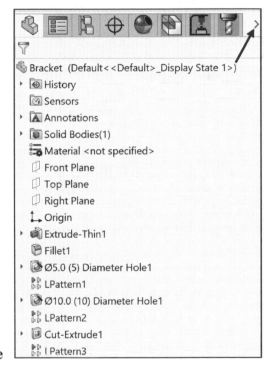

- *FeatureManager design tree* tab.

- *PropertyManager* tab.

- *ConfigurationManager* tab.

- *DimXpertManager* tab.

- *DisplayManager* tab.

- *CAM FeatureManager tree* tab.

- *CAM Operation tree* tab.

- *CAM Tools tree* tab.

Click the direction arrows to expand or collapse the FeatureManager design tree.

CommandManager, FeatureManager, and file folders will vary depending on system set-up and Add-ins.

You learned about creating a new SOLIDWORKS part and opening an existing SOLIDWORKS part along with using the Rollback bar to view the sketches and features.

If you modify a document property from an Overall drafting standard, a modify message is displayed as illustrated.

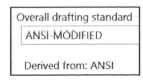

View the provided SOLIDWORKS CSWA Sample Exam folder included in the downloaded materials. The CSWA Sample Exam folder contains a pdf with information on the following: exam details, how to prepare for the exam, how to take the practice exam, taking the exam, sample test questions, test answers, and helpful sites.

SOLIDWORKS Certification Exam Guide & Practice Test

ASSOCIATE Mechanical Design

CSWA: Certified SOLIDWORKS Associate

Go to the 3DEXPERIENCE® Certification Center at https://3dexperience.virtualtester.com/#home to take the free online CSWA sample exam.

The 3DEXPERIENCE® Certification Center is where you are able to login to manage your certificates, take certifications, and make changes to your account settings.

If your school is an academic certification provider, your instructor can allocate free exam credits for the CSWA (Segment 1 & 2), CSWA-SD, CSWA-S, and CSWA-AM certifications. The instructor will require your .edu email address.

Download all needed model files (initial and final) and the SOLIDWORKS CSWA Sample Exam folder from the SDC Publications website www.SDCpublications.com/downloads/978-1-63057-567-0

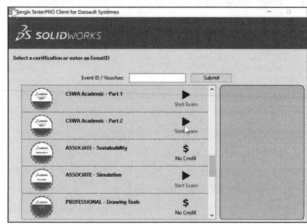

| SOLIDWORKS 2020 |
| SOLIDWORKS 2021 |
| SOLIDWORKS 2022 |
| SOLIDWORKS 2023 |

CHAPTER 2 - CSWA INTRODUCTION AND DRAFTING COMPETENCIES

Chapter Objective

Provide a basic introduction into the curriculum and categories of the Certified SOLIDWORKS Associate (CSWA) exam. Awareness of the exam procedure, process, and required model knowledge. The five exam categories are: Drafting Competencies, Basic Part Creation and Modification, Intermediate Part Creation and Modification, Advanced Part Creation and Modification, and Assembly Creation and Modification.

Introduction

The CSWA Academic exam is provided either in a single 3-hour segment, or 2 - 90-minute segments.

Segment 1 of the CSWA Academic exam is 90 minutes, minimum passing score is 80, with 6 questions. There are two questions in the Basic Part Creation and Modification category, two questions in the Intermediate Part Creation and Modification category and two questions in the Assembly Creation and Modification category.

Segment 2 of the CSWA Academic exam is 90 minutes, minimum passing score is 80 with 8 questions. There are three questions on the CSWA Academic exam in the Drafting Competencies category, three questions in the Advanced Part Creation and Modification category and two questions in the Assembly Creation and Modification category.

The CSWA exam for industry is only provided in a single 3-hour segment. The exam consists of 14 questions in five categories worth a total of 240 points.

All exams cover the same material.

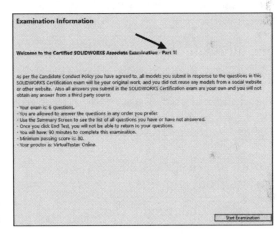

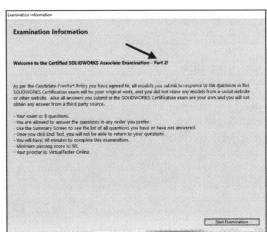

Taking the CSWA Exam (Segment 1 or 2)

Go to the 3DEXPERIENCE®
Certification Center at
https://3dexperience.virtualtester.com/#ho
me.

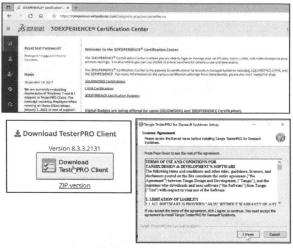

The 3DEXPERIENCE® Certification
Center is where you are able to log in to
manage your certificates, take
certifications, and make changes to your
account settings.

If your school is an academic certification
provider, your instructor can allocate a
free CSWA Segment 1 and Segment 2
exam credit.

The instructor will require your .edu
email address.

Download the TesterPRO Client.

Un-zip the Tester PRO Client.

Read the License Agreement. Agree to
the Candidate Conduct Policy. Click I
Agree.

Install the Tester PRO Client. Click
install. Click Finish.

Select Test Language. Click Continue.

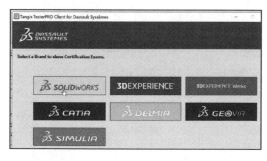

Create an account. If you already have an
account, select "I already have a VirtualTester
UserID and password".

Log in. Click Continue.

Click SOLIDWORKS.

Select either CSWA Academic -
Part 1 or CSWA Academic - Part 2.

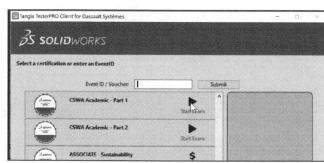

This will depend on the exam credit
that your instructor provided. Click
Start Exam. Select language. Click
Start Examination.

Read the instructions. Agree to the
Confidentiality Agreement and
Candidate Conduct Policy. Click Start Examination.

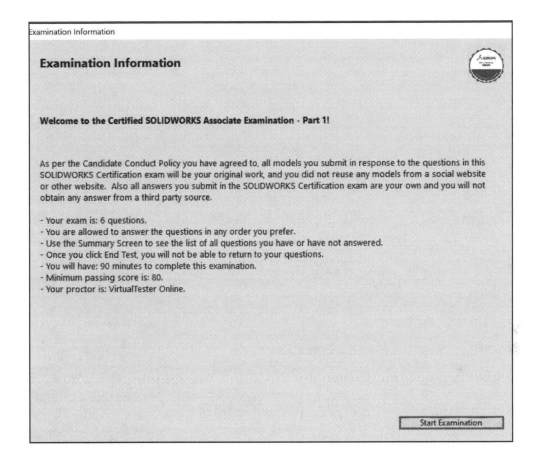

Segment 1: There are two questions in the Basic Part Creation and Modification category, two questions in the Intermediate Part Creation and Modification category and two questions in the Assembly Creation and Modification category.

You are required to build a SOLIDWORKS model with six or more features and to answer a question either on the overall mass, or the location of the Center of mass relative to the default part Origin location.

Segment 1 & 2: Part Creation and Modification requires knowledge of the following:

- Sketch Planes:
  - Front, Top, Right, face, plane, etc.
- Sketch Tools:
  - Line, Circle, Rectangle, Offset Entities, Convert Entities, etc.
- 2D Sketching:
  - Geometric relations and dimensioning
- Extruded Boss/Base feature
- Extruded Cut feature
- Revolved Boss/Base feature

- Mirror, Chamfer, and Fillet feature

- Circular and Linear Pattern feature

- Plane feature

- Mass Properties and Measure tool

- Show/Hide Axes

- Part Modification

- Apply Material

Segment 1 & 2: Assembly Creation and Modification requires knowledge in the following:

- Download an Assembly zip file

- Save zip file to a local hard drive

- Un-zip components

- Create a new Assembly document

- Set document properties

- Insert first component. Fix first component to the origin of the assembly

- Insert all needed components

- Insert needed mates: Coincident, Concentric, Perpendicular, Parallel, Tangent, Distance, Angle, Advanced Distance, Advanced Angle, and Aligned, Anti-Aligned option

- Apply Mass Properties and Measure tool

- Show/Hide Axes

- Create a new Coordinate System

- Modify the assembly

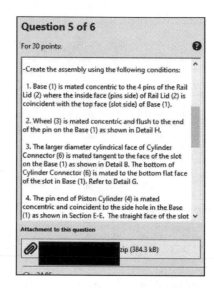

Segment 1

Question 1: Build a basic SOLIDWORKS part. Multiple-choice single answer format. Read the question. Select an answer. Click Next Question. Save all parts.

Your answer should be within 1% of the multiple-choice answer before you move on to the second question.

All answers are in the MMGS unit system. Decimal place 2.

Question 2: Modify question 1. Single answer format. Read the question. Input the answer. Click Next Question.

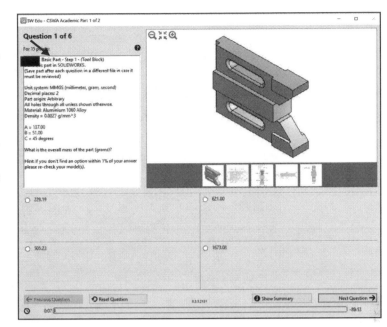

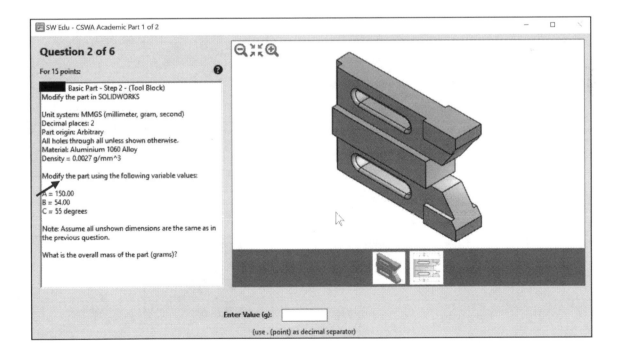

Question 3: Create a new SOLIDWORKS Intermediate part. Multiple-choice single answer format. Read the question. Select an answer. Click Next Question. Your answer should be within 1% of the multiple-choice answer before you move on to the fourth question. Part origin: Arbitrary. All holes through all unless shown otherwise. Decimal place: 2.

Question 4: Modify the SOLIDWORKS Intermediate part. Single answer format. Read the question. Input the answer. Click Next Question.

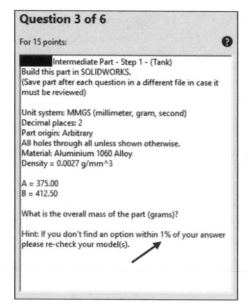

Question 3 of 6

For 15 points: ❷

▇▇▇ Intermediate Part - Step 1 - (Tank)
Build this part in SOLIDWORKS.
(Save part after each question in a different file in case it must be reviewed)

Unit system: MMGS (millimeter, gram, second)
Decimal places: 2
Part origin: Arbitrary
All holes through all unless shown otherwise.
Material: Aluminium 1060 Alloy
Density = 0.0027 g/mm^3

A = 375.00
B = 412.50

What is the overall mass of the part (grams)?

Hint: If you don't find an option within 1% of your answer please re-check your model(s).

Question 4 of 6

For 15 points: ❷

▇▇▇ Intermediate Part - Step 2 - (Tank)
Modify the part in SOLIDWORKS by adding top handle and bottom platform.

Unit system: MMGS (millimeter, gram, second)
Decimal places: 2
Part origin: Arbitrary
All holes through all unless shown otherwise.
Material: Aluminium 1060 Alloy
Density = 0.0027 g/mm^3

1 - Note: Assume all unshown dimensions are the same as in the previous question.

2 - The top platform has 3 cutouts the exact same size and spaced 90 degrees from each other.

What is the overall mass of the part (grams)?

Question 5: Create a SOLIDWORKS assembly. Multiple-choice single answer format. Read the question. Download the zip file. Create an assembly per the instructions. Save the assembly for question 6. Select an answer. Click Next Question. Your answer should be within 1% of the multiple-choice answer before you move on to the final question.

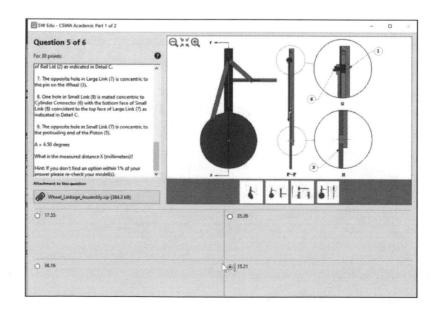

Question 6: Modify the SOLIDWORKS assembly from question 5. Single answer format. Read the question. Input the answer. Click Next Question. All answers are in the MMGS unit system. Decimal place 2. This is the final question in Segment 1.

Segment 2

There are eight questions in Segment 2. Three of the questions are in the Drafting Competencies category. Each question is worth five (5) points.

In the Drafting Competencies category of the exam, you are not required to create or perform an analysis on a part, assembly, or drawing but you are required to have general drafting/drawing knowledge and understanding of various drawing view methods.

The questions are on general drawing views: Projected, Section, Break, Crop, Detail, Alternate Position, etc.

Download all needed model files from the SDC Publication website www.SDCpublications.com/downloads/978-1-63057-567-0.

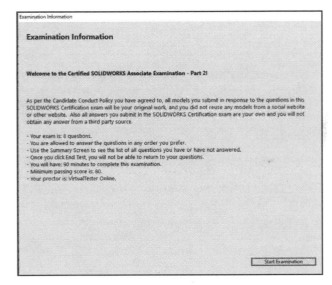

Question 6 of 6

For 30 points:

E30402: Modify this assembly in SOLIDWORKS (Wheel Linkage Assembly)

Unit system: MMGS (millimeter, gram, second)
Decimal places: 2
Assembly origin: Arbitrary

-Using the same assembly created in the previous question modify the following parameters:

A = 16.50 degrees

What is the measured distance X (millimeters)?

Examination Information

Examination Information

Welcome to the Certified SOLIDWORKS Associate Examination - Part 2!

As per the Candidate Conduct Policy you have agreed to, all models you submit in response to the questions in this SOLIDWORKS Certification exam will be your original work, and you did not reuse any models from a social website or other website. Also all answers you submit in the SOLIDWORKS Certification exam are your own and you will not obtain any answer from a third party source.

- Your exam is: 8 questions.
- You are allowed to answer the questions in any order you prefer.
- Use the Summary Screen to see the list of all questions you have or have not answered.
- Once you click End Test, you will not be able to return to your questions.
- You will have: 90 minutes to complete this examination.
- Minimum passing score is: 80.
- Your proctor is: VirtualTester Online.

Start Examination

SOLIDWORKS 2020
SOLIDWORKS 2021
SOLIDWORKS 2022
SOLIDWORKS 2023

Drafting Competencies

Drafting Competencies is one of the five categories on the CSWA exam.

There are three questions (total) on the CSWA exam in the Drafting Competencies category. Each question is worth five (5) points.

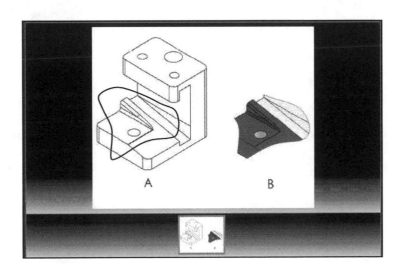

The three questions are in a multiple-choice single answer format. You are allowed to answer the questions in any order you prefer. Use the Summary Screen during the exam to view the list of all questions you have or have not answered.

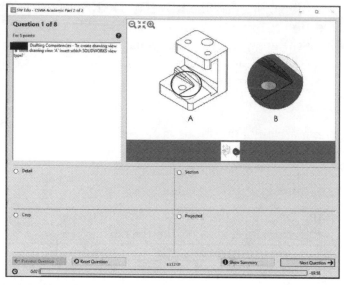

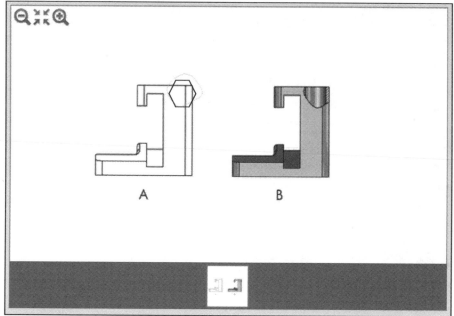

Procedure to Create a Named Drawing view

You need the ability to identify the procedure to create a named drawing view: Standard 3 View, Model View, Projected View, Auxiliary View, Section View, Removed Section, Detail View, Broken-out Section, Break, Crop View and Alternate Position View.

Create a Section view in a drawing by cutting the parent view with a section line. The Section view can be a straight cut section or an offset section defined by a stepped section line. The section line can also include concentric arcs.

Create an Aligned Section view in a drawing through a model, or portion of a model, that is aligned with a selected section line segment. The Aligned Section view is similar to a Section view, but the section line for an aligned section comprises two or more lines connected at an angle.

Create a Crop view, sketch a closed profile such as a circle or spline. The view outside the closed profile disappears as illustrated.

🔆 Crop any drawing view except a Detail view, a view from which a Detail view has been created, or an Exploded view.

Create a Detail view in a drawing to display a portion of a view, usually at an enlarged scale. This detail may be of an orthographic view, a non-planar (isometric) view, a Section view, a Crop view, an Exploded assembly view, or another Detail view.

Section View

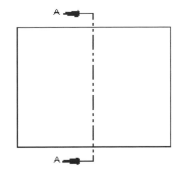
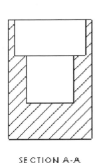

SECTION A-A

Adds a section view, aligned section view, or half section view by cutting the parent view with a section line.

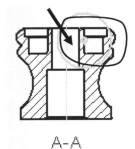
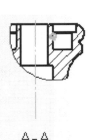

A-A   A-A

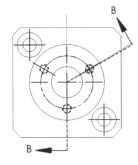
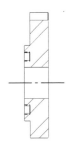

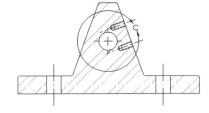
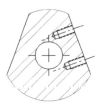

Tutorial: Drawing Named Procedure 2-1

Identify the drawing name view and understand the procedure to create the name view.

1. **View** the illustrated drawing views. The top drawing view is a Break view. The Break view is created by adding a break line to a selected view.

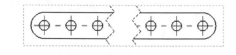

💡 Broken views make it possible to display the drawing view in a larger scale on a smaller size drawing sheet. Reference dimensions and model dimensions associated with the broken area reflect the actual model values.

💡 In views with multiple breaks, the Break line style must be the same.

Tutorial: Drawing Named Procedure 2-2

Identify the drawing name view and understand the procedure to create the name view.

1. **View** the illustrated drawing views. The top drawing view is a Section view of the bottom view. The Section view is created by cutting the parent view with a cutting section line.

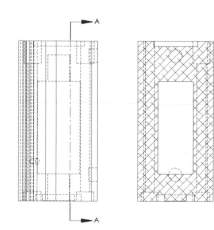

💡 Create a Section view in a drawing by cutting the parent view with a section line. The section view can be a straight cut section or an offset section defined by a stepped section line. The section line can also include Concentric arcs.

Tutorial: Drawing Named Procedure 2-3

Identify the drawing name view and understand the procedure to create the name view.

1. **View** the illustrated drawing views. The view to the right is an Auxilary view of the Front view. Select a reference edge to create an auxiliary view as illustrated.

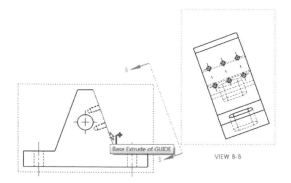

VIEW B-B

💡 An Auxiliary view is similar to a Projected view, but it is unfolded normal to a reference edge in an existing view.

Tutorial: Drawing Named Procedure 2-4

Identify the drawing name view and understand the procedure to create the name view.

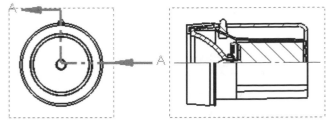

1. **View** the illustrated drawing views. The right drawing view is an Aligned half Section view of the view to the left. The Section view is created by using two lines connected at an angle. Create an Aligned half Section view in a drawing through a model, or portion of a model, that is aligned with a selected section line segment.

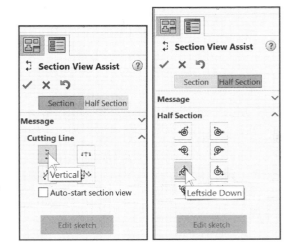

🔆 The Aligned Section view is very similar to a Section View, with the exception that the section line for an aligned half section is comprised of two or more lines connected at an angle.

Tutorial: Drawing Named Procedure 2-5

Identify the drawing name view and understand the procedure to create the name view.

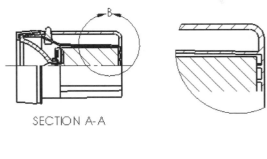

SECTION A-A

1. **View** the illustrated drawing views. The left drawing view is a Detail view of the Section view. The Detail view is created by sketching a circle with the Circle Sketch tool. Click and drag for the location.

The Detail view ⒶA tool provides the ability to add a Detail view to display a portion of a view, usually at an enlarged scale.

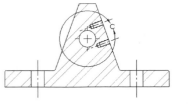

🔆 To create a profile other than a circle, sketch the profile before clicking the Detail view tool. Using a sketch entity tool, create a closed profile around the area to be detailed.

Tutorial: Drawing Named Procedure 2-6

Identify the drawing name view and understand the procedure to create the name view.

1. **View** the illustrated drawing views. The top drawing view is a Broken-out Section view of the bottom drawing view. The Broken-out Section View is part of an existing drawing view, not a separate view. Create the Broken-out Section view with a closed profile, usually by using the Spline Sketch tool. Material is removed to a specified depth to expose inner details.

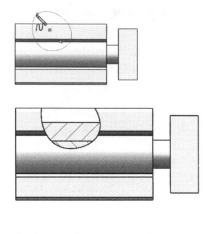

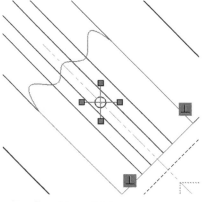

Tutorial: Drawing Named Procedure 2-7

Identify the drawing name view and understand the procedure to create the name view.

1. **View** the illustrated drawing view. The top drawing view is a Crop view. The Crop view is created by a closed sketch profile such as a circle, or spline as illustrated.

The Crop View provides the ability to crop an existing drawing view. You cannot use the Crop tool on a Detail view, a view from which a Detail view has been created, or an Exploded view.

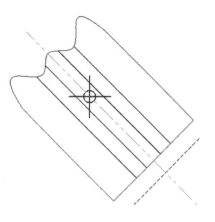

Use the Crop tool to save steps. Example: instead of creating a Section View and then a Detail view, then hiding the unnecessary Section view, use the Crop tool to crop the Section view directly.

In the exam, you are allowed to answer the questions in any order. Use the Summary Screen during the exam to view the list of all questions you have or have not answered.

Tutorial: Drawing Named Procedure 2-8

Identify the drawing name view and understand the procedure to create the name view.

1. **View** the illustrated drawing view. The drawing view is an Alternate Position View. The Alternate Position view tool ⊞ provides the ability to superimpose an existing drawing view precisely on another. The alternate position is displayed with phantom lines.

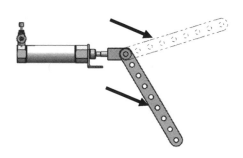

Summary

Drafting Competencies is one of the five categories on the CSWA exam. The CSWA Academic exam is provided either in a single 3-hour segment, or 2 - 90-minute segments.

Segment 2 of the CSWA Academic exam is 90 minutes, minimum passing score is 80 with 8 questions.

Three of the eight questions are in the Drafting Competencies category.

Each question is worth five (5) points. The three questions are in a multiple-choice single answer format.

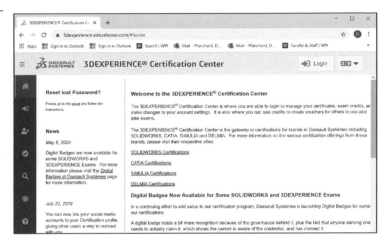

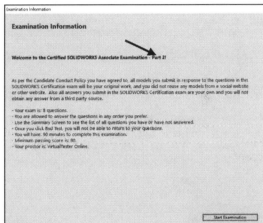

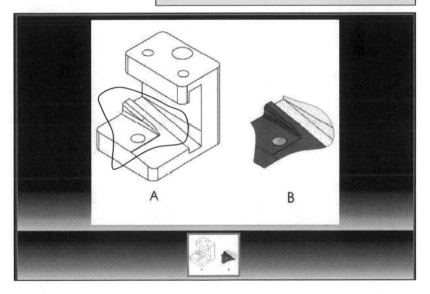

Questions:

1. Identify the illustrated Drawing view.

- A: Projected

- B: Alternative Position

- C: Extended

- D: Aligned Section

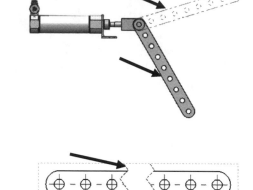

2. Identify the illustrated Drawing view.

- A: Crop

- B: Break

- C: Broken-out Section

- D: Aligned Section

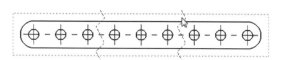

3. Identify the illustrated Drawing view.

- A: Section

- B: Crop

- C: Broken-out Section

- D: Aligned Section

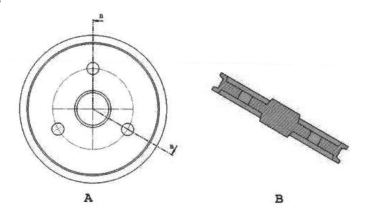

A   B

4. Identify the view procedure. To create the following view, you need to insert a:

- A: Rectangle Sketch tool
- B: Closed Profile: Spline
- C: Open Profile: Circle
- D: None of the above

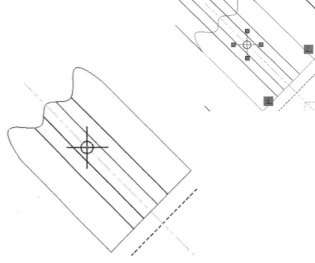

5. Identify the view procedure. To create the following view, you need to insert a:

- A: Open Spline
- B: Closed Spline
- C: 3 Point Arc
- D: None of the above

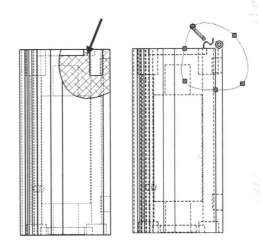

6. Identify the illustrated view type.

- A: Crop
- B: Section
- C: Projected
- D: Detail

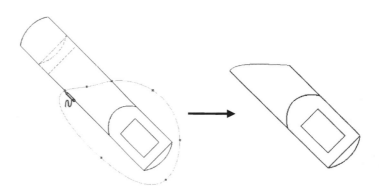

7. To create View B from Drawing View A insert which View type?

- A: Crop
- B: Section
- C: Aligned Section
- D: Projected

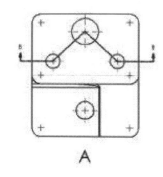

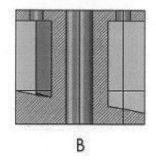

A

B

8. To create View B it is necessary to sketch a closed spline on View A and insert which View type?

- A: Broken out Section
- B: Detail
- C: Section
- D: Projected

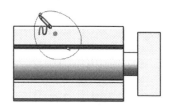

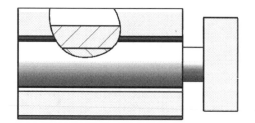

9. To create View B it is necessary to sketch a closed spline on View A and insert which View type?

- A: Horizontal Break
- B: Detail
- C: Section
- D: Broken out Section

A

B

🔆 Screen shots from an older CSWA exam for the Drafting Competencies category. Read each question carefully. Use SOLIDWORKS help if needed.

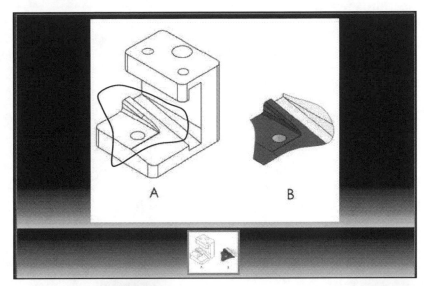

🔆 Zoom in on the part or view if needed.

Alternative Position View:

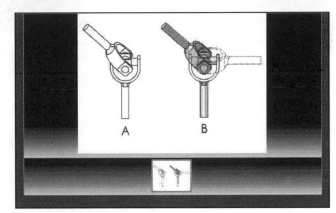

🔆 Screen shots from an older CSWA exam for the *Drafting Competencies* category. Read each question carefully. Use SOLIDWORKS help if needed.

Broken out Section View:

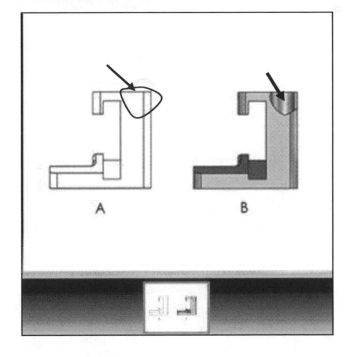

Section View:

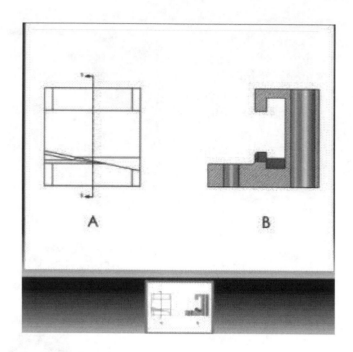

CHAPTER 3 - BASIC PART AND INTERMEDIATE PART CREATION AND MODIFICATION

Chapter Objective

Provide in-depth coverage of the Basic Part Creation and Modification and Intermediate Part Creation and Modification categories on the CSWA exam. Knowledge to create and modify models in these categories from detailed dimensioned illustrations.

Introduction

The main difference between the Basic Part Creation and Modification category and the Intermediate Part Creation and Modification or the Advance Part Creation and Modification category is the complexity of the sketches and the number of dimensions and geometric relations along with an increase in the number of features.

Segment 1 of the CSWA Academic exam has two questions in the Basic Part Creation and Modification category and two questions in the Intermediate Part Creation and Modification category.

Segment 2 of the CSWA Academic exam has no questions in the Basic Part Creation and Modification category and no questions in the Intermediate Part Creation and Modification category.

Note: All answers are in the MMGS unit system. Decimal place 2.

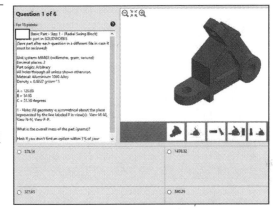

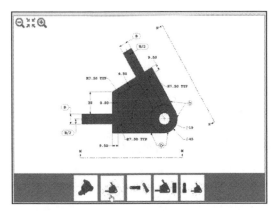

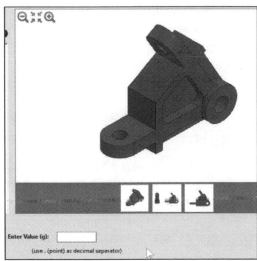

On the completion of the chapter, you will be able to:

- Read and understand an Engineering document used in the CSWA exam.

- Identify the Sketch plane, part Origin location, part dimensions, geometric relations and design intent of the sketch and feature.

- Build a part from a detailed dimensioned illustration using the following SOLIDWORKS tools and features:

  - 2D & 3D sketch tools, Extruded Boss/Base, Extruded Cut, Fillet, Mirror, Revolved Base, Chamfer, Reference geometry, Plane, Axis, Calculate the overall mass of the created part and locate the Center of mass for the created part relative to the Origin.

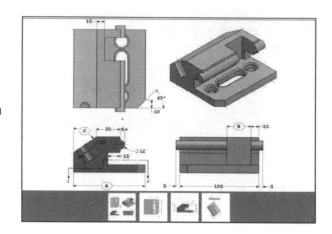

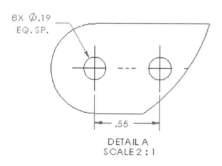

The complexity of the models along with the features progressively increases throughout this chapter to simulate the final types of models that would be provided on the exam.

Engineering Documentation Practices

A 2D drawing view is displayed in the Basic Part Creation and Modification, Intermediate Part Creation and Modification, Advance Part Creation and Modification, and Assembly Creation and Modification categories of the CSWA exam to clarify dimensions and details.

DETAIL A
SCALE 2 : 1

The ability to interpret a 2D drawing view is required.

- Example 1: *8X Ø.19 EQ. SP.* Eight holes with a .19in. diameter are required that are equally (.55in.) spaced.

- Example 2: *R2.50 TYP.* Typical radius of 2.50. The dimension has a two decimal place precision.

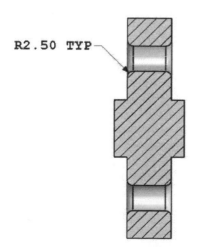

- Example 3: ⊽. The Depth/Deep ⊽ symbol with a 1.50 dimension associated with the hole. The hole Ø.562 has a three decimal place precision.

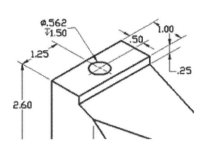

- Example 4: *A+40*. A is provided to you on the CSWA exam. A + 40mm.

💡 N is a Detail view of the M-M Section view.

- Example 5: *ØB*. Diameter of B. B is provided to you on the exam.

- Example 6: ╲. Parallel.

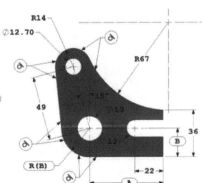

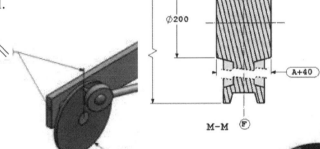

- Example 7: ⊼ The faces are Coincident.

- Example 8: ⌒ Tangent.

💡 Set the correct units in document properties. Set the correct material in the FeatureManager.

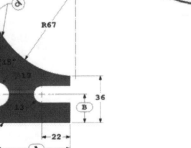

💡 Download all needed model files from the SDC Publication website www.SDCpublications.com/downloads/978-1-63057-567-0.

💡 If your school is an academic certification provider, your instructor can allocate free exam credits for the CSWA (Segment 1 & 2), CSWA-SD, CSWA-S, and CSWA-AM certifications. The instructor will require your .edu email address.

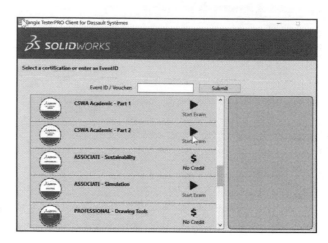

Document Properties

Create a new part.

Set Unit system and precision.

💡 If you don't find your answer (within 1%) in the multiple-choice single answer format section - recheck your model for precision and accuracy.

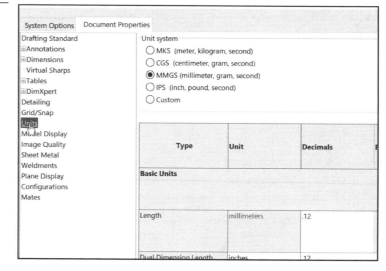

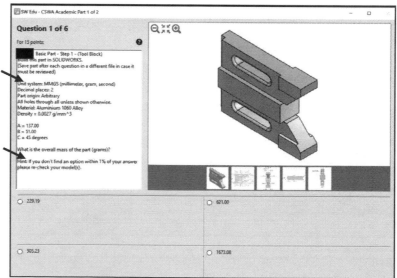

Build a Basic Part from a detailed illustration

Tutorial: Volume/Center of Mass 3-1

Build this model in SOLIDWORKS. Calculate the volume of the part and locate the Center of mass with the provided information.

1. **Create** a new part in SOLIDWORKS.

2. **Build** the illustrated dimensioned model. The model displays all edges on perpendicular planes. Think about the steps to build the model. Insert two features: Extruded Base (Boss-Extrude1) and Extruded Cut (Cut-Extrude1). The part Origin is located in the front left corner of the model. Think about your Base Sketch plane. Keep your Base Sketch simple.

3. **Set** document properties for the model.

4. Create **Sketch1**. Select the Front Plane as the Sketch plane. Sketch1 is the Base sketch. Sketch1 is the profile for the Extruded Base (Boss-Extrude1) feature. Insert the required geometric relations and dimensions.

5. Create the **Extruded Base** feature. Boss-Extrude1 is the Base feature. Blind is the default End Condition in Direction 1. Depth = 2.25in. Identify the extrude direction to maintain the location of the Origin.

6. Create **Sketch2**. Select the Top right face as the Sketch plane for the second feature. Sketch a square. Sketch2 is the profile for the Extruded Cut feature. Insert the required geometric relations and dimensions.

7. Create the **Extruded Cut** feature. Select Through All for End Condition in Direction 1.

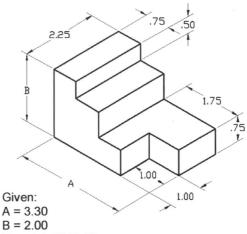

Given:
A = 3.30
B = 2.00
Material: 2014 Alloy
Density = .101 lb/in^3
Units: IPS
Decimal place: 2

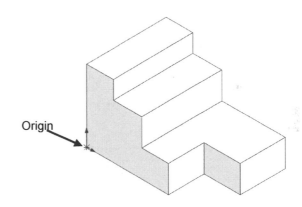

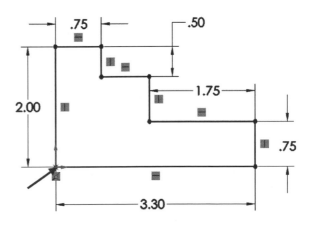

8. **Assign** 2014 Alloy material to the part. Material is required to locate the Center of mass.

9. **Calculate** the volume. The volume = 8.28 cubic inches.

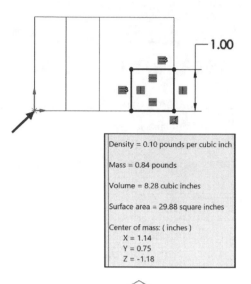

There are numerous ways to build the models in this chapter. A goal is to display different design intents and techniques.

10. **Locate** the Center of mass. The location of the Center of mass is derived from the part Origin.

- X: 1.14 inches

- Y: 0.75 inches

- Z: -1.18 inches

11. **Save** the part. Name it Volume-Center of mass 3-1.

12. **Close** the model.

The principal axes and Center of mass are displayed graphically on the model in the Graphics window.

Tutorial: Volume/Center of Mass 3-2

Build this model in SOLIDWORKS. Calculate the volume of the part and locate the Center of mass with the provided information.

1. **Create** a new part in SOLIDWORKS.

2. **Build** the illustrated dimensioned model. The model displays all edges on perpendicular planes. Think about the steps that are required to build this model. Remember, there are numerous ways to create the models in this chapter.

The CSWA exam is timed. Work efficiently.

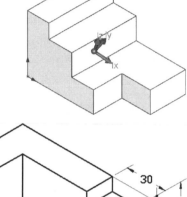

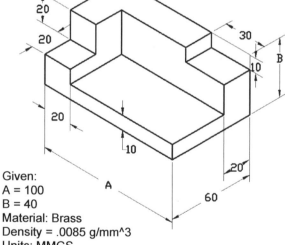

Given:
A = 100
B = 40
Material: Brass
Density = .0085 g/mm^3
Units: MMGS
Decimal place: 2

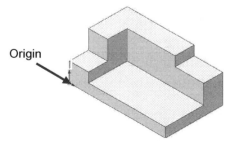

Origin

View the provided Part FeatureManagers. Both FeatureManagers create the same illustrated model. In Option1, there are four sketches and four features (Extruded Base and three Extruded Cuts) that are used to build the model.

In Option2, there are three sketches and three features (Extruded Boss/Base) that are used to build the model. Which FeatureManager is better? In a timed exam, optimize your time and use the least number of features through mirror, pattern, symmetry, etc.

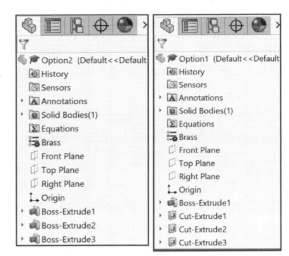

FeatureManager and CommandManager tabs and tree folders will vary depending on system setup and SOLIDWORKS Add-ins.

Create the model using the Option2 Part FeatureManager.

3. **Set** document properties for the model.

4. Create **Sketch1**. Select the Top Plane as the Sketch plane. Sketch a Corner rectangle. Insert the required dimensions.

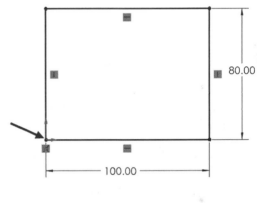

5. Create the **Extruded Base** feature. Boss-Extrude1 is the Base feature. Blind is the default End Condition in Direction 1. Depth = 10mm.

6. Create **Sketch2**. Select the back face of Boss-Extrude1.

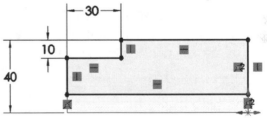

7. Select **Normal To** view. Sketch2 is the profile for the second Extruded Boss/Base feature. Insert the required geometric relations and dimensions as illustrated.

8. Create the second Extruded Boss/Base feature (**Boss-Extrude2**). Blind is the default End Condition in Direction 1. Depth = 20mm. Note the direction of the extrude towards the front of the model.

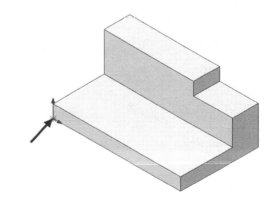

9. Create **Sketch3**. Select the left face of Boss-Extrude1 as the Sketch plane. Sketch3 is the profile for the third Extrude feature. Insert the required geometric relations and dimensions.

10. Create the third Extruded Boss/Base feature (**Boss-Extrude3**). Blind is the default End Condition in Direction 1. Depth = 20mm.

11. **Assign** Brass material to the part.

12. **Calculate** the volume of the model. The volume = 130,000.00 cubic millimeters.

13. **Locate** the Center of mass. The location of the Center of mass is derived from the part Origin.

- X: 43.36 millimeters

- Y: 15.00 millimeters

- Z: -37.69 millimeters

14. **Save** the part. Name it Volume-Center of mass 3-2.

15. **Calculate** the volume of the model using the IPS unit system. The volume = 7.93 cubic inches.

16. **Locate** the Center of mass using the IPS unit system. The location of the Center of mass is derived from the part Origin.

- X: 1.71 inches

- Y: 0.59 inches

- Z: -1.48 inches

17. **Save** the part. Name it Volume-Center of mass 3-2-IPS.

18. **Close** the model.

There are numerous ways to create the models in this chapter. A goal is to display different design intents and techniques.

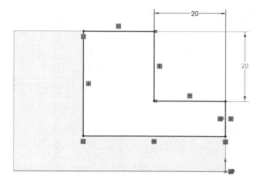

Density = 0.01 grams per cubic millimeter

Mass = 1105.00 grams

Volume = 130000.00 cubic millimeters

Surface area = 23400.00 square millimeters

Center of mass: (millimeters)
  X = 43.46
  Y = 15.00
  Z = -37.69

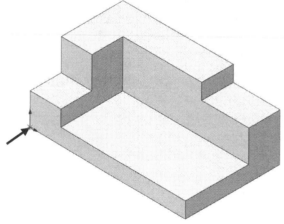

Density = 0.31 pounds per cubic inch

Mass = 2.44 pounds

Volume = 7.93 cubic inches

Surface area = 36.27 square inches

Center of mass: (inches)
  X = 1.71
  Y = 0.59
  Z = -1.48

Tutorial: Mass-Volume 3-3

Build this model in SOLIDWORKS. Calculate the overall mass of the illustrated model with the provided information.

1. **Create** a new part in SOLIDWORKS.

2. **Build** the illustrated model. The model displays all edges on perpendicular planes. Think about the steps required to build the model. Apply the Mirror Sketch tool to the Base sketch. Insert an Extruded Base (Boss-Extrude1) and Extruded-Cut (Cut-Extrude1) feature.

3. **Set** document properties for the model.

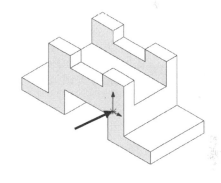

Given:
A = 50, B = 50, C = 120
Material: 6061 Alloy
Density = .0027 g/mm^3
Units: MMGS
Decimal place: 2

🔅 To activate the Mirror Sketch tool, click **Tools**, **Sketch Tools**, **Mirror** from the Main menu. The Mirror PropertyManager is displayed.

4. Create **Sketch1**. Select the Front Plane as the Sketch plane. Apply the Mirror Sketch tool. Select the construction geometry to mirror about as illustrated. Select the Entities to mirror. Insert the required geometric relations and dimensions.

Construction geometry is ignored when the sketch is used to create a feature. Construction geometry uses the same line style as centerlines.

🔅 When you create a new part or assembly, the three default Planes (Front, Right and Top) are aligned with specific views. The Plane you select for the Base sketch determines the orientation of the part.

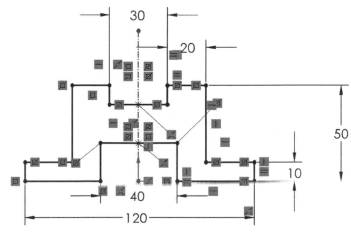

5. Create the **Boss-Extrude1** feature. Boss-Extrude1 is the Base feature. Apply the Mid Plane End Condition in Direction 1 for symmetry. Depth = 50mm.

6. Create **Sketch2**. Select the right face for the Sketch plane. Sketch2 is the profile for the Extruded Cut feature. Insert the required geometric relations and dimensions. Apply construction geometry.

7. Create the **Extruded Cut** feature. Through All is the selected End Condition in Direction 1.

8. **Assign** 6061 Alloy material to the part.

9. **Calculate** the overall mass. The overall mass = 302.40 grams.

10. **Save** the part. Name it Mass-Volume 3-3.

11. **Close** the model.

Tutorial: Mass-Volume 3-4

Build this model in SOLIDWORKS. Calculate the volume of the part and locate the Center of mass with the provided information.

1. **Create** a new part in SOLIDWORKS.

2. **Build** the illustrated model. The model displays all edges on perpendicular planes.

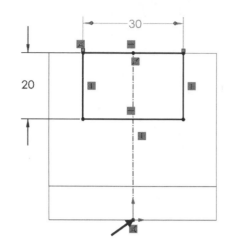

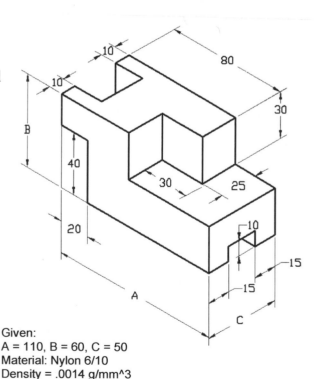

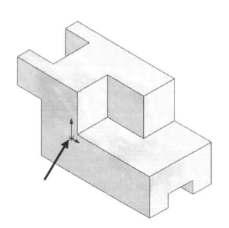

Mass = 302.40 grams

Volume = 112000.00 cubic millimeters

Surface area = 26200.00 square millimeters

Center of mass: (millimeters)
  X = 0.00
  Y = 19.20
  Z = 0.00

Given:
A = 110, B = 60, C = 50
Material: Nylon 6/10
Density = .0014 g/mm^3
Units: MMGS
Decimal Place: 2

View the provided Part FeatureManagers. Both FeatureManagers create the same model. In Option4, there are three sketches and three features that are used to build the model.

In Option3, there are four sketches and four features that are used to build the model. Which FeatureManager is better? In a timed exam, optimize your design time and use the least number of features. Use the Option4 FeatureManager in this tutorial. As an exercise, build the model using the Option3 FeatureManager.

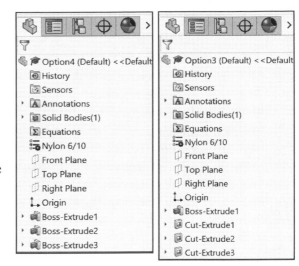

3. **Set** document properties for the model.

4. Create **Sketch1**. Select the Right Plane as the Sketch plane. Sketch1 is the Base sketch. Apply the Mirror Entities Sketch tool. Select the construction geometry to mirror about as illustrated. Select the Entities to mirror. Insert the required geometric relations and dimensions.

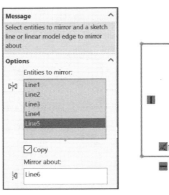

5. Create the **Boss-Extrude1** feature. Boss-Extrude1 is the Base feature. Blind is the default End Condition in Direction 1. Depth = (A - 20mm) = 90mm. Note the direction of the extrude feature.

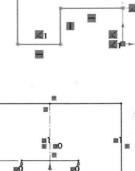

6. Create **Sketch2**. Select the Top face of Boss-Extrude1 for the Sketch plane. Sketch2 is the profile for the second Extruded Boss/Base feature (Boss-Extrude2). Insert the required geometric relations and dimensions.

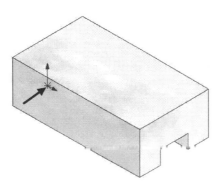

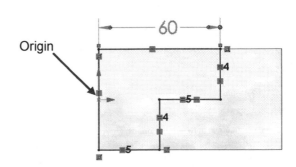

Origin

7. Create the **Boss-Extrude2** feature. Blind is the default End Condition in Direction 1. Depth = 30mm.

8. Create **Sketch3**. Select the left face of Boss-Extrude1 for the Sketch plane. Apply symmetry. Insert the required geometric relations and dimensions. Use construction reference geometry.

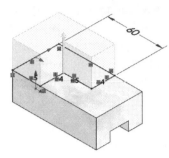

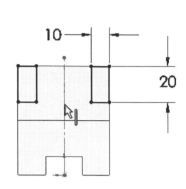

The 20mm dimension for Sketch3 was calculated by: (B - 40mm) = 20mm.

9. Create the **Boss-Extrude3** feature. Blind is the default End Condition in Direction 1. Depth = 20mm. Note the direction of Extrude3.

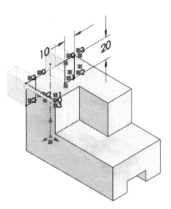

10. **Assign** Nylon 6/10 material to the part.

11. **Calculate** the volume. The volume = 192,500.00 cubic millimeters.

12. **Locate** the Center of mass. The location of the Center of mass is derived from the part Origin.

- X: 35.70 millimeters

- Y: 27.91 millimeters

- Z: -1.46 millimeters

Mass = 269.50 grams

Volume = 192500.00 cubic millimeters

Surface area = 27800.00 square millimeters

Center of mass: (millimeters)
X = 35.70
Y = 27.91
Z = -1.46

13. **Save** the part. Name it Mass-Volume 3-4.

14. **Close** the model.

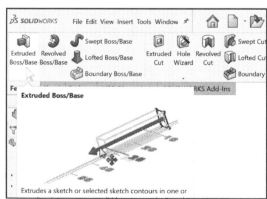

In the previous section, the models that you created displayed all edges on Perpendicular planes and used the Extruded Base, Extruded Boss, or the Extruded Cut feature from the Features toolbar.

In the next section, build models where all edges are not located on Perpendicular planes.

First, let's review a simple 2D Sketch for an Extruded Cut feature.

Tutorial: Simple Cut 3-1

1. **Create** a new part in SOLIDWORKS.

2. **Build** the illustrated model. Start with a 60mm x 60mm x 100mm block. System unit: MMGS. Decimal place: 2. Note the location of the part Origin.

3. Create **Sketch1**. Select the Front Plane as the Sketch plane. Sketch a square as illustrated. Insert the required relations and dimension. The part Origin is located in the bottom left corner of the sketch.

4. Create the **Extruded Base (Boss-Extrude1)** feature. Apply the Mid Plane End Condition in Direction 1. Depth = 100mm.

5. Create **Sketch2**. Select the front face as the Sketch plane. Apply the Line Sketch tool. Sketch a diagonal line. Select the front right vertical midpoint as illustrated. The sketch is fully defined.

6. Create the **Extruded Cut (Cut-Extrude1)** feature. Through All for End Condition in Direction 1 and Direction 2 is selected by default.

7. Apply **2014 Alloy**.

8. **Save** the part. Name it Simple-Cut 3-1.

9. **View** the FeatureManager.

10. **Close** the model.

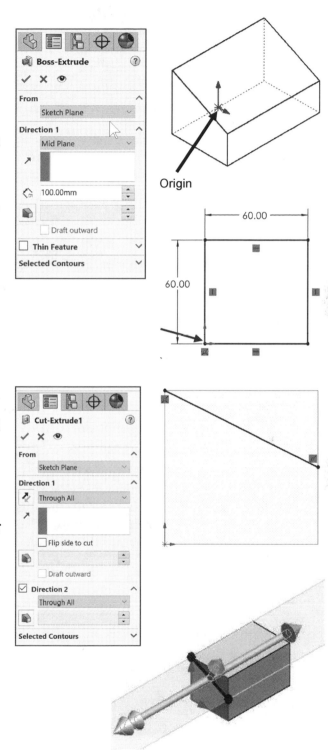

Tutorial: Mass-Volume 3-5

Build this model in SOLIDWORKS.
Calculate the overall mass of the part
and locate the Center of mass with
the provided information.

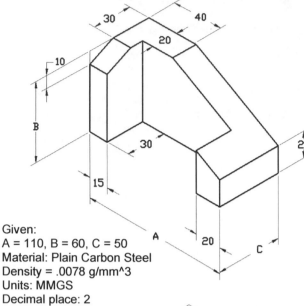

1. **Create** a new part in
   SOLIDWORKS.

2. **Build** the illustrated model. All
   edges of the model are not located
   on Perpendicular planes. Insert an
   Extruded Base (Boss-Extrude1)
   feature and three Extruded Cut
   features to build the model.

Given:
A = 110, B = 60, C = 50
Material: Plain Carbon Steel
Density = .0078 g/mm^3
Units: MMGS
Decimal place: 2

3. **Set** document properties for the
   model. Decimal place: 2.

4. Create **Sketch1**. Select the Front
   Plane as the Sketch plane. Sketch
   a rectangle. Insert the required geometric relations
   and dimensions. The part Origin is located in the
   left bottom corner of the model.

5. Create the **Extruded Base** feature. Blind is the
   default End Condition in Direction 1.
   Depth = 50mm. Note the direction of Extrude1.

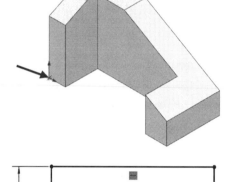

6. Create **Sketch2**. Select the top face of Boss-
   Extrude1 for the Sketch plane. Sketch2 is the
   profile for the first Extruded Cut feature. Insert
   the required relations and dimensions.

7. Create the **Extruded Cut** feature. Select
   Through All for End Condition in Direction 1.

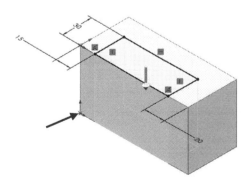

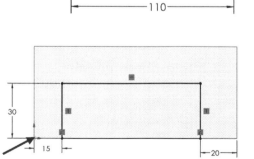

8. Create **Sketch3**. Select the back face of Boss-Extrude1 as the Sketch plane. Sketch a diagonal line. Insert the required geometric relations and dimensions. The Sketch is fully defined.

9. Create the second **Extruded Cut** feature. Through All for End Condition in Direction 1 and Direction 2 is selected by default. Note the direction of the extrude feature.

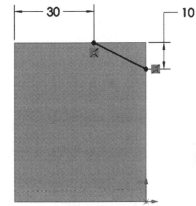

10. Create **Sketch4**. Select the left face of Boss-Extrude1 as the Sketch plane. Sketch a diagonal line. Insert the required geometric relations and dimensions. The sketch is fully defined.

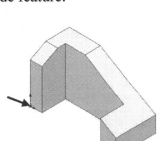

11. Create the third **Extruded Cut** feature. Through All for End Condition in Direction 1 and Direction 2 is selected by default.

12. **Assign** Plain Carbon Steel material to the part.

13. **Calculate** the overall mass. The overall mass = 1130.44 grams.

14. **Locate** the Center of mass. The location of the Center of mass is derived from the part Origin.

- X: 45.24 millimeters
- Y: 24.70 millimeters
- Z: -33.03 millimeters

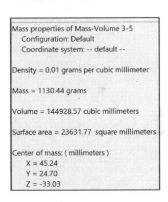

Mass properties of Mass-Volume 3-5
Configuration: Default
Coordinate system: -- default --

Density = 0.01 grams per cubic millimeter

Mass = 1130.44 grams

Volume = 144928.57 cubic millimeters

Surface area = 23631.77 square millimeters

Center of mass: (millimeters)
X = 45.24
Y = 24.70
Z = -33.03

In this category an exam question could read: Build this model. Locate the Center of mass with respect to the part Origin.

- A: X = 45.24 millimeters, Y = 24.70 millimeters, Z = -33.03 millimeters
- B: X = 54.24 millimeters, Y = 42.70 millimeters, Z = 33.03 millimeters
- C: X = 49.24 millimeters, Y = -37.70 millimeters, Z = 38.03 millimeters
- D: X = 44.44 millimeters, Y = -24.70 millimeters, Z = -39.03 millimeters

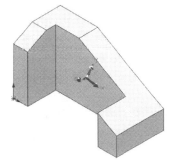

The correct answer is A.

💡 The principal axes and Center of mass are displayed graphically on the model in the Graphics window.

15. **Save** the part. Name it Mass-Volume 3-5.

16. **Close** the model.

Tutorial: Mass-Volume 3-6

Build this model in SOLIDWORKS. Calculate the overall mass of the part and locate the Center of mass with the provided information.

1. **Create** a new part in SOLIDWORKS.

2. **Build** the illustrated model. All edges of the model are not located on Perpendicular planes. Think about the steps required to build the model. Insert two features: Extruded Base (Boss-Extrude1) and Extruded Cut (Cut-Extrude1).

3. **Set** document properties for the model.

4. Create **Sketch1**. Select the Right Plane as the Sketch plane. Apply construction geometry. Insert the required geometric relations and dimensions.

5. Create the **Extruded Base** feature. Boss-Extrude1 is the Base feature. Apply symmetry. Select Mid Plane as the End Condition in Direction 1. Depth = 3.00in.

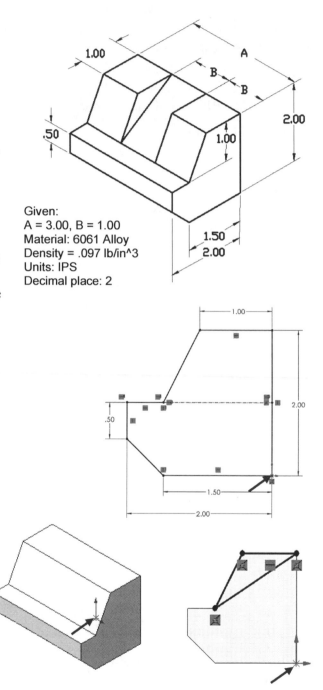

Given:
A = 3.00, B = 1.00
Material: 6061 Alloy
Density = .097 lb/in^3
Units: IPS
Decimal place: 2

6. Create **Sketch2**. Select the Right Plane as the Sketch plane. Select the Line Sketch tool. Insert the required geometric relations. Sketch2 is the profile for the Extruded Cut feature.

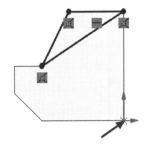

7. Create the **Extruded Cut (Cut-Extrude1)** feature. Apply symmetry. Select Mid Plane as the End Condition in Direction 1. Depth = 1.00in.

8. **Assign** 6061 Alloy material to the part.

9. **Calculate** overall mass. The overall mass = 0.87 pounds.

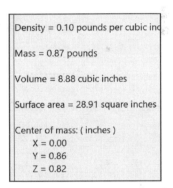

10. **Locate** the Center of mass. The location of the Center of mass is derived from the part Origin.

- X: 0.00 inches

- Y: 0.86 inches

- Z: 0.82 inches

In this category an exam question could read: Build this model. Locate the Center of mass with respect to the part Origin.

- A: X = 0.10 inches, Y = -0.86 inches, Z = -0.82 inches

- B: X = 0.00 inches, Y = 0.86 inches, Z = 0.82 inches

- C: X = 0.15 inches, Y = -0.96 inches, Z = -0.02 inches

- D: X = 1.00 inches, Y = -0.89 inches, Z = -1.82 inches

The correct answer is B.

11. **Save** the part. Name it Mass-Volume 3-6.

12. **Close** the model.

As an exercise, modify the Mass-Volume 3-6 part using the MMGS unit system. Assign Nickel as the material. Calculate the overall mass. The overall mass of the part = 1236.20 grams. Save the part. Name it Mass-Volume 3-6-MMGS.

Density = 0.10 pounds per cubic inch

Mass = 0.87 pounds

Volume = 8.88 cubic inches

Surface area = 28.91 square inches

Center of mass: (inches)
   X = 0.00
   Y = 0.86
   Z = 0.82

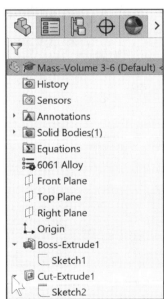

Mass-Volume 3-6 (Default)
- History
- Sensors
- Annotations
- Solid Bodies(1)
- Equations
- 6061 Alloy
- Front Plane
- Top Plane
- Right Plane
- Origin
- Boss-Extrude1
  - Sketch1
- Cut-Extrude1
  - Sketch2

Tutorial: Mass-Volume 3-7

Build this model in SOLIDWORKS. Calculate the overall mass of the part and locate the Center of mass with the provided information.

1. **Create** a new part in SOLIDWORKS.

2. **Build** the illustrated model. All edges of the model are not located on Perpendicular planes. Think about the steps required to build the model. Insert two features: Extruded Base (Boss-Extrude1) and Extruded Cut (Cut-Extrude1).

3. **Set** document properties for the model.

4. Create **Sketch1**. Select the Right Plane as the Sketch plane. Apply the Line Sketch tool. Insert the required geometric relations and dimension. The location of the Origin is in the left lower corner of the sketch.

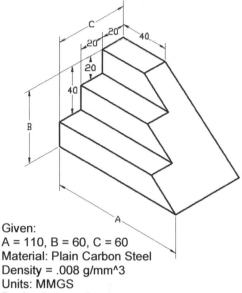

Given:
A = 110, B = 60, C = 60
Material: Plain Carbon Steel
Density = .008 g/mm^3
Units: MMGS
Decimal place: 2

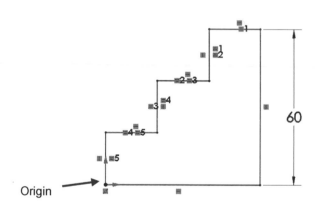

Origin

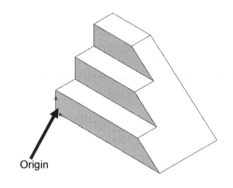

Origin

5. Create the **Extruded Base** feature. Boss-Extrude1 is the Base feature. Blind is the default End Condition in Direction 1. Depth = 110mm.

6. Create **Sketch2**. Select the Front Plane as the Sketch plane. Sketch a diagonal line. Complete the sketch. Sketch2 is the profile for the Extruded Cut feature.

7. Create the **Extruded Cut** feature. Through All for End Condition in Direction 1 and Direction 2 is selected by default.

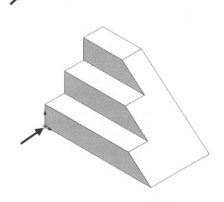

8. **Assign** Plain Carbon Steel material to the part.

9. **Calculate** overall mass. The mass = 1549.60 grams.

10. **Locate** the Center of mass. The location of the Center of mass is derived from the part Origin.

- X: 43.49 millimeters

- Y: 19.73 millimeters

- Z: -35.10 millimeters

11. **Save** the part. Name it Mass-Volume 3-7.

12. **Close** the model.

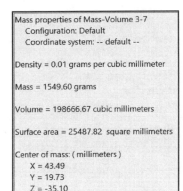

Mass properties of Mass-Volume 3-7
Configuration: Default
Coordinate system: -- default --

Density = 0.01 grams per cubic millimeter

Mass = 1549.60 grams

Volume = 198666.67 cubic millimeters

Surface area = 25487.82 square millimeters

Center of mass: (millimeters)
X = 43.49
Y = 19.73
Z = -35.10

In this category an exam question could read: Build this model. Locate the Center of mass with respect to the part Origin.

- A: X = -43.99 millimeters, Y = 29.73 millimeters, Z = -38.10 millimeters

- B: X = -44.49 millimeters, Y = -19.73 millimeters, Z = 35.10 millimeters

- C: X = 43.49 millimeters, Y = 19.73 millimeters, Z = -35.10 millimeters

- D: X = -1.00 millimeters, Y = 49.73 millimeters, Z = -35.10 millimeters

The correct answer is C.

As an exercise, locate the Center of mass using the IPS unit system, and re-assign copper material. Re-calculate the Center of mass location, with respect to the part Origin. Save the part. Name it Mass-Volume 3-7-IPS.

- X: 1.71 inches

- Y: 0.78 inches

- Z: -1.38 inches

When you create a new part or assembly, the three default Planes (Front, Right and Top) are aligned with specific views. The Plane you select for the Base sketch determines the orientation of the part.

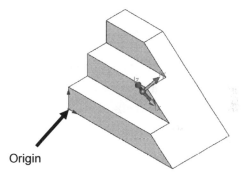

Origin

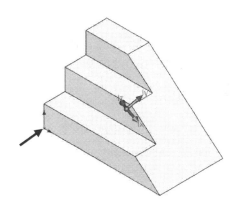

Mass properties of Mass-Volume 3-7-IPS
Configuration: Default
Coordinate system: -- default --

Density = 0.32 pounds per cubic inch

Mass = 3.90 pounds

Volume = 12.12 cubic inches

Surface area = 39.51 square inches

Center of mass: (inches)
X = 1.71
Y = 0.78
Z = -1.38

2D vs. 3D Sketching

Up to this point, the models that you created in this chapter started with a 2D Sketch. Sketches are the foundation for creating features. SOLIDWORKS provides the ability to create either 2D or 3D Sketches. A 2D Sketch is limited to a flat 2D Sketch plane. A 3D sketch can include 3D elements.

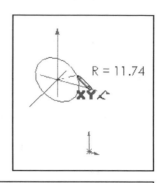

As you create a 3D Sketch, the entities in the sketch exist in 3D space. They are not related to a specific Sketch plane as they are in a 2D Sketch.

You may need to apply a 3D Sketch in the CSWA exam. Below is an example of a 3D Sketch to create a Cut-Extrude feature.

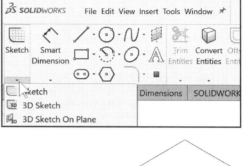

 The complexity of the models increases throughout this chapter to simulate the types of models that are provided on the CSWA exam.

Tutorial 3DSketch 3-1

1. **Create** a new part in SOLIDWORKS.

2. **Build** the illustrated model. Insert two features: Extruded Base and Extruded Cut. Apply the 3D Sketch tool to create the Extruded Cut feature. System units: MMGS. Decimal place: 2.

3. **Set** document properties for the model.

4. Create **Sketch1**. Select the Front Plane as the Sketch plane. Sketch a rectangle. The part Origin is located in the bottom left corner of the sketch. Insert the illustrated geometric relations and dimensions.

5. Create the **Extruded Base (Boss-Extrude1)** feature. Apply symmetry. Select the Mid Plane End Condition in Direction 1. Depth = 100.00mm.

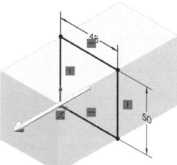

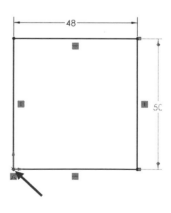

💡 Click **3D Sketch** from the Sketch toolbar. Select the proper Sketch tool.

6. Create **3DSketch1**. Apply the Line Sketch tool. 3DSketch1 is a four point sketch as illustrated. 3DSketch1 is the profile for Extruded Cut feature.

7. Create the **Extruded Cut (Cut-Extrude1)** feature. Select the front right vertical edge as illustrated to remove the material. Edge<1> is displayed in the Direction of Extrusion box.

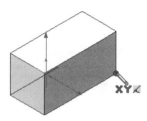

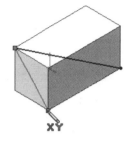
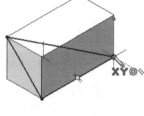

8. **Save** the part. Name it 3DSketch 3-1.

9. **Close** the model.

💡 You can either select the front right vertical edge or the Top face to remove the required material in this tutorial.

Use any of the following tools to create 3D Sketches: Lines, Circles, Rectangles, Arcs, Splines, and Points.

Most relations that are available in 2D Sketching are available in 3D sketching. The exceptions are:

- *Symmetry*

- *Patterns*

- *Offset*

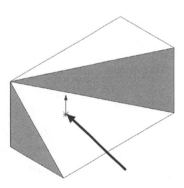
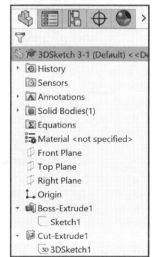
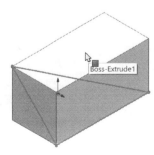

Tutorial: Mass-Volume 3-8

Build this model in SOLIDWORKS. Calculate the volume of the part and locate the Center of mass with the provided information.

1. **Create** a new part in SOLIDWORKS.

2. **Build** the illustrated model. All edges of the model - are not located on Perpendicular planes. Insert two features: Extruded Base (Boss-Extrude1) and Extruded Cut (Cut-Extrude1). Apply a closed four point 3D sketch as the profile for the Extruded Cut feature. The part Origin is located in the lower left front corner of the model.

3. **Set** document properties for the model.

4. Create **Sketch1**. Select the Right Plane as the Sketch plane. Sketch a square. Insert the required geometric relations and dimension.

5. Create the **Extruded Base** feature. Boss-Extrude1 is the Base feature. Blind is the default End Condition in Direction 1. Depth = 4.00in.

6. Create **3DSketch1**. Apply the Line Sketch tool. Create a closed five point 3D sketch as illustrated. 3DSketch1 is the profile for the Extruded Cut feature. Insert the required dimensions.

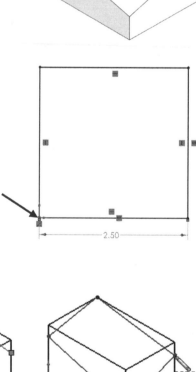

Given:
A = .75, B = 2.50
Material: 2014 Alloy
Density = .10 lb/in^3
Units: IPS
Decimal place: 2

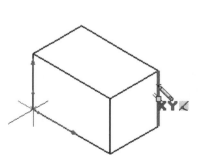
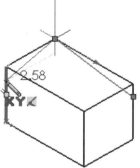
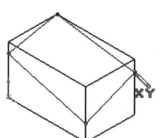

7. Create the **Extruded Cut** feature. Select the front right vertical edge as illustrated. Select Through All for End Condition in Direction 1. Note the direction of the extrude feature.

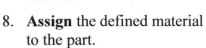
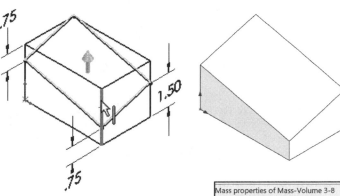
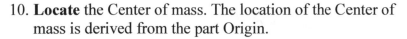

8. **Assign** the defined material to the part.

9. **Calculate** the volume. The volume = 16.25 cubic inches.

Mass properties of Mass-Volume 3-8
  Configuration: Default
  Coordinate system: -- default --

Density = 0.10 pounds per cubic inch

Mass = 1.64 pounds

Volume = 16.25 cubic inches

Surface area = 41.86 square inches

Center of mass: (inches)
  X = 1.79
  Y = 0.85
  Z = -1.35

10. **Locate** the Center of mass. The location of the Center of mass is derived from the part Origin.

- X = 1.79 inches

- Y = 0.85 inches

- Z = -1.35 inches

In this category an exam question could read: Build this model. What is the volume of the part?

- A: 18.88 cubic inches

- B: 19.55 cubic inches

- C: 17.99 cubic inches

- D: 16.25 cubic inches

The correct answer is D.

View the triad location of the Center of mass for the part.

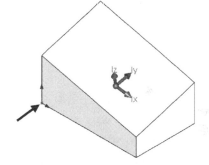

11. **Save** the part. Name it Mass-Volume 3-8.

12. **Close** the model.

As an exercise, calculate the overall mass of the part using the MMGS unit system, and re-assign Nickel as the material. The overall mass of the part = 2263.46 grams. Save the part and name it Mass-Volume 3-8-MMGS.

Density = 0.01 grams per cubic millimeter

Mass = 2263.46 grams

Volume = 266289.79 cubic millimeters

Surface area = 27000.69 square millimeters

Center of mass: (millimeters)
  X = 45.59
  Y = 21.66
  Z = -34.19

🔆 Illustrations will vary depending on your version and system setup.

Chapter 2
Chapter 2 Final Solutions
Chapter 3
Chapter 3 Final Solutions
Chapter 4
Chapter 4 Final Solutions
Chapter 5
Chapter 5 Final Solutions
CSWASampleExam

Tutorial: Mass-Volume 3-9

Build this model in SOLIDWORKS. Calculate the overall mass of the part and locate the Center of mass with the provided information.

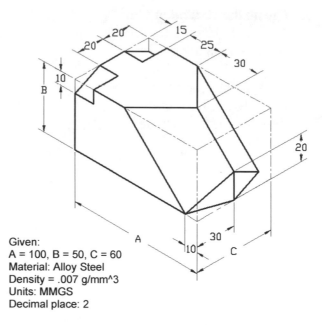

1. **Create** a new part in SOLIDWORKS.

2. **Build** the illustrated model. Insert five sketches and five features to build the model: Extruded Base, three Extruded Cut features and a Mirror feature.

Given:
A = 100, B = 50, C = 60
Material: Alloy Steel
Density = .007 g/mm^3
Units: MMGS
Decimal place: 2

☼ There are numerous ways to build the models in this chapter. A goal is to display different design intents and techniques.

3. **Set** document properties for the model.

4. Create **Sketch1**. Select the Front Plane as the Sketch plane. Sketch a rectangle. Insert the required relations and dimensions. The part Origin is located in the lower left corner of the sketch.

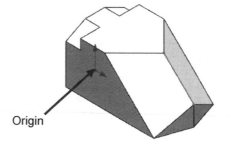

Origin

5. Create the **Extruded Base (Boss-Extrude1)** feature. Apply symmetry. Select the Mid Plane End Condition for Direction 1. Depth = 60mm.

6. Create **Sketch2**. Select the left face of Boss-Extrude1 as the Sketch plane. Insert the required geometric relations and dimensions.

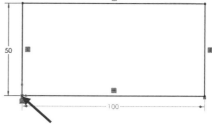

7. Create the first **Extruded Cut** feature. Blind is the default End Condition in Direction 1.

8. Depth = 15mm. Note the direction of the extrude feature.

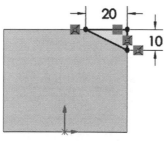

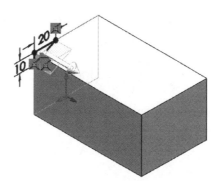

9. Create **Sketch3**. Select the bottom face of Boss-Extrude1 for the Sketch plane. Insert the required geometric relations and dimension.

10. Create the second **Extruded Cut** feature. Blind is the default End Condition in Direction 1. Depth = 20mm.

11. Create **Sketch4**. Select Front Plane as the Sketch plane. Sketch a diagonal line. Sketch4 is the direction of extrusion for the third Extruded Cut feature. Insert the required dimension.

12. Create **Sketch5**. Select the top face of Boss-Extrude1 as the Sketch plane. Sketch5 is the sketch profile for the third Extruded Cut feature. Apply construction geometry. Insert the required geometric relations and dimensions.

13. Create the third **Extruded Cut** feature. Select Through All for End Condition in Direction 1.

14. Select **Sketch4** in the Graphics window for Direction of Extrusion. Line1@Sketch4 is displayed in the Cut-Extrude PropertyManager.

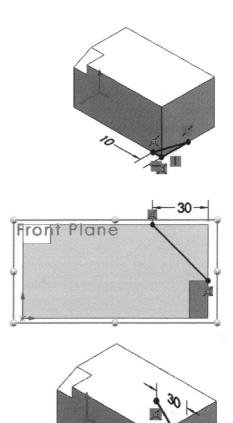

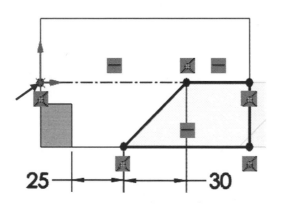

15. Create the **Mirror** feature. Mirror the three Extruded Cut features about the Front Plane. Use the fly-out FeatureManager.

16. **Assign** Alloy Steel material to the part.

17. **Calculate** the overall mass. The overall mass = 1794.10 grams.

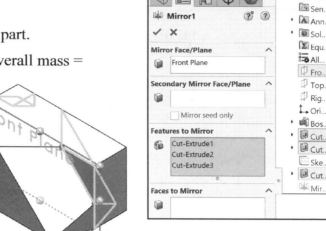

18. **Locate** the Center of mass. The location of the Center of mass is derived from the part Origin.

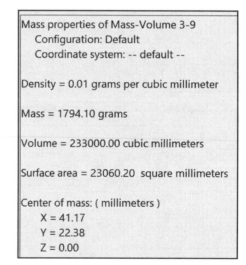

- X = 41.17 millimeters

- Y = 22.38 millimeters

- Z = 0.00 millimeters

View the triad location of the Center of mass for the part.

19. **Save** the part. Name it Mass-Volume 3-9.

20. **Close** the model.

Set document precision from the Document Properties dialog box or from the Dimension PropertyManager. You can also address: Callout value, Tolerance type, and Dimension Text symbols in the Dimension PropertyManager.

You are allowed to answer the questions in any order you prefer. Use the Summary Screen during the CSWA exam to view the list of all questions you have or have not answered.

Mass properties of Mass-Volume 3-9
   Configuration: Default
   Coordinate system: -- default --

Density = 0.01 grams per cubic millimeter

Mass = 1794.10 grams

Volume = 233000.00 cubic millimeters

Surface area = 23060.20 square millimeters

Center of mass: (millimeters)
   X = 41.17
   Y = 22.38
   Z = 0.00

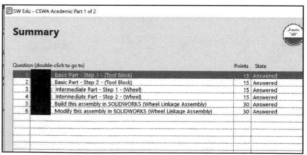

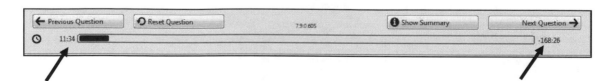

Callout Value

A Callout value is a value that you select in a SOLIDWORKS document. Click a dimension in the Graphics window; the selected dimension is displayed in blue and the Dimension PropertyManager is displayed.

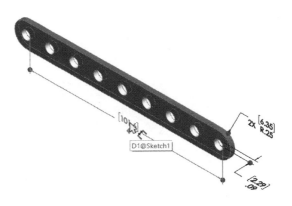

💡 A Callout value is available for dimensions with multiple values in the callout.

Tolerance Type

A Tolerance type is selected from the available drop-down list in the Dimension PropertyManager. The list is dynamic. A few examples of Tolerance type display are listed below:

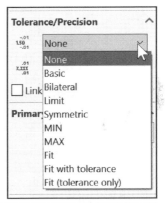

| **Basic** | Adds a box around the dimension text. In geometric dimensioning and tolerancing, **Basic** indicates the theoretically exact value of the dimension. |

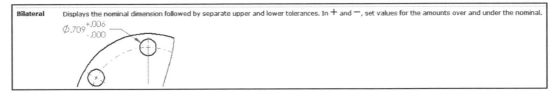

| **Bilateral** | Displays the nominal dimension followed by separate upper and lower tolerances. In + and –, set values for the amounts over and under the nominal. |

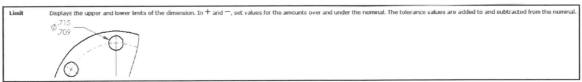

| **Limit** | Displays the upper and lower limits of the dimension. In + and –, set values for the amounts over and under the nominal. The tolerance values are added to and subtracted from the nominal. |

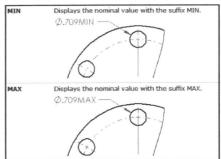

Tutorial: Dimension text 3-1

1. **View** the illustrated model.

2. **Review** the Tolerance, Precision, and Dimension Text.

   a. 2X Ø.190 - Two holes with a diameter of .190. Precision is set to three decimal places.

   b. 2X R.250 - Two corners with a radius of .250. Precision is set to three decimal places.

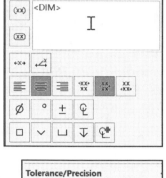

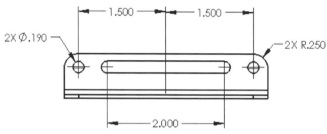

Tutorial: Dimension text 3-2

1. **View** the illustrated model.

2. **Review** the Tolerance, Precision, and Dimension text.

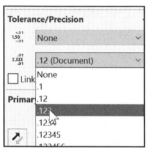

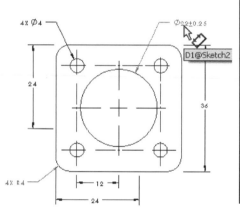

   a. $\varnothing$ 22±0.25 - The primary diameter value of the hole = 22.0mm. Tolerance type: Symmetric. Maximum Variation 0.25mm. Tolerance / Precision is set to two decimal places.

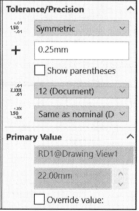

For a Chamfer feature, a second Tolerance/Precision is available.

   b. $36\,^{0}_{-0.5}$ - The primary height = 36mm. Tolerance type: Bilateral. Maximum Variation is 0.0mm. Minimum Variation = -0.5mm. Precision is set to two decimal places. Tolerance is set to one decimal places.

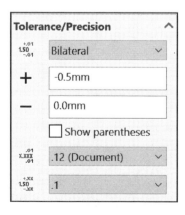

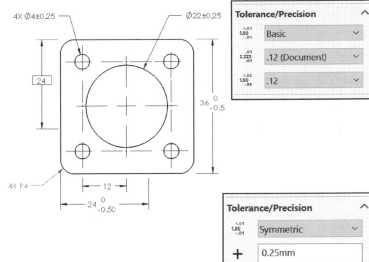

Trailing zeroes are removed according to the ANSI Y.14.5 standard.

c. ┃24┃ - The primary value = 24mm. Tolerance type: Basic. Tolerance / Precision is set to two decimal places.

d. 4X ⌀ 4±0.25 - Four holes with a primary diameter value = 4mm. Tolerance type: Symmetric. Maximum Variation = 0.25mm. Precision/Tolerance is set to two decimal places.

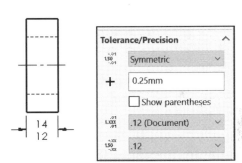

Tutorial: Dimension text 3-3

1. **View** the illustrated model.

2. **Review** the Tolerance, and Precision.

a. ¹⁴₁₂ - The primary value = 12mm. Tolerance type: Limit. Maximum Variation = 2mm. Minimum Variation = 0mm. Tolerance/Precision is set to none.

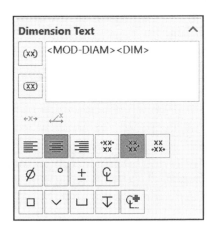

Dimension Text Symbols

Dimension Text symbols are displayed in the Dimension PropertyManager. The Dimension Text box provides eight commonly used symbols and a more button to access the Symbol Library. The eight displayed symbols in the Dimension Text box from left to right are: Diameter, Degree, Plus/Minus, Centerline, Square, Countersink, Counterbore and Depth/Deep.

Review each symbol in the Dimension Text box and in the Symbol library. You are required to understand the meaning of various symbols in a SOLIDWORKS document.

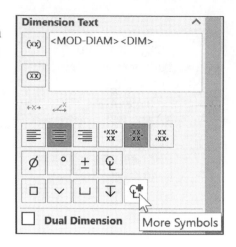

Tutorial: Dimension Text symbols 3-1

1. **View** the illustrated model.

2. **Review** the Dimension Text and document symbols.

   a. 2X Ø3.5 THRU ⌴ Ø6.5⯆3.5 - Two holes with a primary diameter value = 3.5mm, Cbore Ø6.5 with a depth of 3.5.

Tutorial: Dimension Text symbols 3-2

1. **View** the illustrated model.

2. **Review** the Dimension Text and document symbols.

   a. 2X ⌴ Ø5.5 ⯆ 8.8 - Two Cbores with a primary diameter value = 5.5mm with a depth of 8.8.

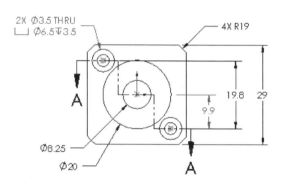

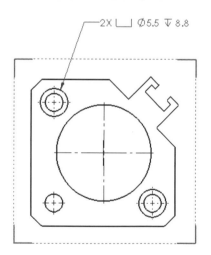

Build Additional Basic Parts

Tutorial: Mass-Volume 3-10

Build this model in SOLIDWORKS. Calculate the overall mass of the part and locate the Center of mass with the provided information.

1. **Create** a new part in SOLIDWORKS.

2. **Build** the illustrated model. Note the Depth/Deep ⊤ symbol with a 1.50 dimension associated with the hole. The hole Ø.562 has a three decimal place precision. Insert three features: Extruded Base (Boss-Extrude1) and two Extruded Cuts. Insert a 3D sketch for the first Extruded Cut feature.

💡 There are numerous ways to build the models in this chapter. A goal is to display different design intents and techniques.

3. **Set** document properties for the model.

4. Create **Sketch1**. Select the Front Plane as the Sketch plane. The part Origin is located in the lower left corner of the sketch. Insert the required geometric relations and dimensions.

5. Create the **Extruded Base (Boss-Extrude1)** feature. Apply symmetry. Select the Mid Plane End Condition in Direction 1. Depth = 2.50in.

6. Create **3DSketch1**. Apply the Line Sketch tool. Create a closed four point 3D sketch. 3DSketch1 is the profile for the first Extruded Cut feature. Insert the required dimensions.

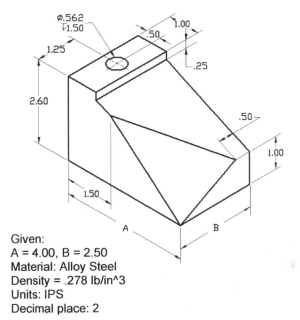

Given:
A = 4.00, B = 2.50
Material: Alloy Steel
Density = .278 lb/in^3
Units: IPS
Decimal place: 2

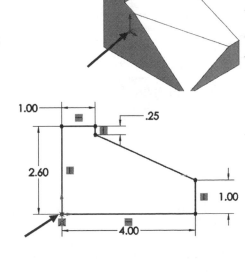

7. Create the first **Extruded Cut** feature. Blind is the default End Conditions. Select the top face as illustrated to be removed. Note the direction of the extrude feature.

8. Create **Sketch2**. Select the top flat face of Boss-Extrude1. Sketch a circle. Insert the required geometric relations and dimensions. The hole diameter Ø.562 has a three decimal place precision.

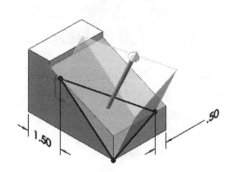

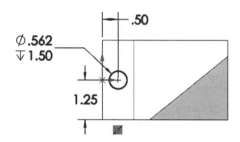

9. Create the second **Extruded Cut** feature. Blind is the default End Condition. Depth = 1.50in. Note: For the exam, you do not need to insert the Depth/Deep ⊽ symbol or note.

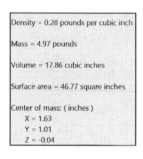

10. **Assign** Alloy Steel material to the part.

11. **Calculate** the overall mass. The overall mass = 4.97 pounds.

12. **Locate** the Center of mass. The location of the Center of mass is derived from the part Origin.

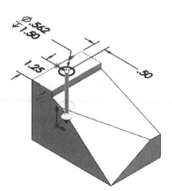

Density = 0.28 pounds per cubic inch

Mass = 4.97 pounds

Volume = 17.86 cubic inches

Surface area = 46.77 square inches

Center of mass: (inches)
  X = 1.63
  Y = 1.01
  Z = -0.04

- X: 1.63 inches

- Y: 1.01 inches

- Z: -0.04 inches

View the triad location of the Center of mass for the part.

13. **Save** the part. Name it Mass-Volume 3-10.

14. **Close** the model.

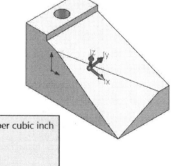

As an exercise, calculate the overall mass of the part using 6061 Alloy.

Modify the "A" dimension from 4.00 to 4.50.

Modify the hole dimension from Ø.562 to Ø.575.

The overall mass of the part = 1.93 pounds.

Save the part. Name it Mass-Volume 3-10A.

Density = 0.10 pounds per cubic inch

Mass = 1.93 pounds

Volume = 19.77 cubic inches

Surface area = 50.66 square inches

Center of mass: (inches)
  X = 1.83
  Y = 0.99
  Z = -0.04

Tutorial: Mass-Volume 3-11

Build this model in SOLIDWORKS. Calculate the overall mass of the part and locate the Center of mass with the provided information.

1. **Create** a new part in SOLIDWORKS.

2. **Build** the illustrated model. Think about the required steps to build this part. Insert four features: Extruded Base, two Extruded Cuts, and a Fillet.

There are numerous ways to build the models in this chapter. A goal is to display different design intents and techniques.

3. **Set** document properties for the model.

4. Create **Sketch1**. Select the Right Plane as the Sketch plane. The part Origin is located in the lower left corner of the sketch. Insert the required geometric relations and dimensions.

5. Create the **Extruded Base (Boss-Extrude1)** feature. Apply symmetry. Select the Mid Plane End Condition for Direction 1. Depth = 4.00in.

6. Create **Sketch2**. Select the top flat face of Boss-Extrude1 as the Sketch plane. Sketch a circle. The center of the circle is located at the part Origin. Insert the required dimension.

7. Create the first **Extruded Cut** feature. Select Through All for End Condition in Direction 1.

8. Create **Sketch3**. Select the front vertical face of Extrude1 as the Sketch plane. Sketch a circle. Insert the required geometric relations and dimensions.

9. Create the second **Extruded Cut** feature. Select Through All for End Condition in Direction 1.

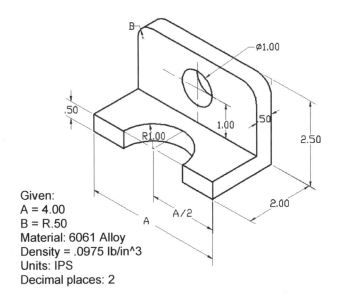

Given:
A = 4.00
B = R.50
Material: 6061 Alloy
Density = .0975 lb/in^3
Units: IPS
Decimal places: 2

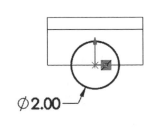

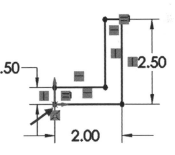

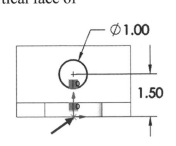

10. Create the **Fillet** feature. Constant radius is selected by default. Fillet the top two edges as illustrated. Radius = .50in.

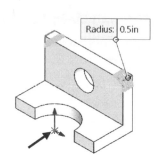

Radius: 0.5in

☀ A Fillet feature removes material. Selecting the correct radius value is important to obtain the correct mass and volume answer in the exam.

11. **Assign** the defined material to the part.

12. **Calculate** the overall mass. The overall mass = 0.66 pounds.

13. **Locate** the Center of mass. The location of the Center of mass is derived from the part Origin.

- X: 0.00 inches
- Y: 0.90 inches
- Z: -1.46 inches

In this category an exam question could read: Build this model. Locate the Center of mass relative to the part Origin.

- A: X = -2.63 inches, Y = 4.01 inches, Z = -0.04 inches
- B: X = 4.00 inches, Y = 1.90 inches, Z = -1.64 inches
- C: X = 0.00 inches, Y = 0.90 inches, Z = -1.46 inches
- D: X = -1.69 inches, Y = 1.00 inches, Z = 0.10 inches

The correct answer is C. Note: Tangent edges and Origin is displayed for educational purposes.

14. **Save** the part. Name it Mass-Volume 3-11.

15. **Close** the model.

As an exercise, calculate the overall mass of the part using the MMGS unit system, and assign 2014 Alloy material to the part. Decimal place: 2.

The overall mass of the part = 310.17 grams. Save the part. Name it Mass-Volume 3-11-MMGS.

Mass = 0.66 pounds

Volume = 6.76 cubic inches

Surface area = 36.99 square inches

Center of mass: (inches)
  X = 0.00
  Y = 0.90
  Z = -1.46

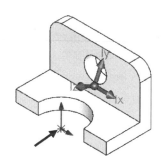

Mass = 310.17 grams

Volume = 110774.26 cubic millimeters

Surface area = 23865.83 square millimeters

Center of mass: (millimeters)
  X = 0.00
  Y = 22.83
  Z = -37.11

Tutorial: Mass-Volume 3-12

Build this model in SOLIDWORKS. Calculate the overall mass of the part and locate the Center of mass with the provided information.

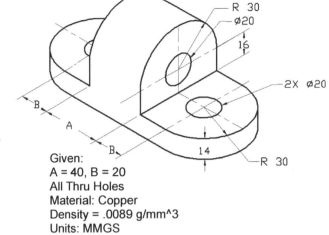

1. **Create** a new part in SOLIDWORKS.

2. **Build** the illustrated model. Insert two features: Extruded Base (Boss-Extrude1) and Extruded Boss (Boss-Extrude2).

Given:
A = 40, B = 20
All Thru Holes
Material: Copper
Density = .0089 g/mm^3
Units: MMGS
Decimal place: 2

3. **Set** document properties for the model.

4. Create **Sketch1**. Select the Top Plane as the Sketch plane. Apply the Centerpoint Straight Slot tool. Insert the required geometric relations and dimensions.

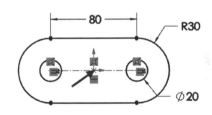

5. Create the **Extruded Base (Boss-Extrude1)** feature. Blind is the default End Condition. Depth = 14mm.

6. Create **Sketch2**. Select the Right Plane as the Sketch plane. Insert the required geometric relations and dimensions.

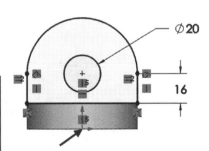

7. Create the **Extruded Boss (Boss-Extrude2)** feature. Apply symmetry. Select the Mid Plane End Condition. Depth = 40mm.

8. **Assign** the defined material to the part.

9. **Calculate** the overall mass. The overall mass = 1605.29 grams.

10. **Locate** the Center of mass. The location of the Center of mass is derived from the part Origin.

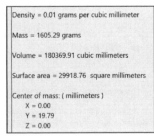

Density = 0.01 grams per cubic millimeter

Mass = 1605.29 grams

Volume = 180369.91 cubic millimeters

Surface area = 29918.76 square millimeters

Center of mass: (millimeters)
    X = 0.00
    Y = 19.79
    Z = 0.00

- X: 0.00 millimeters

- Y: 19.79 millimeters

- Z: 0.00 millimeters

11. **Save** the part. Name it Mass-Volume 3-12.

12. **Close** the model.

☀ There are numerous ways to build the models in this chapter. Optimize your time. The CSWA is a timed exam.

Tutorial: Mass-Volume 3-13

Build this model in SOLIDWORKS. Calculate the volume of the part and locate the Center of mass with the provided information.

1. **Create** a new part in SOLIDWORKS.

2. **Build** the illustrated model. Insert three features: Extruded Base (Boss-Extrude1), Extruded Boss (Boss-Extrude2) and Mirror. Three holes are displayed with an Ø1.00in.

3. **Set** document properties for the model.

4. Create **Sketch1**. Select the Top Plane as the Sketch plane. Apply the Tangent Arc and Line Sketch tool. Insert the required geometric relations and dimensions. Note the location of the Origin.

5. Create the **Extruded Base (Boss-Extrude1)** feature. Blind is the default End Condition. Depth = .50in.

6. Create **Sketch2**. Select the front vertical face of Extrude1 as the Sketch plane. Insert the required geometric relations and dimensions.

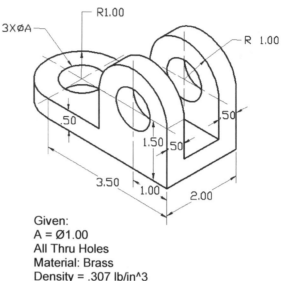

Given:
A = Ø1.00
All Thru Holes
Material: Brass
Density = .307 lb/in^3
Units: IPS
Decimal place: 2

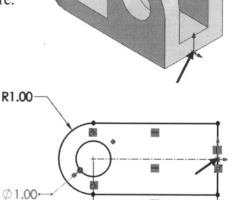

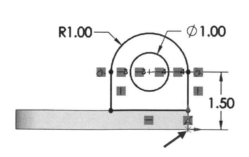

7. Create the **Extruded Boss (Boss-Extrude2)** feature. Blind is the default End Condition in Direction 1. Depth = .50in. Note the direction of the extrude.

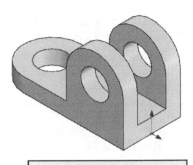

8. Create the **Mirror** feature. Apply Symmetry. Mirror Boss-Extrude2 about the Front Plane.

9. **Assign** the defined material to the part.

10. **Calculate** the volume. The volume = 6.68 cubic inches.

11. **Locate** the Center of mass. The location of the Center of mass is derived from the part Origin.

- X: -1.59 inches

- Y: 0.72 inches

- Z: 0.00 inches

Density = 0.31 pounds per cubic inch

Mass = 2.05 pounds

Volume = 6.68 cubic inches

Surface area = 40.64 square inches

Center of mass: (inches)
  X = -1.59
  Y = 0.72
  Z = 0.00

In this category an exam question could read: Build this model. What is the volume of the model?

- A = 6.19 cubic inches

- B = 7.79 cubic inches

- C = 7.87 cubic inches

- D = 6.68 cubic inches

The correct answer is D.

View the triad location of the Center of mass for the part.

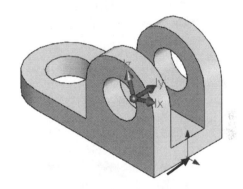

12. **Save** the part. Name it Mass-Volume 3-13.

13. **Close** the model.

As an exercise, calculate the overall mass of the part using the IPS unit system, and assign Copper material to the part. Modify the hole diameters from 1.00in to 1.12in. Decimal place: 2.

The overall mass of the part = 2.05 pounds. Save the part. Name it Mass-Volume 3-13A.

The book is designed to expose the new user to many tools, techniques and procedures. It may not always use the most direct tool or process.

Mass properties of Mass-Volume 3-13A
  Configuration: Default
  Coordinate system: -- default --

Density = 0.32 pounds per cubic inch

Mass = 2.05 pounds

Volume = 6.38 cubic inches

Surface area = 40.00 square inches

Center of mass: (inches)
  X = -1.58
  Y = 0.70
  Z = 0.00

Tutorial: Mass-Volume 3-14

Build this model in SOLIDWORKS. Calculate the overall mass of the part and locate the Center of mass with the provided information.

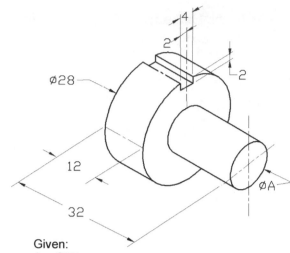

1. **Create** a new part in SOLIDWORKS.

2. **Build** the illustrated model. Insert a Revolved Base feature and Extruded Cut feature to build this part.

3. **Set** document properties for the model.

Given:
A = Ø12
Material: Cast Alloy Steel
Density = .0073 g/mm^3
Units: MMGS
Decimal place: 2

4. Create **Sketch1**. Select the Front Plane as the Sketch plane. Apply the Centerline Sketch tool for the Revolve1 feature. Insert the required geometric relations and dimensions. Sketch1 is the profile for the Revolve1 feature.

5. Create the **Revolved Base** feature. The default angle is 360deg. Select the centerline for the Axis of Revolution.

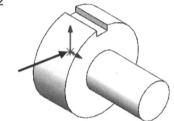

A Revolve feature adds or removes material by revolving one or more profiles around a centerline.

6. Create **Sketch2**. Select the right large circular face of Revolve1 as the Sketch plane. Apply reference construction geometry. Use the Convert Entities and Trim Sketch tools. Insert the required geometric relations and dimensions.

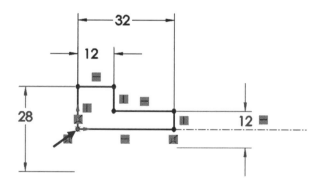

You could also use the 3 Point Arc Sketch tool instead of the Convert Entities and Trim Sketch tools to create Sketch2.

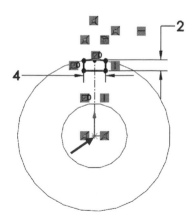

7. Create the **Extruded Cut** feature. Select Through All for End Condition in Direction 1.

8. **Assign** the defined material to the part.

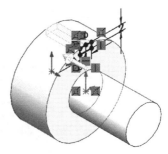

9. **Calculate** the overall mass. The overall mass = 69.77 grams.

10. **Locate** the Center of mass. The location of the Center of mass is derived from the part Origin.

Density = 0.01 grams per cubic millimeter

Mass = 69.77 grams

Volume = 9557.27 cubic millimeters

Surface area = 3069.83 square millimeters

Center of mass: (millimeters)
   X = 9.79
   Y = -0.13
   Z = 0.00

- X = 9.79 millimeters

- Y = -0.13 millimeters

- Z = 0.00 millimeters

11. **Save** the part. Name it Mass-Volume 3-14.

12. **Close** the model.

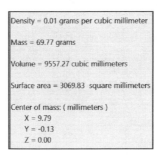

Tutorial: Mass-Volume 3-15

Build this model in SOLIDWORKS. Calculate the overall mass of the part and locate the Center of mass with the provided information.

1. **Create** a new part in SOLIDWORKS.

2. **Build** the illustrated model. Insert two features: Extruded Base (Boss-Extrude1) and Revolved Boss.

3. **Set** document properties for the model.

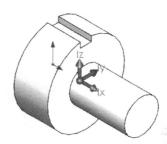

Given:
A = 60, B = 40, C = 8
Material: Cast Alloy Steel
Density = .0073 g/mm^3
Units: MMGS
Decimal place: 2

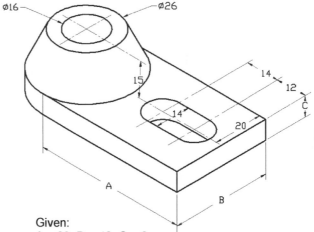

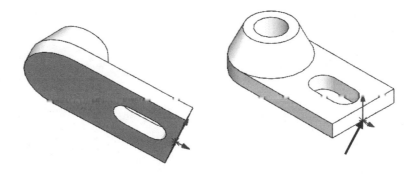

4. Create **Sketch1**. Select the Top Plane as the Sketch plane. Apply construction geometry. Apply the Tangent Arc and Line Sketch tool. Insert the required geometric relations and dimensions.

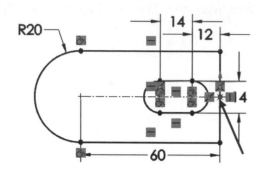

5. Create the **Extruded Base** feature. Blind is the default End Condition. Depth = 8mm.

6. Create **Sketch2**. Select the Front Plane as the Sketch plane. Apply construction geometry for the Revolved Boss feature. Insert the required geometric relations and dimension.

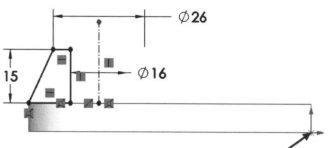

7. Create the **Revolved Boss** feature. The default angle is 360deg. Select the centerline for Axis of Revolution.

8. **Assign** the defined material to the part.

9. **Calculate** the overall mass. The overall mass = 229.46 grams.

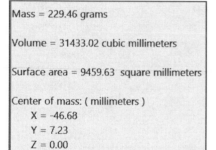

Mass = 229.46 grams

Volume = 31433.02 cubic millimeters

Surface area = 9459.63 square millimeters

Center of mass: (millimeters)
    X = -46.68
    Y = 7.23
    Z = 0.00

10. **Locate** the Center of mass. The location of the Center of mass is derived from the part Origin.

- X = -46.68 millimeters

- Y = 7.23 millimeters

- Z = 0.00 millimeters

In this category an exam question could read: Build this model. What is the overall mass of the part?

- A: 229.46 grams

- B: 249.50 grams

- C: 240.33 grams

- D: 120.34 grams

The correct answer is A.

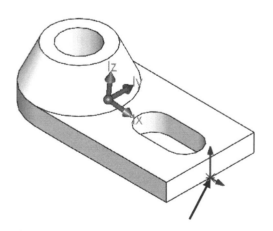

11. **Save** the part. Name it Mass-Volume 3-15.

12. **Close** the model.

Tutorial: Mass-Volume 3-16

Build this model in SOLIDWORKS. Calculate the overall mass of the part and locate the Center of mass with the provided information.

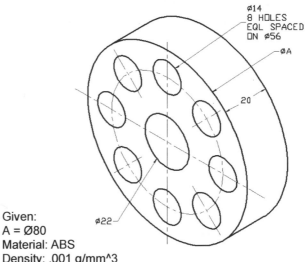

1. **Create** a new part in SOLIDWORKS.

2. **Build** the illustrated model. Insert three features: Extruded Base, Extruded Cut and Circular Pattern. There are eight holes Ø14mm equally spaces on an Ø56mm bolt circle. The center hole = Ø22mm.

Given:
A = Ø80
Material: ABS
Density: .001 g/mm^3
Units: MMGS
Decimal place: 2

3. **Set** document properties for the model.

4. Create **Sketch1**. Select the Front Plane as the Sketch plane. Sketch two circles. The part Origin is located in the center of the sketch. Insert the required geometric relations and dimensions.

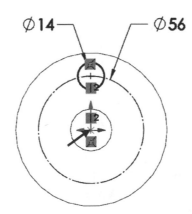

5. Create the **Extruded Base (Boss-Extrude1)** feature. Blind is the default End Condition. Depth = 20mm.

6. Create **Sketch2**. Select the front face as the Sketch plane. Apply construction geometry to locate the seed feature for the Circular Pattern. Insert the required geometric relations and dimensions.

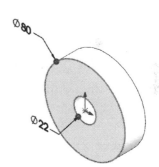

Apply construction reference geometry to assist in creating the sketch entities and geometry that are incorporated into the part. Construction reference geometry is ignored when the sketch is used to create a feature. Construction reference geometry uses the same line style as centerlines.

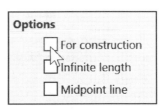

7. Create the **Extruded Cut** feature. Select Through All for End Condition in Direction 1.

8. Create the **Circular Pattern** feature. Create a Circular Pattern of the Cut-Extrude1 feature. Use the View, Temporary Axes command to select the Pattern Axis for the CirPattern1 feature. Instances = 8. Equal spacing is selected by default.

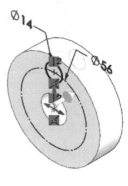

Apply a circular pattern feature to create multiple instances of one or more features that you can space uniformly about an axis.

9. **Assign** the defined material to the part.

10. **Calculate** the overall mass. The overall mass = 69.66 grams.

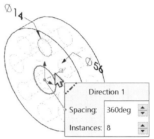

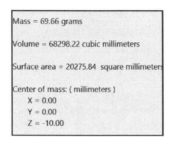

11. **Locate** the Center of mass. The location of the Center of mass is derived from the part Origin.

- X = 0.00 millimeters

- Y = 0.00 millimeters

- Z = -10.00 millimeters

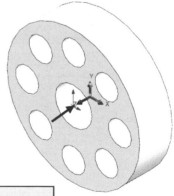

12. **Save** the part. Name it Mass-Volume 3-16.

13. **Close** the model.

As an exercise, select the Top Plane for the Sketch plane to create Sketch1. Recalculate the location of the Center of mass with respect to the part Origin: X = 0.00 millimeters, Y = -10.00 millimeters and Z = 0.00 millimeters. Save the part. Name it Mass-Volume 3-16-TopPlane.

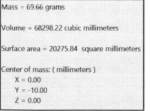

In the next section, the models represent the feature types and complexity that you would see on the exam.

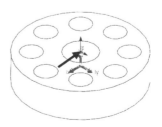

Tutorial: Basic/Intermediate Part 3-1

Build this model in SOLIDWORKS. Calculate the overall mass of the part and locate the Center of mass with the provided information.

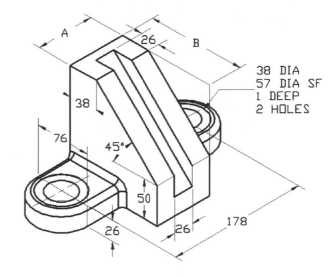

1. **Create** a new part in SOLIDWORKS.

2. **Build** the illustrated model. Think about the various features that create the model. Insert seven features to build this model: Extruded Base, Extruded Cut, Extruded Boss, Fillet, second Extruded Cut, Mirror and a second Fillet. Apply symmetry. Create the left half of the model first, and then apply the Mirror feature.

Given:
A = 76, B = 127
Material: 2014 Alloy
Density: .0028 g/mm^3
Units: MMGS
ALL ROUNDS EQUAL 6MM
Decimal place: 2

💡 There are numerous ways to build the models in this chapter. The goal is to display different design intents and techniques.

3. **Set** document properties for the model. Decimal place: 2.

4. Create **Sketch1**. Select the Front Plane as the Sketch plane. Create the main body of the part. The part Origin is located in the bottom left corner of the sketch. Insert the required geometric relations and dimensions.

5. Create the **Extruded Base** feature. Boss-Extrude1 is the Base feature. Select Mid Plane for End Condition in Direction 1. Depth = 76mm.

6. Create **Sketch2**. Select the top flat face of Extrude1 as the Sketch plane. Create the top cut on the Base feature. Apply construction geometry. Insert the required geometric relations and dimensions.

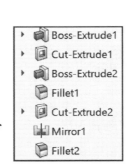

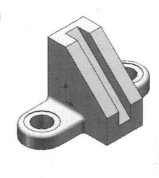

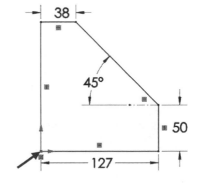

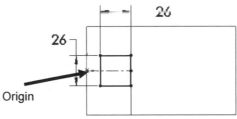

Origin

7. Create the first **Extruded Cut** feature. Select Through All for End Condition in Direction 1. Select the illustrated angled edge for the Direction of Extrusion.

8. Create **Sketch3**. Select the bottom face of Boss-Extrude1 as the Sketch plane. Sketch the first tab with a single hole as illustrated. Insert the required geometric relations and dimensions.

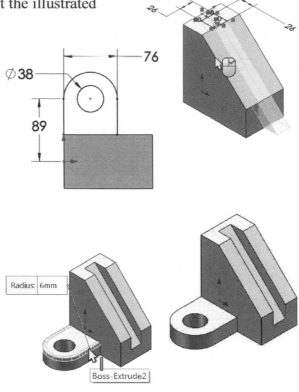

9. Create the **Extruded Boss** feature. Blind is the default End Condition in Direction 1. Depth = 26mm.

10. Create the first **Fillet** feature. Fillet the top edge of the left tab. Radius = 6mm. Constant Size fillet is selected by default.

11. Create **Sketch4**. Select the top face of Extrude3 as the Sketch plane. Sketch a circle. Insert the required dimension.

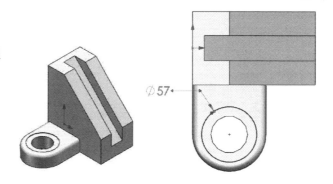

12. Create the second **Extruded Cut** feature. Blind is the default End Condition in Direction 1. Depth = 1mm. The model displayed an Ø57mm Spot Face hole with a 1mm depth.

13. Create the **Mirror** feature. Mirror about the Front Plane. Mirror the Cut-Extrude2, Fillet1, and Boss-Extrude2 feature.

14. Create the second Constant Size **Fillet** feature. Fillet the top inside edge of the left tab and the top inside edge of the right tab. Radius = 6mm.

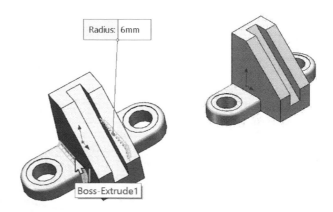

15. **Assign** the defined material to the part.

16. **Calculate** the overall mass of the part. The overall mass = 3437.29 grams.

17. **Locate** the Center of mass. The location of the Center of mass is derived from the part Origin.

- X = 49.21 millimeters

- Y = 46.88 millimeters

- Z = 0.00 millimeters

18. **Save** the part. Name it Part-Modeling 3-1.

19. **Close** the model.

In this category, an exam question could read: Build this model. What is the overall mass of the part?

- A: 3944.44 grams

- B: 4334.29 grams

- C: 3437.29 grams

- D: 2345.69 grams

The correct answer is C.

As an exercise, modify all ALL ROUNDS from 6mm to 8mm. Modify the material from 2014 Alloy to 6061 Alloy.

Modify the Sketch1 angle from 45deg to 30deg. Modify the Extrude3 depth from 26mm to 36mm. Recalculate the location of the Center of mass with respect to the part Origin.

- X = 49.76 millimeters

- Y = 34.28 millimeters

- Z = 0.00 millimeters

20. **Save** the part. Name it Part-Modeling 3-1-Modify.

Tangent edges and Origin is displayed for educational purposes.

When you create a new part or assembly, the three default Planes (Front, Right and Top) are aligned with specific views. The Plane you select for the Base sketch determines the orientation of the part.

Mass = 3437.29 grams

Volume = 1227602.22 cubic millimeters

Surface area = 101091.11 square millimeters

Center of mass: (millimeters)
X = 49.21
Y = 46.88
Z = 0.00

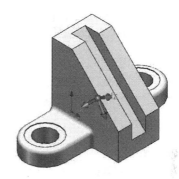

Mass = 3024.12 grams

Volume = 1120042.82 cubic millimeters

Surface area = 92861.98 square millimeters

Center of mass: (millimeters)
X = 49.76
Y = 34.28
Z = 0.00

Tutorial: Basic/Intermediate Part 3-2

Build this model in SOLIDWORKS. Calculate the overall mass of the part and locate the Center of mass with the provided information.

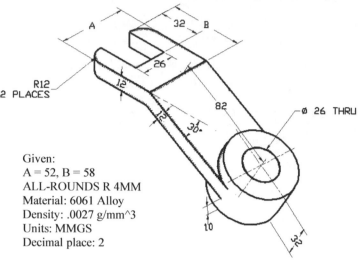

1. **Create** a new part in SOLIDWORKS.

Given:
A = 52, B = 58
ALL-ROUNDS R 4MM
Material: 6061 Alloy
Density: .0027 g/mm^3
Units: MMGS
Decimal place: 2

2. **Build** the illustrated model. Think about the various features that create the part. Insert seven features and a plane to build this part: Extruded-Thin1, Boss-Extrude1, Cut-Extrude1, Cut-Extrude2 and three Fillets. Apply reference construction planes to build the circular features.

3. **Set** document properties for the model.

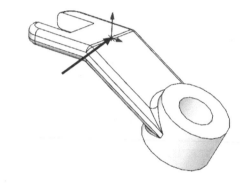

4. Create **Sketch1**. Select the Front Plane as the Sketch plane. Apply construction geometry as the reference line for the 30deg angle. Insert the required geometric relations and dimensions. Note the location of the Origin.

5. Create the **Extrude-Thin1** feature. This is the Base feature. Apply symmetry. Select Mid Plane for End Condition in Direction 1 to maintain the location of the Origin. Depth = 52mm. Thickness = 12mm.

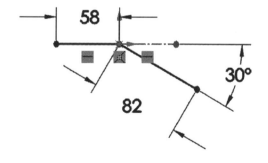

Use the Thin Feature option to control the extrude thickness, not the Depth.

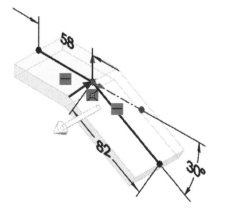

6. Create **Plane1**. Plane1 is the Sketch plane for the Extruded Boss (Boss-Extrude1) feature. Select the midpoint and the top face as illustrated. Plane1 is located in the middle of the top and bottom faces. Select Parallel Plane at Point for option.

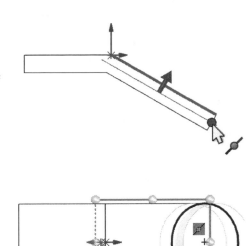

🔆 Create Plane1 to use the Depth dimension of 32mm.

7. Create **Sketch2**. Select Plane1 as the Sketch plane. Use the Normal To view tool. Sketch a circle to create the Extruded Boss feature. Insert the required geometric relations.

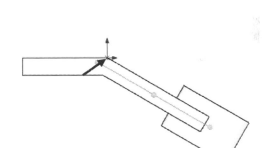

🔆 The Normal To view tool rotates and zooms the model to the view orientation normal to the selected plane, planar face, or feature.

8. Create the **Extruded Boss** feature. Apply Symmetry. Select Mid Plane for End Condition in Direction 1. Depth = 32mm.

9. Create **Sketch3**. Select the top circular face of Boss-Extrude1 as the Sketch plane. Sketch a circle. Insert the required geometric relation and dimension.

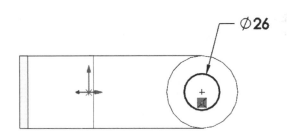

There are numerous ways to create the models in this chapter. A goal is to display different design intents and techniques.

10. Create the first **Extruded Cut** feature. Select Through All for End Condition in Direction 1.

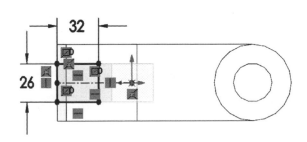

11. Create **Sketch4**. Select the top face of Extrude-Thin1 as the Sketch plane. Apply construction geometry. Insert the required geometric relations and dimensions.

12. Create the second **Extruded Cut** feature. Select Through All for End Condition in Direction 1.

13. Create the **Fillet1** feature. Fillet the left and right edges of Extrude-Thin1 as illustrated. Radius = 12mm.

14. Create the **Fillet2** feature. Fillet the top and bottom edges of Extrude-Thin1 as illustrated. Radius = 4mm.

15. Create the **Fillet3** feature. Fillet the rest of the model, six edges as illustrated. Radius = 4mm.

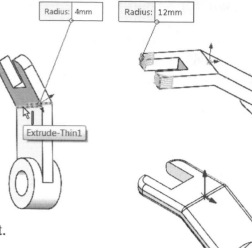

16. **Assign** the defined material to the part.

17. **Calculate** the overall mass of the part. The overall mass = 300.65 grams.

18. **Locate** the Center of mass. The location of the Center of mass is derived from the part Origin.

- X: 34.26 millimeters

- Y: -29.38 millimeters

- Z: 0.00 millimeters

19. **Save** the part. Name it Part-Modeling 3-2.

20. **Close** the model.

```
Mass = 300.65 grams

Volume = 111352.30 cubic millimeters

Surface area = 23765.33  square millimeters

Center of mass: ( millimeters )
    X = 34.26
    Y = -29.38
    Z = 0.00
```

As an exercise, modify the Fillet2 and Fillet3 radius from 4mm to 2mm. Modify the Fillet1 radius from 12m to 10mm. Modify the material from 6061 Alloy to ABS.

Modify the Sketch1 angle from 30deg to 45deg. Modify the Extrude depth from 32mm to 38mm. Recalculate the location of the Center of mass with respect to the part Origin.

- X = 27.62 millimeters

- Y = -40.44 millimeters

- Z = 0.00 millimeters

21. **Save** the part. Name it Part-Modeling 3-2-Modify.

In the exam; you are allowed to answer the questions in any order. Use the Summary Screen during the exam to view the list of all questions you have or have not answered.

```
Mass = 123.60 grams

Volume = 121173.81 cubic millimeters

Surface area = 25622.46  square millimeters

Center of mass: ( millimeters )
    X = 27.62
    Y = -40.44
    Z = 0.00
```

Tutorial: Basic/Intermediate Part 3-3

Build this model in SOLIDWORKS. Calculate the volume of the part and locate the Center of mass with the provided information.

1. **Create** a new part in SOLIDWORKS.

2. **Build** the illustrated model. Think about the various features that create this model. Insert five features and a plane to build this part: Extruded Base, two Extruded Bosses, Extruded Cut and a Rib. Insert a reference plane to create the Boss-Extrude2 feature.

3. **Set** document properties for the model.

4. Create **Sketch1**. Select the Top Plane as the Sketch plane. Sketch a center rectangle. Use the horizontal construction line as the Plane1 reference. Insert the required relations and dimensions.

5. Create the **Extruded Base** feature. Blind is the default End Condition in Direction 1. Depth = 1.00in. Note the extrude direction is downward.

💡 Create planes to aid in the modeling for the exam. Use planes to sketch, to create a section view, for a neutral plane in a draft feature, and so on.

💡 The created plane is displayed 5% larger than the geometry on which the plane is created, or 5% larger than the bounding box. This helps reduce selection problems when planes are created directly on faces or from orthogonal geometry.

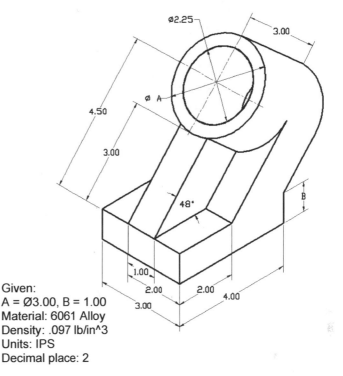

Given:
A = Ø3.00, B = 1.00
Material: 6061 Alloy
Density: .097 lb/in^3
Units: IPS
Decimal place: 2

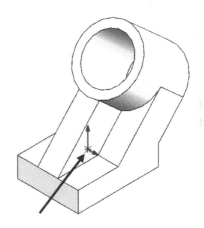

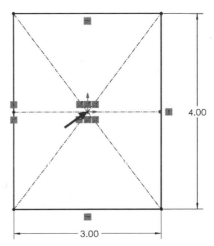

6. Create **Plane1**. Plane1 is the Sketch plane for the Extruded Boss feature. Show Sketch1. Select the horizontal construction line in Sketch1 and the top face of Boss-Extrude1. Angle = 48deg.

🔆 Click **View, Sketches** or **View, Hide/Show, Sketches** from the Main menu to displayed sketches in the Graphics window.

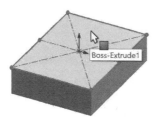 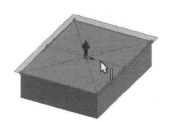

🔆 The Normal To view tool rotates and zooms the model to the view orientation normal to the selected plane, planar face, or feature.

7. Create **Sketch2**. Select Plane1 as the Sketch plane. Create the Extruded Boss profile. Insert the required geometric relations and dimension. Note: Dimension to the front **top edge** of Boss-Extrude1 as illustrated.

8. Create the first **Extruded Boss** feature. Select the Up To Vertex End Condition in Direction 1. Select the back top right vertex point as illustrated.

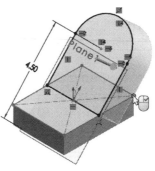

9. Create **Sketch3**. Select the back angled face of Boss-Extrude2 as the Sketch plane. Sketch a circle. Insert the required geometric relations.

10. Create the third **Extruded Boss** feature. Blind is the default End Condition in Direction 1. Depth = 3.00in.

11. Create **Sketch4**. Select the front face of Boss-Extrude3 as the Sketch plane. Sketch a circle. Sketch4 is the profile for the Extruded Cut feature. Insert the required geometric relation and dimension.

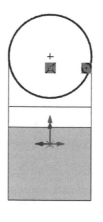 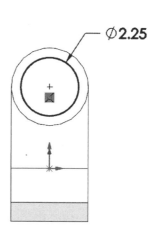

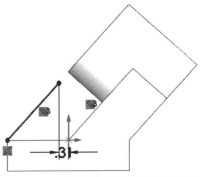

🔆 The part Origin is displayed in blue.

12. Create the **Extruded Cut** feature. Select Through All for End Condition in Direction 1.

13. Create **Sketch5**. Select the Right Plane as the Sketch plane. Insert a Parallel relation to partially define Sketch5. Sketch5 is the profile for the Rib feature. Sketch5 does not need to be fully defined. Sketch5 locates the end conditions based on existing geometry.

14. Create the **Rib** feature. Thickness = 1.00in.

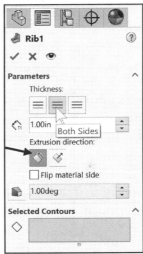

🔆 The Rib feature is a special type of extruded feature created from open or closed sketched contours. The Rib feature adds material of a specified thickness in a specified direction between the contour and an existing part. You can create a rib feature using single or multiple sketches.

15. **Assign** 6061 Alloy material to the part.

16. **Calculate** the volume. The volume = 30.65 cubic inches.

17. **Locate** the Center of mass. The location of the Center of mass is derived from the part Origin.

- X: 0.00 inches
- Y: 0.73 inches
- Z: -0.86 inches

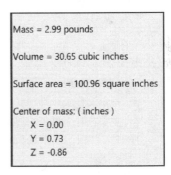

Mass = 2.99 pounds

Volume = 30.65 cubic inches

Surface area = 100.96 square inches

Center of mass: (inches)
  X = 0.00
  Y = 0.73
  Z = -0.86

18. **Save** the part. Name it Part-Modeling 3-3.

19. **Close** the model.

As an exercise, modify the Rib1 feature from 1.00in to 1.25in.

Modify the Extrude depth from 3.00in to 3.25in.

Modify the material from 6061 Alloy to Copper.

Modify the Plane1 angle from 48deg to 30deg.

Recalculate the volume of the part. The new volume = 26.94 cubic inches.

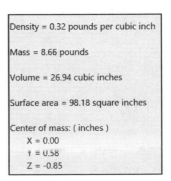

Density = 0.32 pounds per cubic inch

Mass = 8.66 pounds

Volume = 26.94 cubic inches

Surface area = 98.18 square inches

Center of mass: (inches)
  X = 0.00
  Y = 0.58
  Z = -0.85

20. **Save** the part. Name it Part-Modeling 3-3-Modify.

Tutorial: Basic/Intermediate Part 3-4

Build this model in
SOLIDWORKS. Calculate the
volume of the part and locate the
Center of mass with the provided
information.

1. **Create** a new part in
   SOLIDWORKS.

2. **Build** the illustrated model.
   Apply symmetry. Think about
   the various features that create
   the part. Insert six features:
   Extruded Base, two Extruded
   Cuts, Mirror, Extruded Boss,
   and a third Extruded Cut.

3. **Set** document properties for the
   model.

4. Create **Sketch1**. Select the Top
   Plane as the Sketch plane.
   Apply symmetry. The part
   Origin is located in the center
   of the rectangle. Insert the
   required relations and dimensions.

5. Create the **Extruded Base** (Boss-Extrude1)
   feature. Blind is the default End Condition in
   Direction 1. Depth = .50in.

6. Create **Sketch2**. Select the top face of Boss-
   Extrude1 for the Sketch plane. Sketch a circle.
   Insert the required relations and dimensions.

7. Create the first **Extruded Cut** feature. Select
   Through All as End Condition in Direction1.

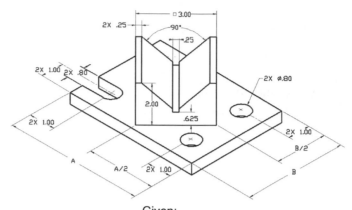

Given:
A = 6.00, B = 4.50
Material: 2014 Alloy
Plate thickness = .50
Units: IPS
Decimal place: 2

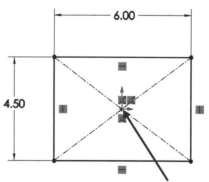

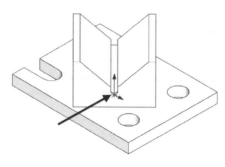

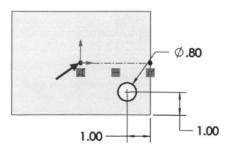

8. Create **Sketch3**. Select the top face of Boss-Extrude1 for the Sketch plane. Insert the required geometric relations and dimensions.

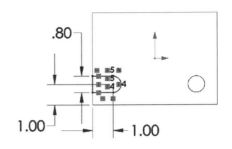

🔅 Click **View**, **Temporary** axes to view the part temporary axes in the Graphics window.

9. Create the second **Extruded Cut** feature. Select Through All as End Condition in Direction1.

10. Create the **Mirror** feature. Mirror the two Extruded Cut features about the Front Plane.

11. Create **Sketch4**. Select the top face of Boss-Extrude1 as the Sketch plane. Apply construction geometry to center the sketch. Insert the required relations and dimensions.

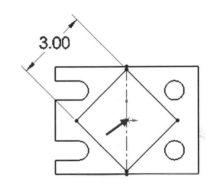

12. Create the **Extruded Boss** feature. Blind is the default End Condition in Direction 1. Depth = 2.00in.

13. Create **Sketch5**. Select the front face of Boss-Extrude as illustrated for the Sketch plane. Sketch5 is the profile for the third Extruded Cut feature. Apply construction geometry. Insert the required dimensions and relations.

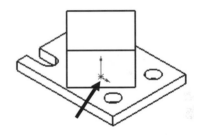

14. Create the third **Extruded Cut** feature. Through All is selected for End Condition in Direction 1 and Direction 2.

15. **Assign** 2014 Alloy material to the part.

16. **Calculate** the volume of the part. The volume = 25.12 cubic inches.

17. **Locate** the Center of mass. The location of the Center of mass is derived from the part Origin.

- X: 0.06 inches

- Y: 0.80 inches

- Z: 0.00 inches

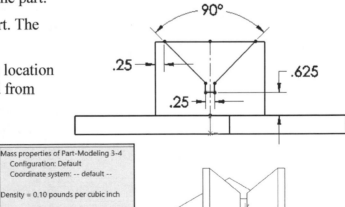

Mass properties of Part-Modeling 3-4
  Configuration: Default
  Coordinate system: -- default --

Density = 0.10 pounds per cubic inch

Mass = 2.54 pounds

Volume = 25.12 cubic inches

Surface area = 88.33 square inches

Center of mass: (inches)
  X = 0.06
  Y = 0.80
  Z = 0.00

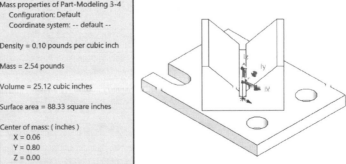

18. **Save** the part. Name it Part-Modeling 3-4.

19. **Close** the model.

Summary

Basic Part Creation and Modification and Intermediate Part Creation and Modification are two of the five categories on the CSWA exam.

There are two questions on the CSWA Academic exam (Part 1) in the *Basic Part Creation and Modification* category. One question is in a multiple-choice single answer format and the other question (Modification of the model) is in the fill in the blank format. Each question is worth fifteen (15) points for a total of thirty (30) points.

The main difference between the *Basic Part Creation and Modification* category and the *Intermediate Part Creation and Modification* or the *Advance Part Creation and Modification* category is the complexity of the sketches and the number of dimensions and geometric relations along with an increase in the number of features.

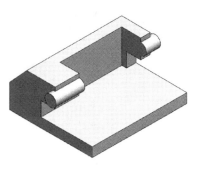

During the CSWA exam, SOLIDWORKS provides various model views. Click on the additional view icons during the exam to better understand the part and provided information. Read each question carefully. Identify the dimensions, center of mass, units and location of the Origin. Apply needed material.

If your school is an academic certification provider, your instructor can allocate free exam credits for Segment 1 & 2. The instructor will require your .edu email address.

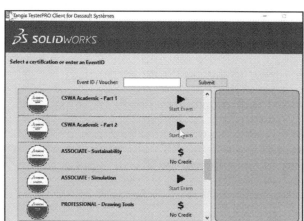

Questions

Question 1:

In Tutorial, Volume/Center of mass 3-2 you built the model using the FeatureManager that had three features vs. four features in the FeatureManager.

Calculate the overall mass of the part, volume, and locate the Center of mass with the provided information using the Option1 FeatureManager.

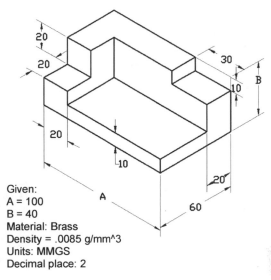

Given:
A = 100
B = 40
Material: Brass
Density = .0085 g/mm^3
Units: MMGS
Decimal place: 2

Question 2:

In Tutorial, Mass/Volume 3-4 you built the model using the FeatureManager that had three features vs. four features.

Calculate the overall mass of the part, volume, and locate the Center of mass with the provided information using the Option3 FeatureManager.

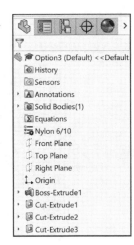

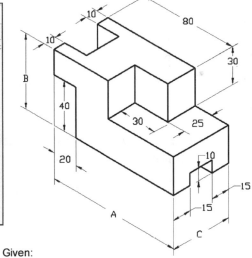

Given:
A = 110, B = 60, C = 50
Material: Nylon 6/10
Density = .0014 g/mm^3
Units: MMGS
Decimal place: 2

Question 3:

In Tutorial, Basic/Intermediate 3-4 you built the illustrated model. Modify the plate thickness from .50in to .25in. Modify the Sketch5 angle from 90deg to 75deg. Re-assign the material from 2014 Alloy to 6061 Alloy.

Calculate the overall mass of the part, volume, and locate the Center of mass with the provided information.

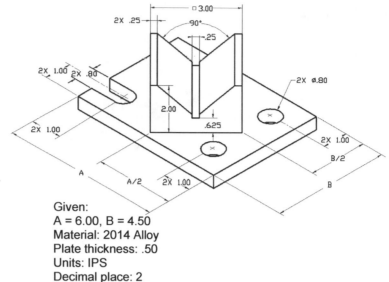

Given:
A = 6.00, B = 4.50
Material: 2014 Alloy
Plate thickness: .50
Units: IPS
Decimal place: 2

Question 4:

Build this model: Set document properties, identify the correct Sketch planes, apply the correct Sketch and Feature tools and apply material.

Calculate the overall mass of the part, volume, and locate the Center of mass with the provided illustrated information.

- Material: 6061 Alloy
- Units: MMGS
- Decimal place: 2
- All Thru Holes

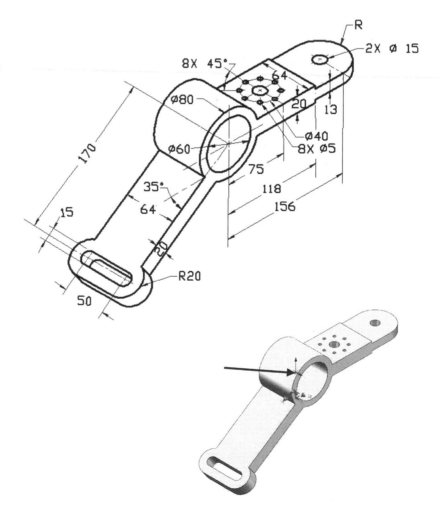

Question 5:

Build this model. Set document properties and identify the correct Sketch planes. Apply the correct Sketch and Feature tools, and apply material.

Calculate the overall mass of the part, volume, and locate the Center of mass with the provided information.

- Material: 6061 Alloy

- Units: MMGS

- Decimal place: 2

- All Thru Holes

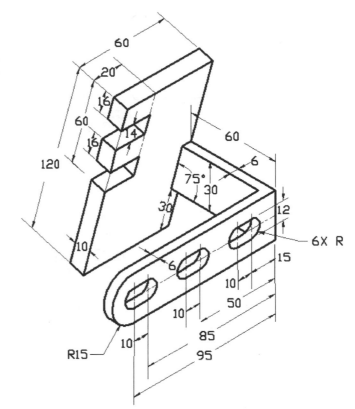

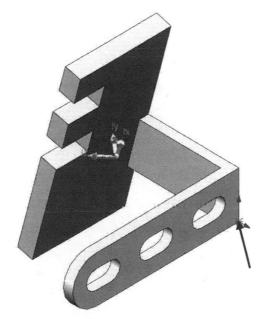

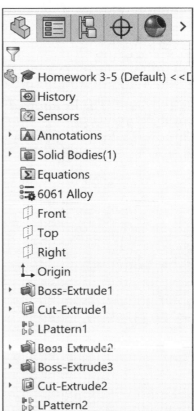

Question 6:

Build this model. Set document properties and identify the correct Sketch planes. Apply the correct Sketch and Feature tools, and apply material.

Calculate the overall mass of the part with the provided information. Note: The Origin is arbitrary.

- Material: Copper
- Units: MMGS
- A = 100
- B = 80
- Decimal place: 2

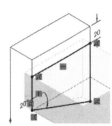

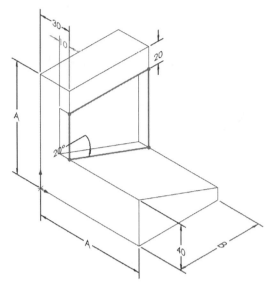

Question 7:

Build this model. Set document properties and identify the correct Sketch planes. Apply the correct Sketch and Feature tools, and apply material.

Calculate the overall mass of the part with the provided information. The location of the Origin is arbitrary.

- Material: 6061
- Units: MMGS
- Decimal place: 2
- A = 16.00
- B = 40.00
- Side A is perpendicular to side B
- C = 18.00

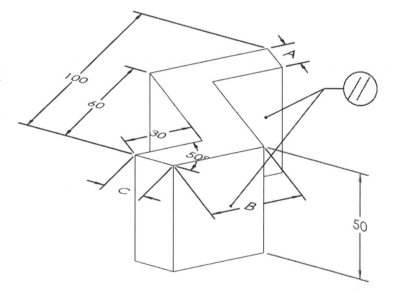

Question 8:

Build this model in SOLIDWORKS. Set document properties and identify the correct Sketch planes. Apply the correct Sketch and Feature tools, and apply material.

Calculate the overall mass of the part and volume with the provided information.

- Material: 6061

- Units: IPS

- Decimal place: 2

- View the provided drawing views for details.

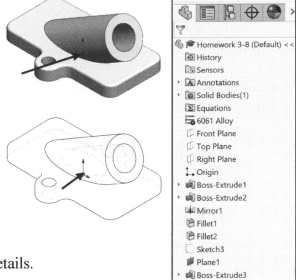

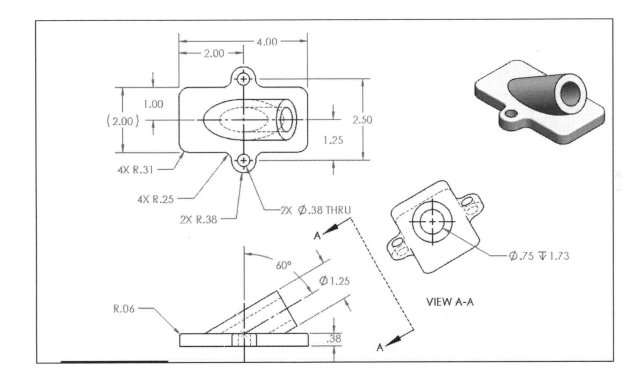

Question 9:

Build this model in SOLIDWORKS. Set document properties and identify the correct Sketch planes. Apply the correct Sketch and Feature tools, and apply material.

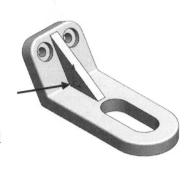

Calculate the overall mass of the part and volume with the provided information.

- Material: 6061
- Units: IPS
- Decimal place: 2
- View the provided drawing views for details

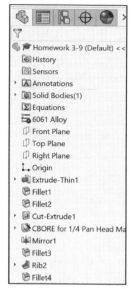

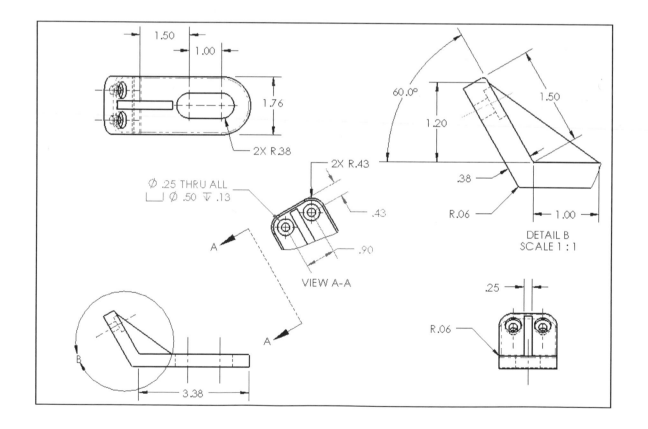

Question 10:

Build the illustrated model.

Calculate the overall mass in grams. Calculate the volume of the part in cubic millimeters.

- Decimal place: 2
- Material: AISI 304
- Units: MMGS
- All Holes ⊽ 25mm
- All Rounds 5mm
- All Holes Ø4mm

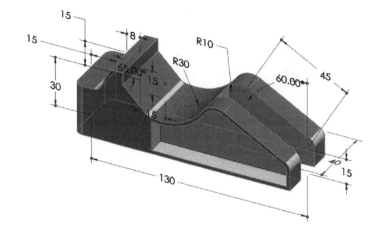

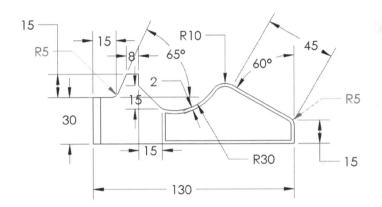

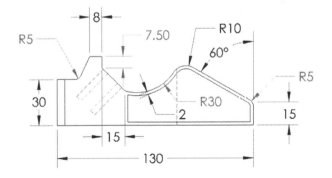

Front views

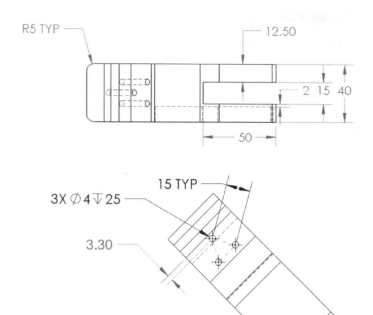

Top and Auxiliary view

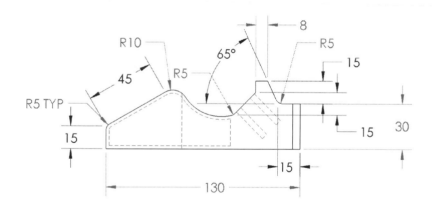

Calculate the mass:

A = 888.48grams

B = 990.50grams

C = 788.48grams

D = 820.57grams

Back view

If you don't find your answer (within 1%) in the multiple-choice single answer format section, recheck your solid model for precision and accuracy. It could be as simple as missing a few fillets.

Calculate the volume:

A = 102259.43 cubic millimeters

B = 133359.47 cubic millimeters

C = 111059.43 cubic millimeters

D = 125059.49 cubic millimeters

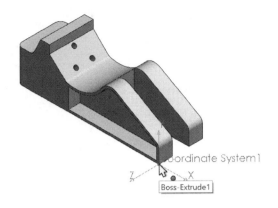

Question 10A:

Create a new coordinate system.

The new coordinate system location is at the front right bottom point (vertex) of the model.

Enter the Center of Mass:

X = **-80.39 millimeters**

Y = **-15.93 millimeters**

Z = **-22.65 millimeters**

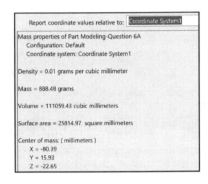

Question 10B:

Modify the model.

Calculate the overall mass in grams. Calculate the volume of the part in cubic millimeters.

- Modify all fillets (rounds) to 7mm.

- Modify the overall length to 140mm.

- Modify material to 1060 alloy.

Enter the mass:

309.75 grams

Enter the volume:

114721.22 cubic millimeters

If you don't find your answer (within 1%) in the multiple-choice single answer format section, recheck your solid model for precision and accuracy. It could be as simple as missing a few fillets.

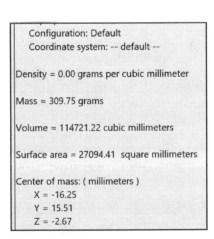

Question 11:

Build the illustrated model.

Calculate the overall mass in grams. Calculate the volume of the part in cubic millimeters.

- Decimal place: 2

- Material: 1060 Alloy

- Units: MMGS

- TYP $\emptyset 12$

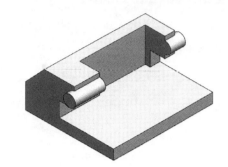

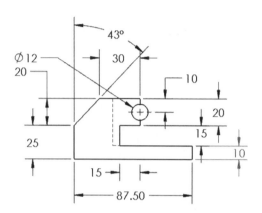

Front view

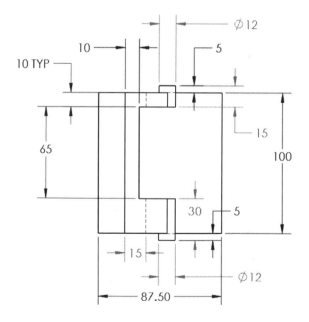

Top view

Calculate the mass:

A = 600.92 grams

B = 509.92 grams

C = 701.93 grams

D = 599.34 grams

Calculate the volume:

A = 188860.93 cubic millimeters
B = 196660.93 cubic millimeters
C = 198880.65 cubic millimeters
D = 230021.67 cubic millimeters

Question 11A:

Modify the model. Calculate the overall mass in grams. Calculate the volume of the part in cubic millimeters.

- Modify material to Plain Carbon Steel.

- Modify TYP Hole diameter from TYP ϕ 12 to TPY ϕ 10.

Enter the mass:

1465.70 grams

Enter the volume:

187910.60 cubic millimeters

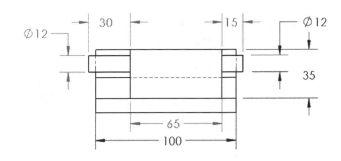

Right view

Configuration: Default
Coordinate system: -- default --

Density = 0.00 grams per cubic millimeter

Mass = 509.92 grams

Volume = 188860.93 cubic millimeters

Surface area = 32545.06 square millimeters

Center of mass: (millimeters)
  X = 31.39
  Y = 16.55
  Z = 1.37

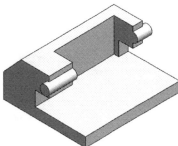

Configuration: modified
Coordinate system: -- default --

Density = 0.01 grams per cubic millimeter

Mass = 1465.70 grams

Volume = 187910.60 cubic millimeters

Surface area = 32373.16 square millimeters

Center of mass: (millimeters)
  X = 31.29
  Y = 16.45
  Z = 1.33

Question 11B:

Create a new coordinate system with the provided illustration.

The new coordinate system location is at the front left bottom point (vertex) of the model.

Enter the Center of Mass:

X = 31.29 millimeters

Y = 16.45 millimeters

Z = -48.67 millimeters

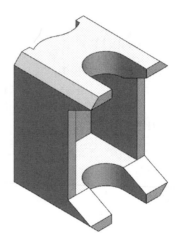

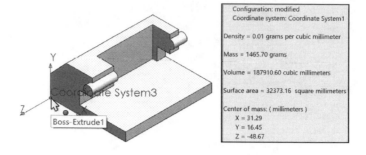

Question 12:

Build the illustrated model.

Calculate the overall mass in grams. Calculate the volume of the part in cubic millimeters.

- Decimal place: 2

- Material: Plain Carbon Steel

- Units: MMGS

- The part is not **symmetrical** about the Front Plane

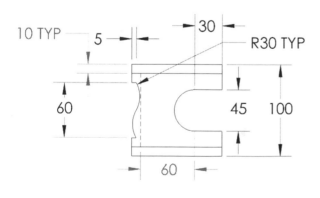

Top view

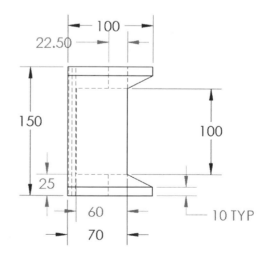

Front view

Calculate the mass:

A = 4411.5 grams

B = 4079.32 grams

C = 4234.30 grams

D = 5322.00 grams

Calculate the volume:

A = 522989.22 cubic millimeters

B = 555655.11 cubic millimeters

C = 511233.34 cubic millimeters

D = 655444.00 cubic millimeters

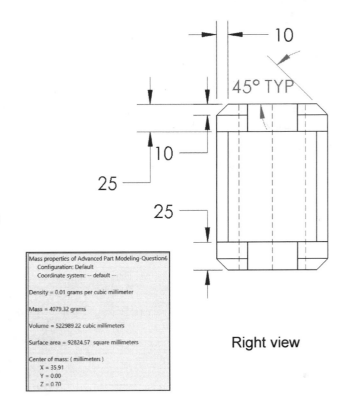

Mass properties of Advanced Part Modeling-Question6
    Configuration: Default
    Coordinate system: -- default --

Density = 0.01 grams per cubic millimeter

Mass = 4079.32 grams

Volume = 522989.22 cubic millimeters

Surface area = 92824.57 square millimeters

Center of mass: (millimeters)
    X = 35.91
    Y = 0.00
    Z = 0.70

Right view

Question 12A:

Create a new coordinate system.

Center a new coordinate system with the provided illustration. The new coordinate system location is at the back right bottom point (vertex) of the model.

Enter the Center of Mass:

X = -64.09 millimeters

Y = 75.00 millimeters

Z = 40.70 millimeters

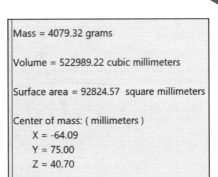

Boss-Extrude1

Mass = 4079.32 grams

Volume = 522989.22 cubic millimeters

Surface area = 92824.57 square millimeters

Center of mass: (millimeters)
    X = -64.09
    Y = 75.00
    Z = 40.70

Question 13:

Build the provided model.

Calculate the overall mass in grams. Calculate the volume of the part in cubic millimeters.

Open Homework problem 3-13 from the Homework chapter exercises. Use the rollback bar to obtain features and dimensions. Think about the various ways that this model can be built.

- Decimal place: 2

- Material: 1060 Alloy

- Units: MMGS

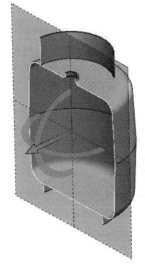

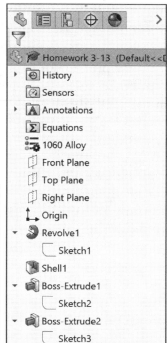

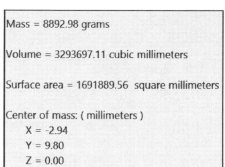

Mass = 8892.98 grams

Volume = 3293697.11 cubic millimeters

Surface area = 1691889.56 square millimeters

Center of mass: (millimeters)
  X = -2.94
  Y = 9.80
  Z = 0.00

💡 Screen shots from an older CSWA exam for a Basic/Intermediate part.

Click on the additional views to understand the part and to provide information. Read each question carefully.

Understand the dimensions, center of mass and units. Apply needed materials.

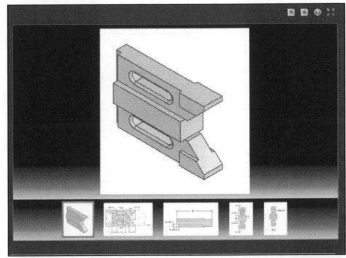

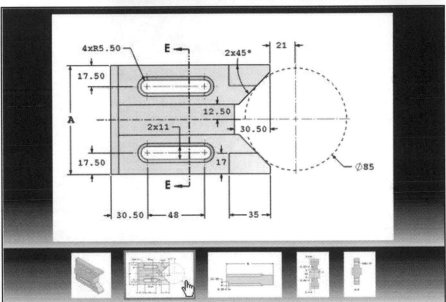

Screen shots from the exam

Zoom in on the part or view if needed.

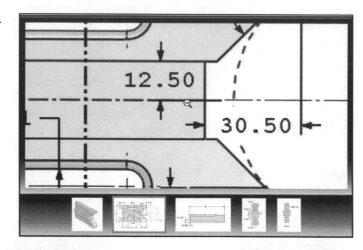

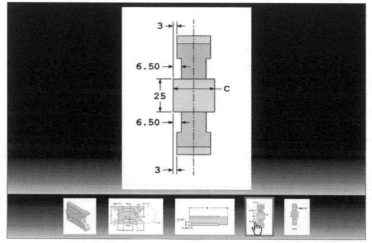

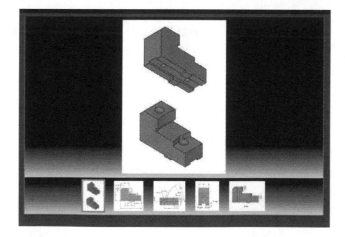

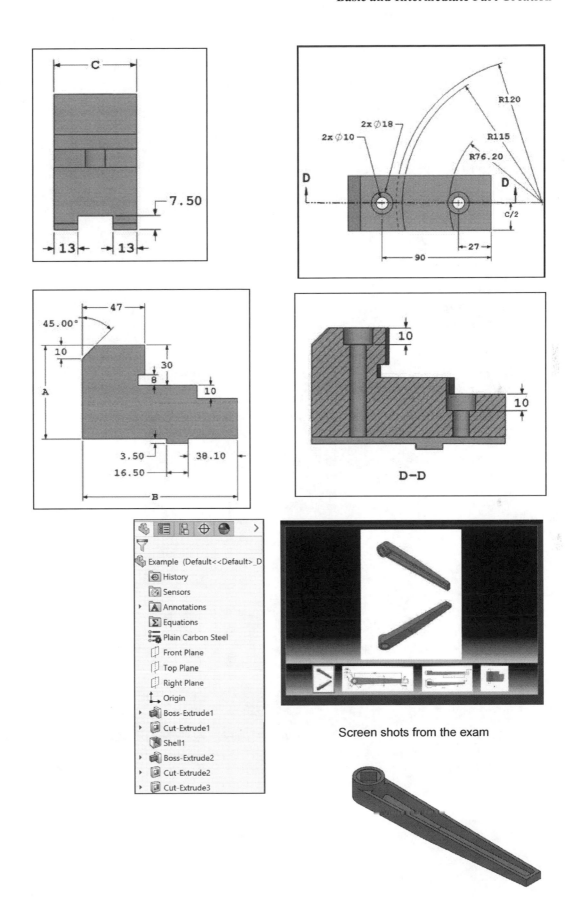

Screen shots from the exam

💡 Screen shots from an older CSWA exam for a Basic/Intermediate part.

Click on the additional views to understand the part and to provide information. Read each question carefully.

Understand the dimensions, center of mass and units. Apply needed materials.

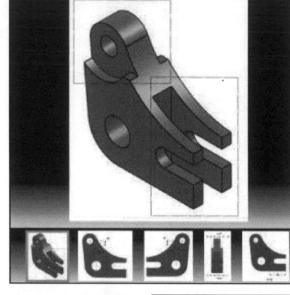

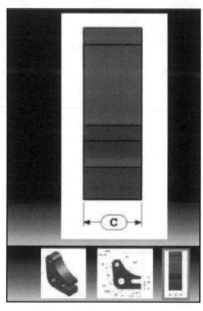

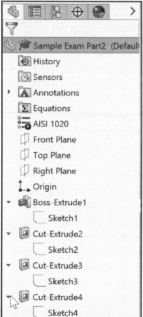

Screen shot from the new sample CSWA exam for a Basic/Intermediate part.

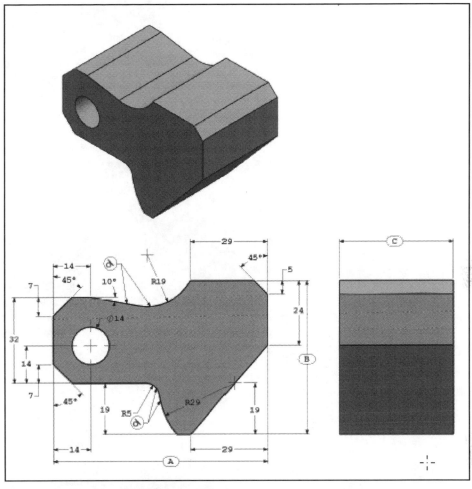

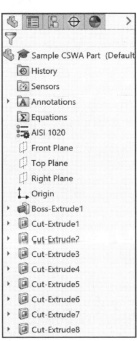

Screen shot from the new sample CSWA exam for a Basic/Intermediate part.

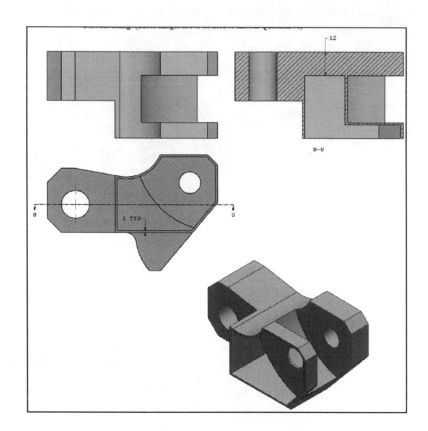

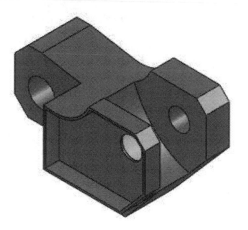

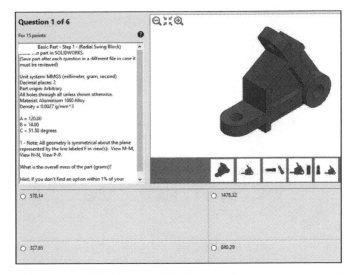

Question 1 of 6

For 15 points:

Basic Part - Step 1 - (Radial Swing Block)

...this part in SOLIDWORKS.
(Save part after each question in a different file in case it must be reviewed)

Unit system: MMGS (millimeter, gram, second)
Decimal places: 2
Part origin: Arbitrary
All holes through all unless shown otherwise.
Material: Aluminium 1060 Alloy
Density = 0.0027 g/mm^3

A = 120.00
B = 14.00
C = 51.50 degrees

1 - Note: All geometry is symmetrical about the plane represented by the line labeled F in view(s): View M-M, View N-N, View P-P.

What is the overall mass of the part (grams)?

Hint: If you don't find an option within 1% of your

○ 578.14   ○ 1478.32

○ 327.65   ○ 840.29

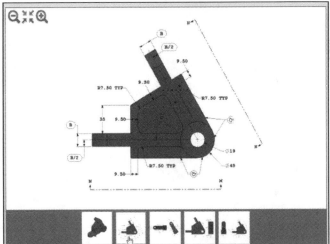

Enter Value (g): [   ]

(use . (point) as decimal separator)

Screen shot from an exam

Notes:

CHAPTER 4 - ADVANCED PART CREATION AND MODIFICATION

Objectives

Provide in-depth coverage of the Advanced Part Creation and Modification category on the CSWA exam. Knowledge to create and modify models in this category from detailed dimensioned illustrations.

Introduction

The main difference between the Advanced Part Creation and Modification and the Basic Part Creation and Modification or the Intermediate Part Creation and Modification category is the complexity of the sketches and the number of dimensions and geometric relations along with an increased number of features.

There are three questions on the CSWA exam (part 2) in this category. The first question is in a multiple-choice single answer format and the other two questions (Modification of the model) are in the fill in the blank format.

Each question is worth fifteen (15) points for a total of forty-five (45) points.

You are required to build a model with six or more features and to answer a question either on the overall mass, volume, or the location of the Center of mass for the created model relative to the default part Origin location. You are then requested to modify the model and answer fill in the blank format questions.

On the completion of the chapter, you will be able to:

- Read and understand an Engineering document used in the CSWA exam.

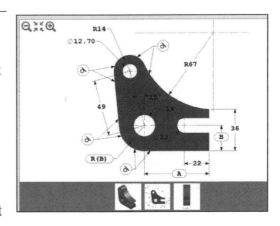

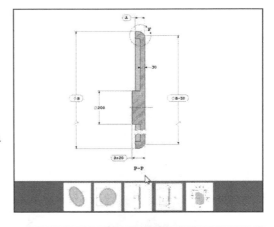

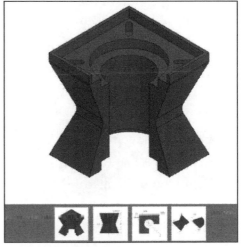

Screen shots from an exam

- Identify the Sketch plane, part Origin location, part dimensions, geometric relations and design intent of the sketch and feature.

- Build an advanced part from a detailed dimensioned illustration using the following tools and features:

  - 2D & 3D Sketch tools, Extruded Boss/Base, Extruded Cut, Fillet, Chamfer, Mirror, Revolved Boss/Base, Linear & Circular Pattern, Plane, Axis, and Revolved Cut.

- Locate the Center of mass relative to the part Origin.

- Create a new coordinate system.

- Locate the Center of mass relative to a created Coordinate system.

In the *Advanced Part Creation and Modification* category, there are two dimension modification questions based on the first (multiple choice) question. You should be within 1% of the multiple-choice answer before you go on to the modification single answer section.

Download all needed model files (initial and final) and the SOLIDWORKS CSWA Sample Exam folder from the SDC Publications website www.SDCpublications.com/downloads/978-1-63057-567-0

Build an Advanced Part from a detailed dimensioned illustration

Tutorial: Advanced Part 4-1

An exam question in this category could read: Build this part in SOLIDWORKS. Calculate the overall mass and locate the Center of mass of the illustrated model.

1. **Create** a new part in SOLIDWORKS.

2. **Build** the illustrated model. Insert seven features: Extruded Base, two Extruded Bosses, two Extruded Cuts, a Chamfer and a Fillet.

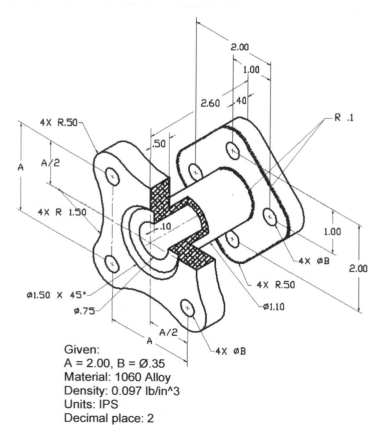

Given:
A = 2.00, B = Ø.35
Material: 1060 Alloy
Density: 0.097 lb/in^3
Units: IPS
Decimal place: 2

The complexity of the models along with the features progressively increases throughout this chapter to simulate the final types of models that would be provided on the exam.

Think about the steps that you would take to build the illustrated part. Identify the location of the part Origin. Start with the back base flange. Review the provided dimensions and annotations in the part illustration.

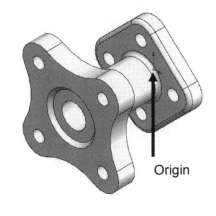

Origin

🔅 The main difference between the *Advanced Part Creation and Modification* and the *Basic Part Creation and Modification* or the *Intermediate Part Creation and Modification* category is the complexity of the sketches and the number of dimensions and geometric relations along with an increased number of features.

3. **Set** document properties for the model.

4. Create **Sketch1**. Sketch1 is the Base sketch. Select the Front Plane as the Sketch plane. Apply construction geometry. Sketch a horizontal and vertical centerline. Sketch four circles. Insert an Equal relation. Insert a Symmetric relation about the vertical and horizontal centerlines. Sketch two top angled lines and a tangent arc. Apply the Mirror Sketch tool. Complete the sketch. Insert the required geometric relations and dimensions.

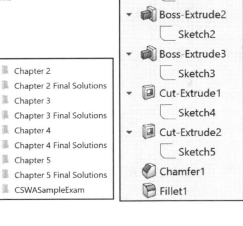

- Boss-Extrude1
  - Sketch1
- Boss-Extrude2
  - Sketch2
- Boss-Extrude3
  - Sketch3
- Cut-Extrude1
  - Sketch4
- Cut-Extrude2
  - Sketch5
- Chamfer1
- Fillet1

🔅 In a Symmetric relation, the selected items remain equidistant from the centerline, on a line perpendicular to the centerline. Sketch entities to select: a centerline and two points, lines, arcs or ellipses.

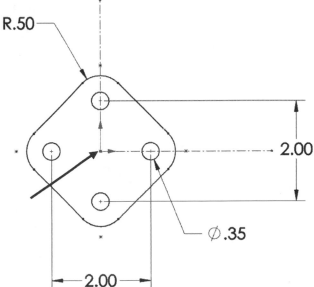

R.50

2.00

Ø.35

2.00

In the exam, you are allowed to answer the questions in any order. Use the Summary Screen during the exam to view the list of all questions you have or have not answered.

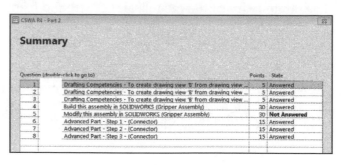

The Sketch Fillet tool rounds the selected corner at the intersection of two sketch entities, creating a tangent arc.

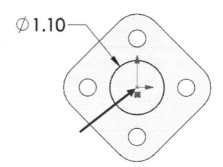

5. Create the **Extruded Base** feature. Boss-Extrude1 is the Base feature. Blind is the default End Condition in Direction 1. Depth = .40in.

6. Create **Sketch2**. Select the front face of Boss-Extrude1 as the Sketch plane. Sketch a circle. Insert the required geometric relation and dimension.

7. Create the first **Extruded Boss** feature. Blind is the default End Condition in Direction 1. The Extrude feature is the tube between the two flanges. Depth = 1.70in. Note: 1.70in = 2.60in - (.50in + .40in).

$\emptyset 1.10$

The complexity of the models along with the features progressively increases throughout this chapter to simulate the final types of parts that could be provided on the CSWA exam.

When you create a new part or assembly, the three default Planes (Front, Right and Top) are aligned with specific views. The Plane you select for the Base sketch determines the orientation of the part.

Illustrations may vary depending on your version, setup, and Add-ins.

There are no Surfacing or Boundary feature questions on the CSWA exam.

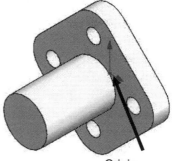

Origin

8. Create **Sketch3**. Select the front circular face of Boss-Extrude2 as the Sketch plane. Sketch a horizontal and vertical centerline. Sketch the top two circles. Insert an Equal and Symmetric relation between the two circles. Mirror the top two circles about the horizontal centerline. Insert dimensions to locate the circles from the Origin. Apply either the 3 Point Arc or the Centerpoint Arc Sketch tool. The center point of the Tangent Arc is aligned with a Vertical relation to the Origin. Complete the sketch.

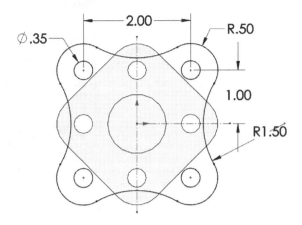

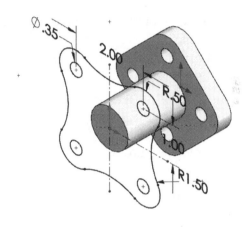

☀ Use the Centerpoint Arc Sketch tool to create an arc from a: centerpoint, a start point, and an end point.

☀ Apply the Tangent Arc Sketch tool to create an arc, tangent to a sketch entity.

The Arc PropertyManager controls the properties of a sketched Centerpoint Arc, Tangent Arc, and 3 Point Arc.

9. Create the second **Extruded Boss** feature. Blind is the default End Condition in Direction 1. Depth = .50in.

10. Create **Sketch4**. Select the front face of the Extrude feature as the Sketch plane. Sketch a circle. Insert the required geometric relation and dimension.

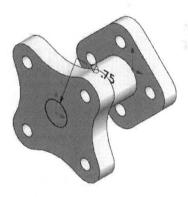

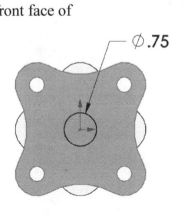

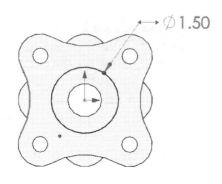

11. Create the first **Extruded Cut** feature. Select the Through All End Condition for Direction 1.

12. Create **Sketch5**. Select the front face of the Extrude feature as the Sketch plane. Sketch a circle. Insert the required geometric relation and dimension.

13. Create the second **Extruded Cut** feature. Blind is the default End Condition for Direction1.
Depth = .10in.

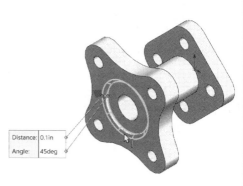

14. Create the **Chamfer** feature. In order to have the outside circle 1.50in, select the inside edge of the sketched circle. Create an Angle Distance chamfer.
Distance = .10in.
Angle = 45deg.

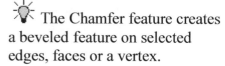

🔆 The Chamfer feature creates a beveled feature on selected edges, faces or a vertex.

15. Create the **Fillet** feature. Fillet the two edges as illustrated. Radius = .10in.

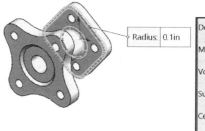

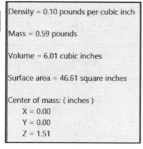

16. **Assign** 1060 Alloy material to the part. Material is required to calculate the overall mass of the part.

17. **Calculate** the overall mass. The overall mass = 0.59 pound.

18. **Locate** the Center of mass. The location of the Center of mass is relative to the part Origin.

- X: 0.00

- Y: 0.00

- Z: 1.51

19. **Save** the part. Name it Advanced Part 4-1.

20. **Close** the model.

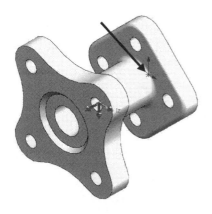

Tutorial: Advanced Part 4-2

An exam question in this category could read: Build this part in SOLIDWORKS. Calculate the overall mass and locate the Center of mass of the illustrated model.

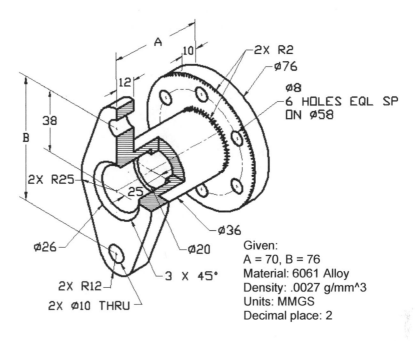

Given:
A = 70, B = 76
Material: 6061 Alloy
Density: .0027 g/mm^3
Units: MMGS
Decimal place: 2

1. **Create** a new part in SOLIDWORKS.

2. **Build** the illustrated dimensioned model. Insert eight features: Extruded Base, Extruded Cut, Circular Pattern, two Extruded Bosses, Extruded Cut, Chamfer and Fillet.

Think about the steps that you would take to build the illustrated part. Review the provided information. Start with the six hole flange.

🔅 Tangent edges are displayed for educational purposes.

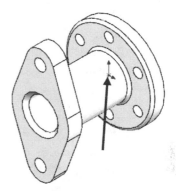

3. **Set** the document properties for the model.

4. Create **Sketch1**. Sketch1 is the Base sketch. Select the Front Plane as the Sketch plane. Sketch two circles. Insert the required geometric relations and dimensions.

5. Create the **Extruded Base** feature. Blind is the default End Condition in Direction 1. Depth = 10mm. Note the direction of the extrude feature to maintain the Origin location.

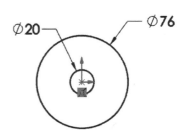

6. Create **Sketch2**. Select the front face of Boss-Extrude1 as the Sketch plane. Sketch2 is the profile for first Extruded Cut feature. The Extruded Cut feature is the seed feature for the Circular Pattern. Apply construction reference geometry. Insert the required geometric relations and dimensions.

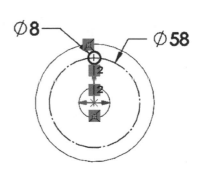

7. Create the **Extruded Cut** feature. Cut-Extrude1 is the first bolt hole. Select Through All for End Condition in Direction 1.

8. Create the **Circular Pattern** feature. Default Angle = 360deg. Number of instances = 6. Select the center axis for the Pattern Axis box.

💡 The Circular Pattern PropertyManager is displayed when you pattern one or more features about an axis.

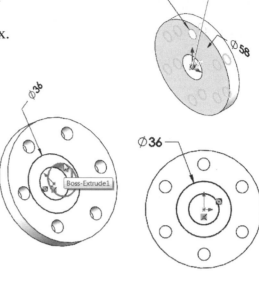

9. Create **Sketch3**. Select the front face of the Extrude feature as the Sketch plane. Sketch two circles. Insert a Coradial relation on the inside circle. The two circles share the same centerpoint and radius. Insert the required dimension.

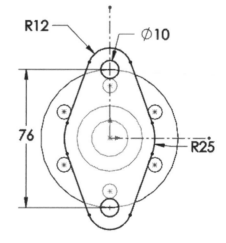

10. Create the first **Extruded Boss (Boss-Extrude2)** feature. The Boss-Extrude2 feature is the connecting tube between the two flanges. Blind is the default End Condition in Direction 1. Depth = 48mm.

11. Create **Sketch4**. Select the front circular face of Extrude3 as the Sketch plane. Sketch a horizontal and vertical centerline from the Origin. Sketch the top and bottom circles symmetric about the horizontal centerline. Dimension the distance between the two circles and their diameter. Create the top centerpoint arc with the centerpoint Coincident to the top circle. The start point and the end point of the arc are horizontal. Sketch the two top angled lines symmetric about the vertical centerline. Apply symmetry. Mirror the two lines and the centerpoint arc about the horizontal centerline. Insert the left and right tangent arcs with a centerpoint Coincident with the Origin. Complete the sketch.

12. Create the second **Extruded Boss** (Boss-Extrude3) feature. Blind is the default End Condition in Direction 1. Depth = 12mm.

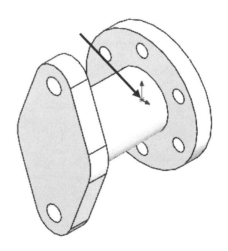

13. Create **Sketch5**. Select the front face of the Extrude feature as the Sketch plane. Sketch a circle. The part Origin is located in the center of the model. Insert the required dimension.

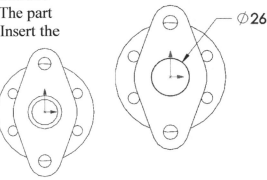

14. Create the second **Extruded Cut** feature. Blind is the default End Condition in Direction 1. Depth = 25mm.

15. Create the **Chamfer** feature. Chamfer1 is an Angle distance chamfer. Chamfer the inside edge of the Extrude feature as illustrated. Distance = 3mm. Angle = 45deg.

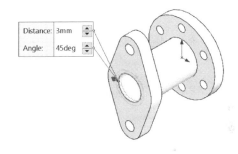

16. Create the **Fillet** feature. Fillet the two edges of Extrude1. Radius = 2mm.

17. **Assign** 6061 Alloy material to the part.

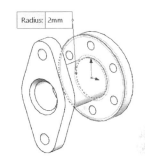

18. **Calculate** the overall mass of the part. The overall mass = 276.97 grams.

19. **Locate** the Center of mass. The location of the Center of mass is relative to the part Origin.

- X: 0.00 millimeters

- Y: 0.00 millimeters

- Z: 21.95 millimeters

20. **Save** the part. Name it Advanced Part 4-2.

21. **Close** the model.

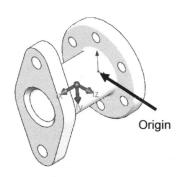

In the Advanced Part Modeling category, an exam question could read: Build this part in SOLIDWORKS. Locate the Center of mass with respect to the part Origin.

- A: X = 0.00 millimeters, Y = 0.00 millimeters, Z = 21.95 millimeters

- B: X = 21.95 millimeters, Y = 10.00 millimeters, Z = 0.00 millimeters

- C: X = 0.00 millimeters, Y = 0.00 millimeters, Z = -27.02 millimeters

- D: X= 1.00 millimeters, Y = -1.01 millimeters, Z = -0.04 millimeters

The correct answer is A.

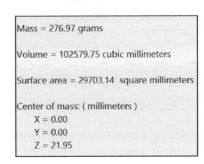

Mass = 276.97 grams

Volume = 102579.75 cubic millimeters

Surface area = 29703.14 square millimeters

Center of mass: (millimeters)
X = 0.00
Y = 0.00
Z = 21.95

Tutorial: Advanced Part 4-3

An exam question in this category could read: Build this part in SOLIDWORKS. Calculate the volume and locate the Center of mass of the illustrated model.

1. **Create** a new part in SOLIDWORKS.

2. **Build** the illustrated dimensioned model. Insert five sketches, five features and a Reference plane: Extruded Base, Plane1, Extruded Boss, Extruded Cut, Fillet and Extruded Cut.

Think about the steps that you would take to build the illustrated part. Insert a Reference plane to create the Extruded Boss feature. Create Sketch2 for Plane1. Plane1 is the Sketch plane for Sketch3. Sketch3 is the profile for Boss-Extrude2.

3. **Set** document properties for the model.

4. Create **Sketch1**. Sketch1 is the Base sketch. Select the Top Plane as the Sketch plane. Sketch a rectangle. Insert the required geometric relations and dimensions.

5. Create the **Extruded Base (Boss-Extrude1)** feature. Blind is the default End Condition in Direction 1. Depth = .700in.

6. Create **Sketch2**. Select the top face of Boss-Extrude1 as the Sketch plane. Sketch a diagonal line as illustrated. Plane1 is the Sketch plane for Sketch3. Sketch3 is the sketch profile for Boss-Extrude2. The Origin is located in the bottom left corner of the sketch. Complete the sketch.

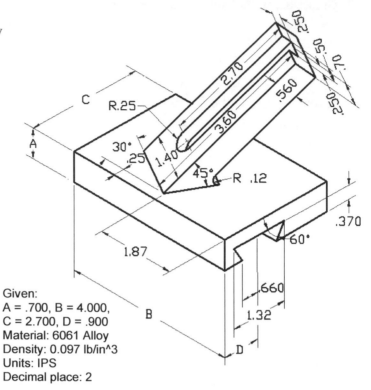

Given:
A = .700, B = 4.000,
C = 2.700, D = .900
Material: 6061 Alloy
Density: 0.097 lb/in^3
Units: IPS
Decimal place: 2

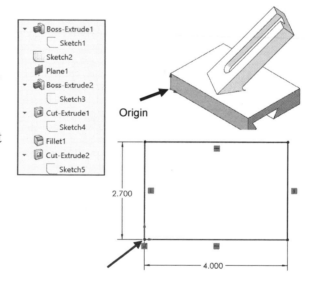

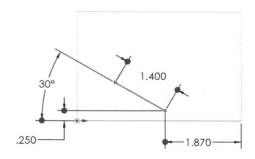

7. Create **Plane1**. Show Sketch2. Select the top face of Boss-Extrude1 and Sketch2. Sketch2 and face<1> are the Reference Entities. Angle = 45 deg.

💡 Activate the Plane PropertyManager. Click **Plane** from the Reference Geometry Consolidated toolbar, or click **Insert**, **Reference Geometry**, **Plane** from the Menu bar.

💡 View Sketch2. Click **View**, **Hide/Show**, **Sketches** from the Menu bar.

8. Create **Sketch3**. Select Plane1 as the Sketch plane. Select the Line Sketch tool. Use Sketch2 as a reference for the width dimension of the rectangle. Insert the required geometric relations and dimension. Sketch3 is the sketch profile for Boss-Extrude2.

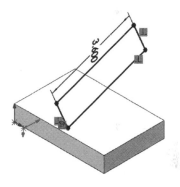

9. Create the **Extruded Boss** feature. Boss-Extrude2 is located on Plane1. Blind is the default End Condition in Direction 1. Depth = .560in.

10. Create **Sketch4**. Select the top angle face of Boss-Extrude2 as the Sketch plane. Sketch4 is the profile for the first Extruded Cut feature. Apply a Mid point relation with the Centerline Sketch tool. Insert a Parallel, Symmetric, Perpendicular, and Tangent relation. Insert the required dimensions.

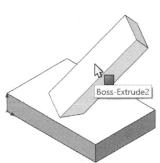

11. Create the first **Extruded Cut** feature. Blind is the default End Condition. Depth = .250in.

💡 There are numerous ways to build the models in this chapter. A goal is to display different design intents and techniques.

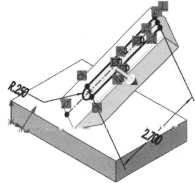

12. Create the **Fillet** feature. Fillet the illustrated edge. Edge<1> is displayed in the Items To Fillet box. Radius = .12in.

13. Create **Sketch5**. Select the right face of Boss-Extrude1 as the Sketch plane. Insert the required relations and dimensions.

14. Create the second **Extruded Cut** feature. Select Through All as the End Condition in Direction 1.

15. **Assign** 6061 Alloy material to the part.

16. **Calculate** the volume of the part. The volume = 8.19 cubic inches.

17. **Locate** the Center of mass. The location of the Center of mass is relative to the part Origin.

- X: 2.08 inches

- Y: 0.79 inches

- Z: -1.60 inches

18. **Save** the part. Name it Advanced Part 4-3.

19. **Close** the model.

As an exercise, apply the MMGS unit system to the part. Modify the material from 6061 Alloy to ABS. Modify the Plane1 angle from 45deg to 30deg. Decimal place: 2.

Calculate the total mass of the part and the location of the Center of mass relative to the part Origin. Save the part. Name it Advanced Part 4-3 MMGS System.

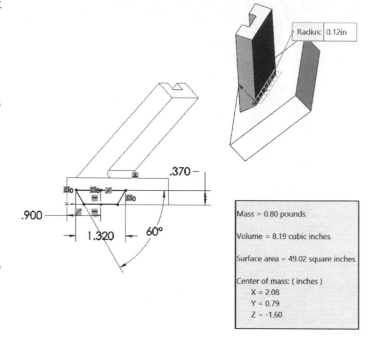

Radius: 0.12in

.370

.900

1.320   60°

Mass = 0.80 pounds

Volume = 8.19 cubic inches

Surface area = 49.02 square inches

Center of mass: (inches)
   X = 2.08
   Y = 0.79
   Z = -1.60

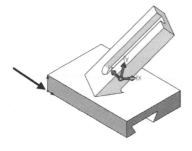

Mass = 134.58 grams

Volume = 131941.14 cubic millimeters

Surface area = 30575.45 square millimeters

Center of mass: (millimeters)
   X = 53.89
   Y = 16.98
   Z = -42.47

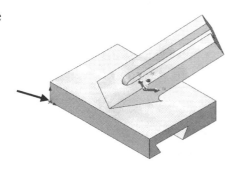

Tutorial: Advanced Part 4-4

An exam question in this category could read: Build this part in SOLIDWORKS. Calculate the volume and locate the Center of mass of the illustrated model.

1. **Create** a new part in SolidWorks.

2. **Build** the illustrated dimensioned model. Create the part with eleven sketches, eleven features and a Reference plane: Extruded Base, Plane1, two Extruded Bosses, two Extruded Cuts, Extruded Boss, Extruded Cut, Extruded-Thin, Mirror, Extruded Cut and Extruded Boss.

Think about the steps that you would take to build the illustrated part. Create the rectangular Base feature. Create Sketch2 for Plane1. Insert Plane1 to create the Extruded Boss feature: Boss-Extrude2. Plane1 is the Sketch plane for Sketch3. Sketch3 is the sketch profile for Boss-Extrude2.

3. **Set** document properties for the model.

4. Create **Sketch1**. Sketch1 is the Base sketch. Select the Top Plane as the Sketch plane. Sketch a rectangle. Insert the required geometric relations and dimensions. Note the location of the Origin.

5. Create the **Extruded Base (Boss-Extrude1)** feature. Blind is the default End Condition in Direction 1. Depth = .500in.

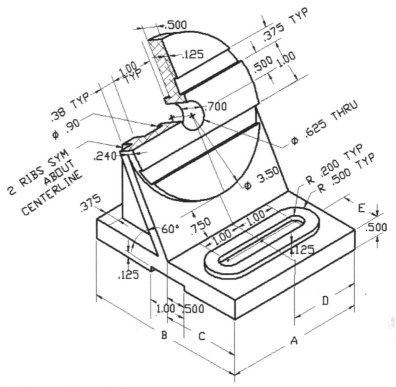

Given:
A = 3.500, B = 4.200, C = 2.000, D =1.750, E = 1.000
Material: 6061 Alloy
Density: 0.097 lb/in^3
Units: IPS
Decimal place: 2

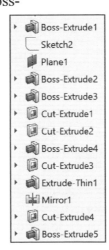

- Boss-Extrude1
  - Sketch2
  - Plane1
- Boss-Extrude2
- Boss-Extrude3
- Cut-Extrude1
- Cut-Extrude2
- Boss-Extrude4
- Cut-Extrude3
- Extrude-Thin1
- Mirror1
- Cut-Extrude4
- Boss-Extrude5

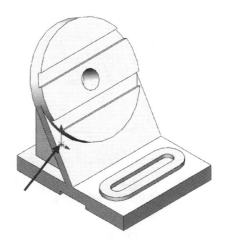

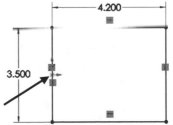

6. Create **Sketch2**. Sketch2 is the sketch profile for Plane1. Select the top face of Extrude1 as the Sketch plane. Sketch a centerline. Show Sketch2.

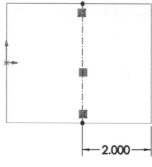

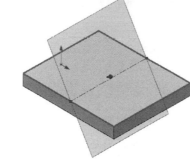

7. Create **Plane1**. Select the top face of Boss-Extrude1 and Sketch2. Face<1> and Line1@Sketch2 are displayed in the Selections box. Angle = 60deg.

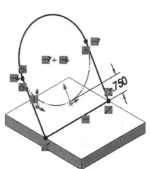

8. Create **Sketch3**. Select Plane1 as the Sketch plane. Sketch3 is the sketch profile for the Extrude feature. Utilize the Convert Entities Sketch tool to convert the Sketch2 line to Plane1. Sketch two equal vertical lines Collinear with the left and right edges. Sketch a construction circle with a diameter Coincident to the left and right vertical lines. Create an 180deg tangent arc between the two vertical lines. Insert the required geometric relations and dimensions. Complete the sketch. Utilize the First arc condition from the Leaders tab in the Dimension PropertyManager to minimum the dimension to the bottom of the circle, Sketch2.

🔆 Insert a Construction circle when dimensions are reference to a minimum or maximum arc condition.

9. Create the **Extruded Boss** (Boss-Extrude2) feature. Blind is the default End Condition. Depth = .260in. Note: .260in = (.500in - .240in). The extrude direction is towards the back.

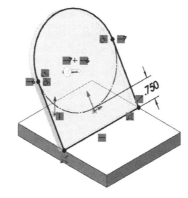

10. Create **Sketch4**. Select the right angled face of the Extrude feature as the Sketch plane. Wake-up the center point of the tangent Arc. Sketch a circle. The circle is Coincident and Coradial to the Extrude feature.

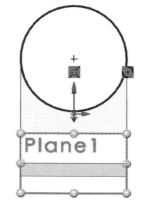

11. Create the second **Extruded Boss** (Boss-Extrude3) feature. Blind is the default End Condition in Direction 1. Depth = .240in.

12. Create **Sketch5**. Sketch5 is the profile for the Extruded Cut feature. Select the right angle face of the Extrude feature as the Sketch plane. Apply the Convert Entities and Trim Sketch tools. Insert the required geometric relations and dimensions.

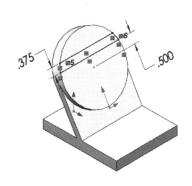

13. Create the first **Extruded Cut** feature. Blind is the default End Condition. Depth = .125in.

14. Create **Sketch6**. Select the right angle face of Boss-Extrude3 as the Sketch plane. Apply the Convert Entities and Trim Sketch tools. Insert the required geometric relations and dimensions.

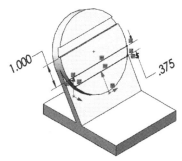

15. Create the second **Extruded Cut** feature. Blind is the default End Condition. Depth = .125in.

16. Create **Sketch7**. Select the left angled face of Boss-Extrude2 as the Sketch plane. Sketch a circle. Insert the required geometric relation and dimension.

17. Create the third **Extruded Boss (Boss-Extrude4)** feature. Blind is the default End Condition. Depth = .200in. Note: .200in – (.700in - .500in).

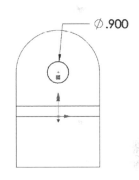

18. Create **Sketch8**. Select the flat circular face of the Extrude feature as illustrated as the Sketch plane. Sketch a circle. Insert the required dimension.

19. Create the third **Extruded Cut** feature. Select Through All for End Condition in Direction 1.

20. Create **Sketch9**. Select the left flat top face of Extrude1 as the Sketch Plane. Sketch a line parallel to the front edge as illustrated. Insert the required geometric relations and dimensions.

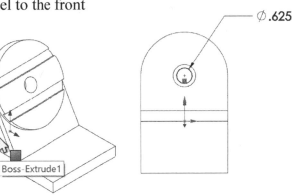

Boss-Extrude1

💡 The book is designed to expose the new user to many tools, techniques and procedures. It may not always use the most direct tool or process.

💡 When you create a new part or assembly, the three default Planes (Front, Right and Top) are aligned with specific views. The Plane you select for the Base sketch determines the orientation of the part.

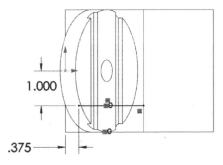

21. Create the **Extrude-Thin1** feature. Extrude-Thin1 is the left support feature. Select Up To Surface for End Condition in Direction 1. Select face<1> for direction as illustrated. Thickness = .38in. Select One-Direction.

22. Create the **Mirror** feature. Mirror the Extrude-Thin1 feature about the Front Plane.

23. Create **Sketch10**. Select the bottom front flat face of Boss-Extrude1 as the Sketch plane. Sketch10 is the profile for the fourth Extruded Cut feature. Insert the required geometric relations and dimensions.

24. Create the fourth **Extruded Cut** feature. Select Through All for End Condition in Direction 1.

25. Create **Sketch11**. Select the top face of Boss-Extrude1 as the Sketch plane. Apply construction geometry. Sketch11 is the profile for the Boss-Extrude5 feature. Insert the required geometric relations and dimensions.

26. Create the **Extruded Boss** feature. Blind is the default End Condition in Direction 1. Depth = .125in.

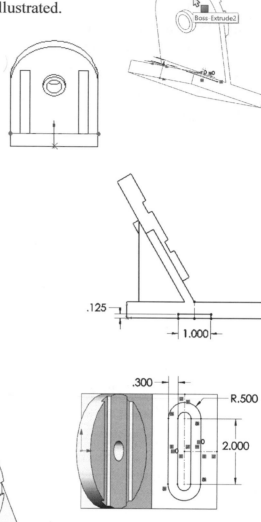

Click on the additional views during the CSWA exam to better understand the part and provided information. Read each question carefully. Identify the dimensions, center of mass and units. Apply needed material.

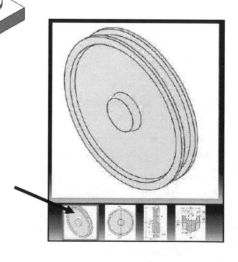

27. **Assign** 6061 Alloy material to the part.

28. **Calculate** the volume of the part. The volume = 14.05 cubic inches.

29. **Locate** the Center of mass. The location of the Center of mass is relative to the part Origin.

- X: 1.59 inches

- Y: 1.19 inches

- Z: 0.00 inches

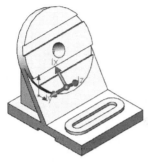

Mass properties of Advanced Part 4-4
Configuration: Default
Coordinate system: -- default --

Density = 0.10 pounds per cubic inch

Mass = 1.37 pounds

Volume = 14.05 cubic inches

Surface area = 79.45 square inches

Center of mass: (inches)
X = 1.59
Y = 1.19
Z = 0.00

30. **Save** the part. Name it Advanced Part 4-4.

31. **Close** the model.

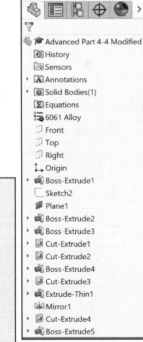

In the Advanced Part Modeling category, an exam question could read: Build this model. Calculate the volume of the part.

- A: 14.05 cubic inches

- B: 18.66 cubic inches

- C: 17.44 cubic inches

- D: 12.71 cubic inches

The correct answer is A.

As an exercise, modify A from 3.50in to 3.60in. Modify B from 4.20in to 4.10in. Modify the Plane1 angle from 60deg to 45deg. Modify the system units from IPS to MMGS. Decimal place: 2.

Calculate the mass and locate the Center of mass. The mass = 597.09 grams.

- X: 34.27 millimeters

- Y: 26.70 millimeters

- Z: 0.00 millimeters

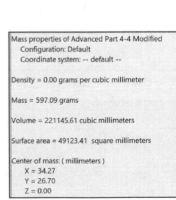

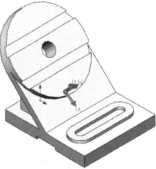

Mass properties of Advanced Part 4-4 Modified
Configuration: Default
Coordinate system: -- default --

Density = 0.00 grams per cubic millimeter

Mass = 597.09 grams

Volume = 221145.61 cubic millimeters

Surface area = 49123.41 square millimeters

Center of mass: (millimeters)
X = 34.27
Y = 26.70
Z = 0.00

32. **Save** the part. Name it Advanced Part 4-4 Modified.

Calculate the Center of Mass Relative to a Created Coordinate System Location

In the Simple Part Modeling chapter, you located the Center of mass relative to the default part Origin. In the Advanced Part Modeling category, you may need to locate the Center of mass relative to a created coordinate system location. The exam model may display a created coordinate system location. Example:

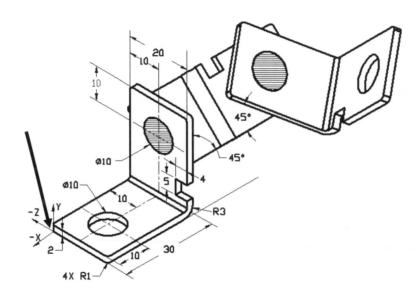

The SOLIDWORKS software displays positive values for (X, Y, Z) coordinates for a reference coordinate system. The CSWA exam displays either a positive or negative sign in front of the (X, Y, Z) coordinates to indicate direction as illustrated, (-X, +Y, -Z).

The following section reviews creating a Coordinate System location for a part.

Tutorial: Coordinate location 4-1

Use the Mass Properties tool to calculate the Center of mass for a part. Decimal place: 2. Unit system: MMGS.

1. **Open** the Plate-3-Point part from the SOLIDWORKS CSWA Folder\Chapter 4 location. View the location of the part Origin.

2. **Locate** the Center of mass. The location of the Center of mass is relative to the part Origin.

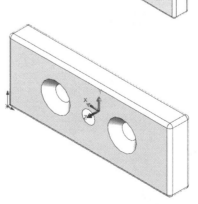

- X = 28.01 millimeters

- Y = 11.01 millimeters

- Z = -3.49 millimeters

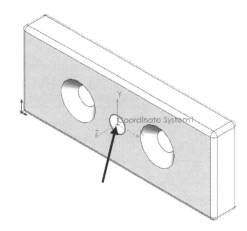

Create a new coordinate system location as illustrated. Locate the new coordinate system location at the center of the center hole. Decimal place: 2. The center of the CBORE hole is under defined. Note: It appears center on the Base-Extrude.

3. Right-**click** the **front face** of Base-Extrude.

4. Click **Sketch** from the Context toolbar.

5. Click the **edge** of the center hole as illustrated.

6. Click **Convert Entities** from the Sketch toolbar. The center point for the new coordinate location is displayed.

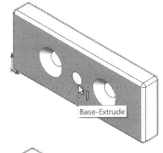

7. **Exit** the sketch. Sketch4 is displayed.

8. Click the **Coordinate System** tool from the Consolidated Reference Geometry toolbar. The Coordinate System PropertyManager is displayed.

9. Click the **center point** of the center hole in the Graphics window. Point2@Sketch4 is displayed in the Selections box as the Origin.

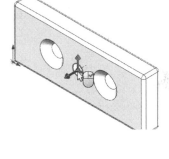

10. Click **OK** from the Coordinate System PropertyManager. Coordinate System1 is displayed.

11. **View** the new coordinate location at the center of the center hole.

View the Mass Properties of the part with the new coordinate location.

12. Click the **Mass Properties** tool from the Evaluate tab.

13. Select **Coordinate System1** from the Output box. The Center of mass relative to the new location is located at the following coordinates: X = .21 millimeters, Y = .31 millimeters, Z = -3.49 millimeters.

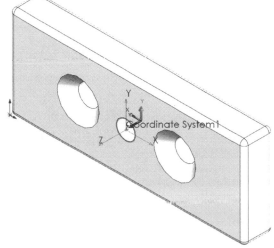

☀ To reverse the direction of an axis, click its **Reverse Axis Direction** button in the Coordinate System PropertyManager.

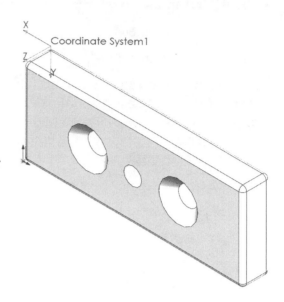

Tutorial: Coordinate location 4-2

Create a new coordinate system location. Locate the new coordinate system at the top back point as illustrated. The center of the CBORE hole is under defined. It appears center on the Base-Extrude.

1. **Open** the Plate-1 part from the SOLIDWORKS CSWA Folder\Chapter 4 location.

2. **View** the location of the part Origin.

3. Drag the **Rollback bar** under the Base-Extrude feature in the FeatureManager.

4. Click the **Coordinate System** tool from the Consolidated Reference Geometry toolbar. The Coordinate System PropertyManager is displayed.

5. Click the **back left vertex** as illustrated.

6. Click the **top back horizontal** edge as illustrated. Do not select the midpoint.

7. Click the **back left vertical** edge as illustrated.

8. Click **OK** from the Coordinate System PropertyManager. Coordinate System1 is displayed in the FeatureManager and in the Graphics window.

9. Drag the **Rollback bar** to the bottom of the FeatureManager.

10. **Calculate** the Center of mass relative to the new coordinate system.

11. Select **Coordinate System1**. The Center of mass relative to the new location is located at the following coordinates:

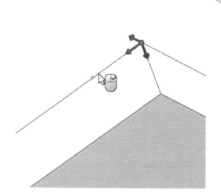

- X = -28.01 millimeters

- Y = 10.99 millimeters

- Z = 3.51 millimeters

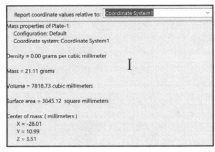

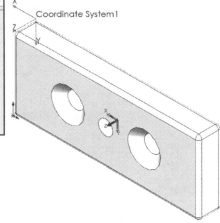

12. **Close** the model.

Tutorial: Advanced Part 4-5

An exam question in this category could read: Build this part in SOLIDWORKS. Calculate the overall mass and locate the Center of mass of the illustrated model.

1. **Create** a new part in SOLIDWORKS.

2. **Build** the illustrated dimensioned model. Insert thirteen features: Extrude-Thin1, Fillet, two Extruded Cuts, Circular Pattern, two Extruded Cuts, Mirror, Chamfer, Extruded Cut, Mirror, Extruded Cut and Mirror.

Think about the steps that you would take to build the illustrated part. Review the provided information. The depth of the left side is 50mm. The depth of the right side is 60mm.

🔅 There are numerous ways to build the models in this chapter. A goal is to display different design intents and techniques.

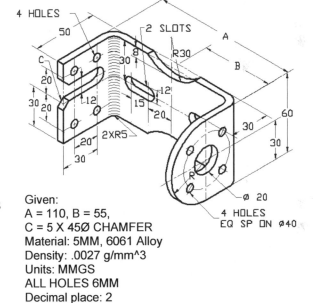

Given:
A = 110, B = 55,
C = 5 X 45Ø CHAMFER
Material: 5MM, 6061 Alloy
Density: .0027 g/mm^3
Units: MMGS
ALL HOLES 6MM
Decimal place: 2

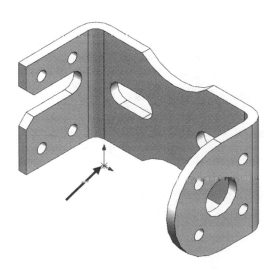

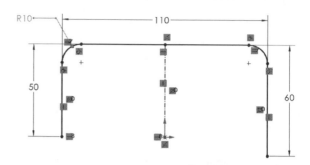

💡 If the inside radius = 5mm and the material thickness = 5mm, then the outside radius = 10mm.

3. **Set** document properties for the model.

4. Create **Sketch1**. Sketch1 is the Base sketch. Select the Top Plane as the Sketch plane. Apply the Line and Sketch Fillet Sketch tools. Apply construction geometry. Insert the required geometric relations and dimensions.

5. Create the **Extrude-Thin1** feature. Extrude-Thin1 is the Base feature. Apply symmetry in Direction 1. Depth = 60mm. Thickness = 5mm. Check the Auto-fillet corners box. Radius = 5mm.

💡 The Auto-fillet corners option creates a round at each edge where lines meet at an angle.

6. Create the **Fillet** feature. Fillet1 is a full round fillet. Fillet the three illustrated faces: top, front and bottom.

7. Create **Sketch2**. Select the right face as the Sketch plane. Wake-up the centerpoint. Sketch a circle. Insert the required relation and dimension.

8. Create the first **Extruded Cut** feature. Select Up To Next for the End Condition in Direction 1.

💡 The Up To Next End Condition extends the feature from the sketch plane to the next surface that intercepts the entire profile. The intercepting surface must be on the same part.

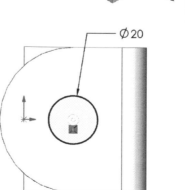

9. Create **Sketch3**. Select the right face as the Sketch plane. Create the profile for the second Extruded Cut feature. This is the seed feature for CirPattern1. Apply construction geometry to locate the center point of Sketch3. Insert the required relations and dimensions.

10. Create the second **Extruded Cut** feature. Select Up To Next for the End Condition in Direction 1.

11. Create the **Circular Pattern** feature. Number of Instances = 4. Default angle = 360deg.

12. Create **Sketch4**. Select the left outside face of Extrude-Thin1 as the Sketch plane. Apply the Line and Tangent Arc Sketch tool to create Sketch4. Insert the required geometric relations and dimensions.

13. Create the third **Extruded Cut** feature. Select Up To Next for End Condition in Direction 1. The Slot on the left side of Extrude-Thin1 is created.

14. Create **Sketch5**. Select the left outside face of Extrude-Thin1 as the Sketch plane. Sketch two circles. Insert the required geometric relations and dimensions.

15. Create the fourth **Extruded Cut** feature. Select Up To Next for End Condition in Direction 1.

There are numerous ways to create the models in this chapter. A goal is to display different design intents and techniques.

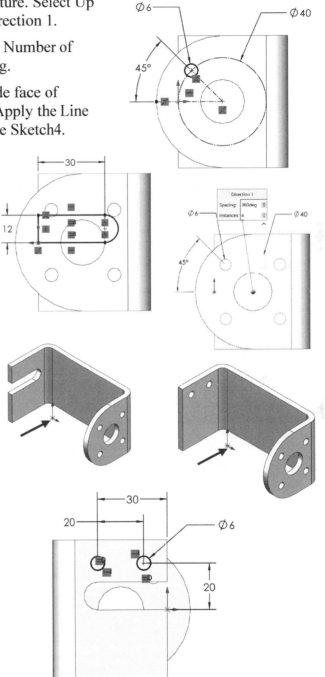

16. Create the first **Mirror** feature. Mirror the top two holes about the Top Plane.

17. Create the **Chamfer** feature. Create an Angle distance chamfer. Chamfer the selected edges as illustrated. Distance = 5mm. Angle = 45deg.

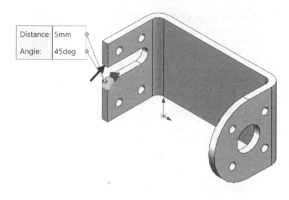

18. Create **Sketch6**. Select the front face of Extrude-Thin1 as the Sketch plane. Insert the required geometric relations and dimensions.

19. Create the fifth **Extruded Cut** feature. Select Thought All for End Condition in Direction 1.

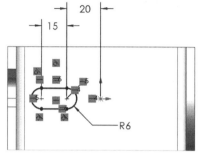

20. Create the second **Mirror** feature. Mirror Extrude5 about the Right Plane.

21. Create **Sketch7**. Select the front face of Extrude-Thin1 as the Sketch plane. Apply the 3 Point Arc Sketch tool. Apply the min First Arc Condition option. Insert the required geometric relations and dimensions.

22. Create the last **Extruded Cut** feature. Through All is the End Condition in Direction 1 and Direction 2.

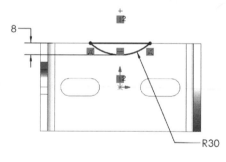

23. Create the third **Mirror** feature. Mirror the Extrude feature about the Top Plane as illustrated.

24. **Assign** the material to the part.

25. **Calculate** the overall mass of the part. The overall mass = 132.45 grams.

26. **Locate** the Center of mass relative to the part Origin:

- X: 1.83 millimeters

- Y: -0.27 millimeters

- Z: -35.38 millimeters

27. **Save** the part. Name it Advanced Part 4-5.

28. **Close** the model.

Mass properties of Advanced Part 4-5
Configuration: Default
Coordinate system: -- default --

Density = 0.00 grams per cubic millimeter

Mass = 132.45 grams

Volume = 49055.56 cubic millimeters

Surface area = 24219.80 square millimeters

Center of mass: (millimeters)
X = 1.83
Y = -0.27
Z = -35.38

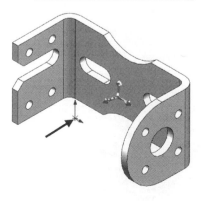

All questions on the exam are in a multiple-choice single answer or fill in the blank format. In the Advanced Part Modeling category, an exam question could read: Build this model. Calculate the overall mass of the part with the provided information.

- A: 139.34 grams

- B: 155.19 grams

- C: 132.45 grams

- D: 143.91 grams

The correct answer is C.

Mass properties of Advanced Part 4-5
  Configuration: Default
  Coordinate system: -- default --

Density = 0.00 grams per cubic millimeter

Mass = 132.45 grams

Volume = 49055.56 cubic millimeters

Surface area = 24219.80 square millimeters

Center of mass: (millimeters)
  X = 1.83
  Y = -0.27
  Z = -35.38

☀ Use the Options button in the Mass Properties dialog box to apply custom settings to units.

Tutorial: Advanced Part 4-5A

An exam question in this category could read: Build this part in SOLIDWORKS. Locate the Center of mass. Note the coordinate system location of the model as illustrated.

Where do you start? Build the model as you did in the Tutorial: Advanced Part 4-5. Create Coordinate System1 to locate the Center of mass.

1. **Open** Advanced Part 4-5 from your SOLIDWORKS folder.

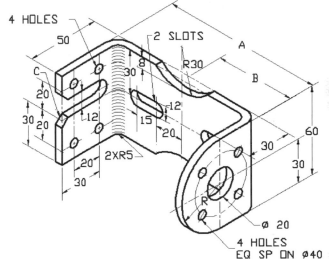

A = 110, B = 55, C = 5 X 45Ø CHAMFER
Material: 5MM, 6061 Alloy
Density: .0027 g/mm^3
Units: MMGS
ALL HOLES 6MM
Decimal place: 2

Create the illustrated coordinate system location.

2. Show **Sketch2** from the FeatureManager design tree.

3. Click the **center point** of Sketch2 in the Graphics window as illustrated.

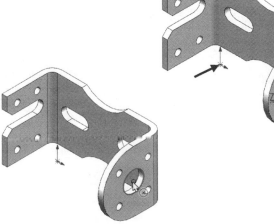

4. Click the **Coordinate System** tool from the Consolidated Reference Geometry toolbar. The Coordinate System PropertyManager is displayed. Point2@Sketch2 is displayed in the Origin box.

5. Click **OK** from the Coordinate System PropertyManager. Coordinate System1 is displayed.

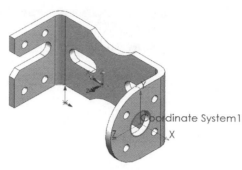

6. **Locate** the Center of mass based on the location of the illustrated coordinate system. Select Coordinate System1.

- X: -53.17 millimeters

- Y: -0.27 millimeters

- Z: -15.38 millimeters

7. **Save** the part. Name it Advanced Part 4-5A.

8. **Close** the model.

```
Mass = 132.45 grams

Volume = 49055.56 cubic millimeters

Surface area = 24219.80  square millimeters

Center of mass: ( millimeters )
    X = -53.17
    Y = -0.27
    Z = -15.38
```

Tutorial: Advanced Part 4-5B

Build this part in SOLIDWORKS. Locate the Center of mass. View the location of the coordinate system. The coordinate system is located at the left front point of the model.

Build the illustrated model as you did in the Tutorial: Advanced Part 4-5. Create Coordinate System1 to locate the Center of mass for the model.

1. **Open** Advance Part 4-5 from your SOLIDWORKS folder.

Create the illustrated coordinate system.

2. Click the **vertex** as illustrated for the Origin location.

4 HOLES
50
2 SLOTS
A
C
R30
B
20
8
30
30 20
12
12
15
20
2XR5
20
30
R
60
30
30
Ø 20
4 HOLES
EQ SP ON Ø40

Given:
A = 110, B = 55,
C = 5 X 45Ø CHAMFER
Material: 5MM, 6061 Alloy
Density: .0027 g/mm^3
Units: MMGS
ALL HOLES 6MM
Decimal place: 2

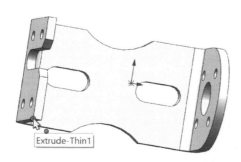

Extrude-Thin1

Coordinate System1

3. Click the **Coordinate System** tool from the Consolidated Reference Geometry toolbar. The Coordinate System PropertyManager is displayed. Vertex<1> is displayed in the Origin box.

4. Click the **bottom horizontal edge** as illustrated. Edge<1> is displayed in the X Axis Direction box. Make any direction modification if needed.

5. Click the **left back vertical edge** as illustrated. Edge<2> is displayed in the Y Axis Direction box. Make any direction modification if needed.

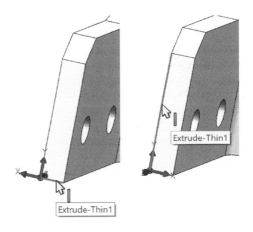

6. Click **OK** from the Coordinate System PropertyManager. Coordinate System1 is displayed.

7. **Locate** the Center of mass based on the location of the illustrated coordinate system. Select Coordinate System1.

- X: 56.83 millimeters

- Y: -29.73 millimeters

- Z: 35.38 millimeters

8. **Save** the part. Name it Advanced Part 4-5B.

9. **Close** the model.

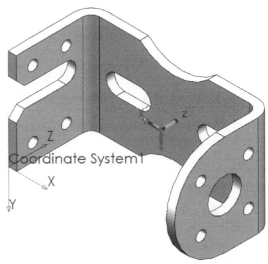

In the Advanced Part Modeling category, an exam question could read: Build this model. Locate the Center of mass.

- A: X = 56.83 millimeters, Y = -29.73 millimeters, Z = 35.38 millimeters

- B: X = 1.80 millimeters, Y = -0.27 millimeters, Z = -35.54 millimeters

- C: X = -59.20 millimeters, Y = -0.27 millimeters, Z = -15.54 millimeters

- D: X= -1.80 millimeters, Y = 1.05 millimeters, Z = -0.14 millimeters

The correct answer is A.

Mass properties of Advanced Part 4-5B
    Configuration: Default
    Coordinate system: Coordinate System1

Density = 0.00 grams per cubic millimeter

Mass = 132.45 grams

Volume = 49055.56 cubic millimeters

Surface area = 24219.80 square millimeters

Center of mass: (millimeters)
    X = 56.83
    Y = -29.73
    Z = 35.38

Tutorial: Advanced Part 4-6

An exam question in this category could read: Build this part in SOLIDWORKS. Calculate the overall mass and locate the Center of mass of the illustrated model.

1. **Create** a new part in SOLIDWORKS.

2. **Build** the illustrated dimensioned model. Insert twelve features and a Reference plane: Extrude-Thin1, two Extruded Bosses, Extruded Cut, Extruded Boss, Extruded Cut, Plane1, Mirror and five Extruded Cuts.

Think about the steps that you would take to build the illustrated part. Create an Extrude-Thin1 feature as the Base feature.

3. **Set** document properties for the model. Review the given information.

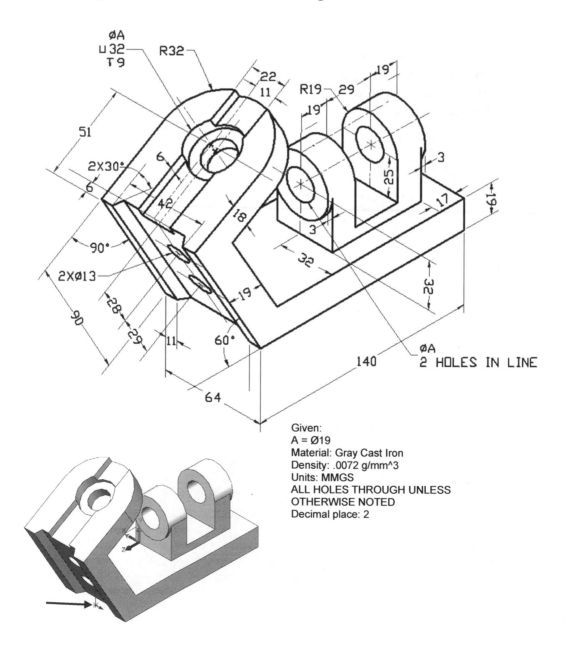

Given:
A = Ø19
Material: Gray Cast Iron
Density: .0072 g/mm^3
Units: MMGS
ALL HOLES THROUGH UNLESS
OTHERWISE NOTED
Decimal place: 2

4. Create **Sketch1**. Sketch1 is the Base sketch. Select the Right Plane as the Sketch plane. Apply construction geometry. Insert the required geometric relations and dimensions. Sketch1 is the profile for Extrude-Thin1. Note the location of the Origin.

5. Create the **Extrude-Thin1** feature. Apply symmetry. Select Mid Plane as the End Condition in Direction 1. Depth = 64mm. Thickness = 19mm.

6. Create **Sketch2**. Select the top narrow face of Extrude-Thin1 as the Sketch plane. Sketch three lines: two vertical and one horizontal and a tangent arc. Insert the required geometric relations and dimensions.

7. Create the **Boss-Extrude1** feature. Blind is the default End Condition in Direction 1. Depth = 18mm.

8. Create **Sketch3**. Select the Right Plane as the Sketch plane. Sketch a rectangle. Insert the required geometric relations and dimensions. Note: 61mm = (19mm - 3mm) x 2 + 29mm.

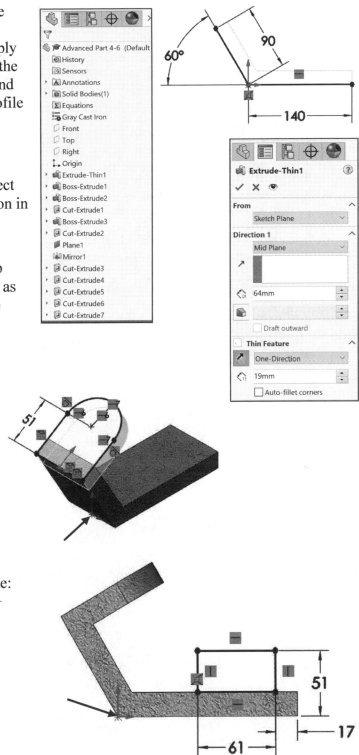

9. Create the **Boss-Extrude2** feature. Select Mid Plane for End Condition in Direction 1. Depth = 38mm. Note: 2 x R19.

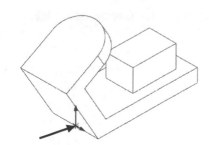

10. Create **Sketch4**. Select the Right Plane as the Sketch plane. Sketch a vertical centerline from the top midpoint of the sketch. The centerline is required for Plane1. Plane1 is a Reference plane. Sketch a rectangle symmetric about the centerline. Insert the required relations and dimensions. Sketch4 is the profile for Extrude3.

11. Create the first **Extruded Cut** feature. Extrude in both directions. Select Through All for End Condition in Direction 1 and Direction 2.

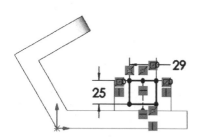

12. Create **Sketch5**. Select the inside face of the Extrude feature for the Sketch plane. Sketch a circle from the top midpoint. Sketch a construction circle. Construction geometry is required for future features. Complete the sketch.

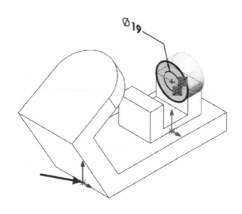

13. Create the **Extruded Boss** (Boss-Extrude3) feature. Blind is the default End Condition. Depth = 19mm.

14. Create **Sketch6**. Select the inside face for the Sketch plane. Show Sketch5. Select the construction circle in Sketch5. Apply the Convert Entities Sketch tool.

15. Create the second **Extruded Cut** feature. Select the Up To Next End Condition in Direction 1.

There are numerous ways to create the models in this chapter. A goal is to display different design intents and techniques.

Tangent edges and origin are displayed for educational purposes.

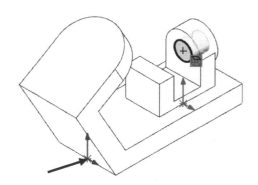

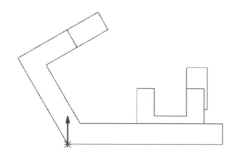

16. Create **Plane1**. Apply symmetry. Create Plane1 to mirror Cut-Extrude2 and Boss-Extrude3. Create a Parallel Plane at Point. Select the midpoint of Sketch4, and Face<1> as illustrated. Point1@Sketch4 and Face<1> is displayed in the Selections box.

17. Create the **Mirror** feature. Mirror Cut-Extrude2 and Boss-Extrude3 about Plane1.

The Mirror feature creates a copy of a feature, (or multiple features), mirrored about a face or a plane. You can select the feature or you can select the faces that comprise the feature.

18. Create **Sketch7**. Select the top front angled face of Extrude-Thin1 as the Sketch plane. Apply the Centerline Sketch tool. Insert the required geometric relations and dimensions.

19. Create the third **Extruded Cut** feature. Select Through All for End Condition in Direction 1. Select the edge for the vector to extrude.

Click on the additional views during the CSWA exam to better understand the part and provided information. Read each question carefully. Identify the dimensions, center of mass and units. Apply needed material.

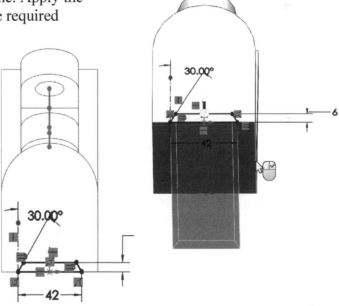

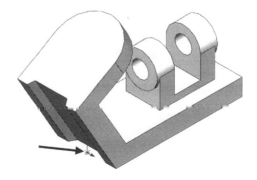

20. Create **Sketch8**. Select the top front angled face of Extrude-Thin1 as the Sketch plane. Sketch a centerline. Sketch two vertical lines and a horizontal line. Select the top arc edge. Apply the Convert Entities Sketch tool. Apply the Trim Sketch tool to remove the unwanted arc geometry. Insert the required geometric relations and dimension.

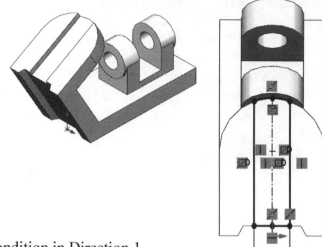

21. Create the fourth **Extruded Cut** feature. Blind is the default End Condition in Direction 1. Depth = 6mm.

22. Create **Sketch9**. Create a Cbore with Sketch9 and Sketch10. Select the top front angled face of Extrude-Thin1 as illustrated for the Sketch plane. Extrude8 is the center hole in the Extrude-Thin1 feature. Sketch a circle. Insert the required geometric relations and dimension.

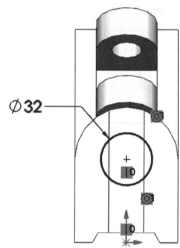

23. Create the fifth **Extrude Cut** feature. Blind is the default End Condition. Depth = 9mm. This is the first feature for the Cbore.

24. Create **Sketch10**. Select the top front angled face of Extrude-Thin1 as the Sketch plane. Sketch a circle. Insert the required geometric relation and dimension. A = Ø19.

25. Create the sixth **Extruded Cut** feature. Select the Up To Next End Condition in Direction 1. The Cbore is complete.

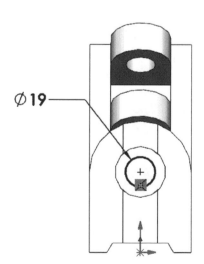

☀ In the exam you are allowed to answer the questions in any order. Use the Summary Screen during the exam to view the list of all questions you have or have not answered.

26. Create **Sketch11**. Select the front angle face of the Extrude feature for the Sketch plane. Sketch two circles. Insert the required geometric relations and dimensions.

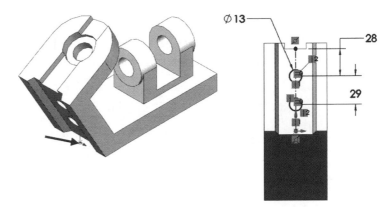

27. Create the last **Extruded Cut** feature. Select the Up To Next End Condition in Direction 1.

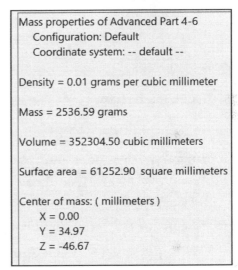

The FilletXpert manages, organizes and reorders constant radius fillets.

The FilletXpert automatically calls the FeatureXpert tool when it has trouble placing a fillet on the specified geometry.

28. **Assign** the material to the part.

29. **Calculate** the overall mass of the part. The overall mass = 2536.59 grams.

30. **Locate** the Center of mass relative to the part Origin:

- X: 0.00 millimeters

- Y: 34.97 millimeters

- Z: -46.67 millimeters

31. **Save** the part.

32. **Name** it Advanced Part 4-6.

The book is designed to expose the new user to many tools, techniques and procedures. It may not always use the most direct tool or process.

Mass properties of Advanced Part 4-6
  Configuration: Default
  Coordinate system: -- default --

Density = 0.01 grams per cubic millimeter

Mass = 2536.59 grams

Volume = 352304.50 cubic millimeters

Surface area = 61252.90 square millimeters

Center of mass: (millimeters)
  X = 0.00
  Y = 34.97
  Z = -46.67

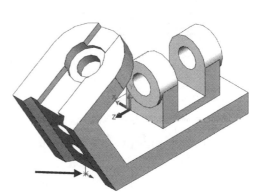

Tutorial: Advanced Part 4-6A

An exam question in this category could read: Build this part in SOLIDWORKS. Locate the Center of mass for the illustrated coordinate system.

Where do you start? Build the illustrated model as you did in the Tutorial: Advanced Part 4-6. Create Coordinate System1 to locate the Center of mass for the model.

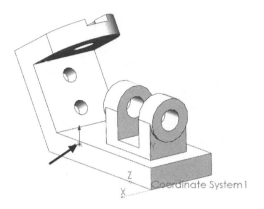

1. **Open** Advanced Part 4-6 from your folder.

Create the illustrated Coordinate system.

2. Click the **Coordinate System** tool from the Consolidated Reference Geometry toolbar. The Coordinate System PropertyManager is displayed.

3. Click the **bottom right vertex** of Extrude-Thin1 as illustrated. Vertex<1> is displayed in the Origin box.

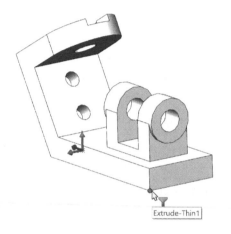

4. Click **OK** from the Coordinate System PropertyManager. Coordinate System1 is displayed.

5. **Locate** the Center of mass based on the location of the illustrated coordinate system. Select Coordinate System1.

- X: -32.00 millimeters

- Y: 34.97 millimeters

- Z: 93.33 millimeters

6. **Save** the part.

7. **Name** it Advanced Part 4-6A.

8. **View** the Center of mass with the default coordinate system.

9. **Close** the model.

Mass properties of Advanced Part 4-6
   Configuration: Default
   Coordinate system: Coordinate System1

Density = 0.01 grams per cubic millimeter

Mass = 2747.98 grams

Volume = 352304.50 cubic millimeters

Surface area = 61252.90 square millimeters

Center of mass: (millimeters)
   X = -32.00
   Y = 34.97
   Z = 93.33

Tutorial: Advanced Part 4-7

An exam question in this category could read: Build this part in SOLIDWORKS. Calculate the overall mass and locate the Center of mass of the illustrated model.

🔆 If needed, see additional model views and dimensions at the end of this tutorial.

1. **Create** a new part in SOLIDWORKS.

2. **Build** the illustrated dimensioned model. Insert thirteen features: Extruded Base, nine Extruded Cuts, two Extruded Bosses and a Chamfer. Note: The center point of the top hole is located 30mm from the top right edge.

Think about the steps that you would take to build the illustrated part. Review the centerlines that outline the overall size of the part.

3. **Set** document properties for the model.

4. Create **Sketch1**. Sketch1 is the Base sketch. Select the Right Plane as the Sketch plane. Sketch a rectangle. Insert the required geometric relations and dimensions. The part Origin is located in the bottom left corner of the sketch.

5. Create the **Extruded Base** feature. Blind is the default End Condition in Direction 1. Depth = 50mm. Boss-Extrude1 is the Base feature.

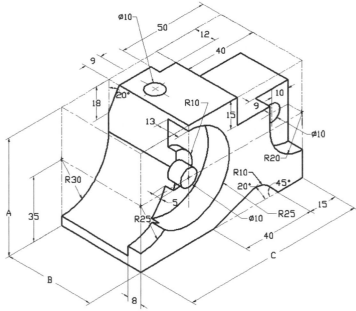

Given:
A = 63, B = 50, C = 100
Material: Copper
Units: MMGS
Density: .0089 g/mm^3
Top hole is 20mm from the top front edge.
All HOLES THROUGH ALL
Decimal place: 2

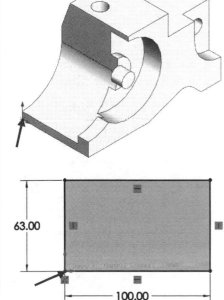

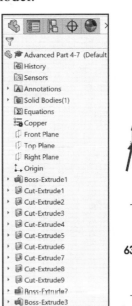

6. Create **Sketch2**. Select the right face of Extrude1 as the Sketch plane. Sketch a 90deg tangent arc. Sketch three lines to complete the sketch. Insert the required geometric relations and dimensions.

7. Create the first **Extruded Cut** feature. Offset the extrude feature. Select the Offset Start Condition. Offset value = 8.0mm. Blind is the default End Condition. Depth = 50mm.

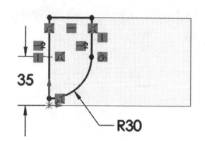

The default Start Condition in the Extrude PropertyManager is Sketch Plane. The Offset start condition starts the extrude feature on a plane that is offset from the current Sketch plane.

8. Create **Sketch3**. Select the right face as the Sketch plane. Create the Extrude profile. Insert the required geometric relations and dimensions.

9. Create the second **Extruded Cut** feature. Select Through All for End Condition in Direction 1.

10. Create **Sketch4**. Select the top face of the Extrude feature as the Sketch Plane. Select the top edge to reference the 10mm dimension. Insert the required geometric relations and dimensions.

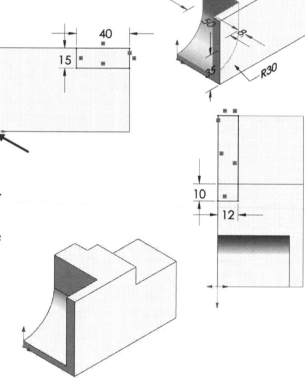

11. Create the third **Extruded Cut** feature. Select Through All for End Condition in Direction 1.

12. Create **Sketch5**. Select the right face of the Extrude feature as the Sketch Plane. Apply construction geometry. Sketch a 90deg tangent arc. Sketch three lines to complete the sketch. Insert the required geometric relations and dimensions.

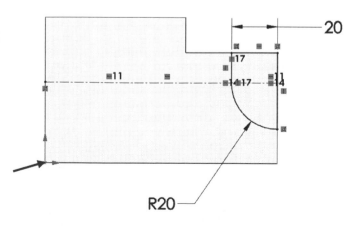

13. Create the fourth **Extruded Cut** feature. Blind is the default End Condition. Depth = 9mm.

14. Create **Sketch6**. Select the right face of the Extrude feature as the Sketch plane. Sketch a circle. Insert the required dimensions.

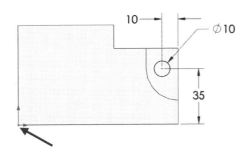

15. Create the fifth **Extruded Cut** feature. Select Through All for End Condition in Direction 1.

16. Create **Sketch7**. The top hole is 20mm from the top front edge. Select the top face of Extrude1 as the Sketch Plane. Sketch a circle. Insert the required dimensions and relations.

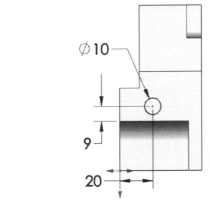

17. Create the sixth **Extruded Cut** feature. Select Through All for End Condition in Direction 1.

There are numerous ways to create the models in this chapter. A goal of this text is to display different design intents and techniques.

18. Create **Sketch8**. Select the right face of Extrude1 as the Sketch plane. Insert a tangent arc as illustrated. Complete the sketch. Insert the required relations and dimensions.

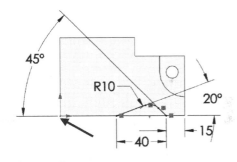

19. Create the seventh **Extruded Cut** feature. Apply symmetry. Select the Through All End Condition in Direction 1 and Direction 2.

20. Create **Sketch9**. Select the right face of Extrude1 as the Sketch plane. Select Hidden Lines Visible. Sketch two construction circles centered about the end point of the arc. Apply the 3 Point Arc Sketch tool. Complete the sketch. Insert the required relations and dimensions.

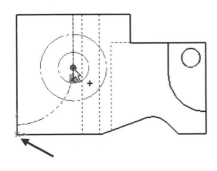

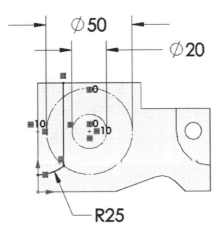

21. Create the eight **Extruded Cut** feature. Select the Through All End Condition in Direction 1 and Direction 2. Note the direction of the extrude feature from the illustration.

22. Create **Sketch10**. Select the right face of Extrude1 as the Sketch plane. Sketch a circle centered at the end point of the arc as illustrated. Apply the Trim Entities Sketch tool. Display Sketch9. Complete the sketch. Insert the required geometric relations.

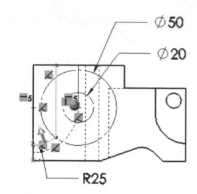

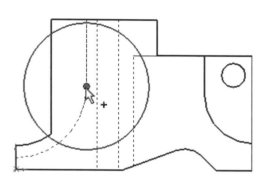
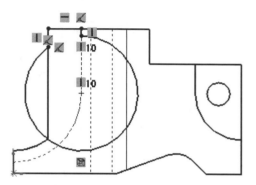

23. Create the ninth **Extruded Cut** feature. Blind is the default End Condition in Direction 1. Depth = 13mm. The feature is displayed.

24. Create **Sketch11**. Select the right face of the Extrude feature as the Sketch plane. Select the construction circle from Sketch9, the left arc, and the left edge as illustrated. Apply the Convert Entities Sketch tool. Apply the Trim Sketch tool. Insert the required relations.

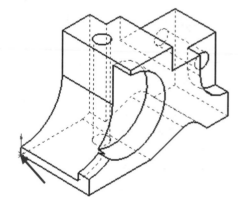

25. Create the **Extruded Boss** feature. Blind is the default End Condition in Direction 1. Depth = 5.00mm.

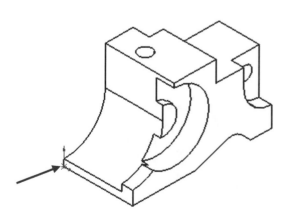
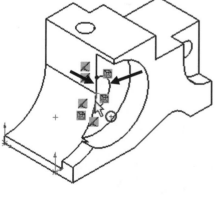

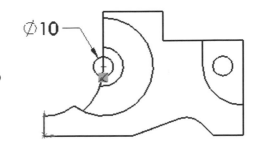

26. Create **Sketch12**. Select the right face of the Extrude feature as the Sketch plane. Sketch a circle. Insert the require relation and dimension.

27. Create the **Extruded Boss** feature. Select the Up To Surface End Condition in Direction 1. Select the right face of Extrude1 for Direction 1.

28. Create the **Chamfer** feature. Chamfer the left edge as illustrated. Distance = 18mm. Angle = 20deg.

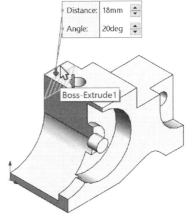

29. **Assign** material to the part.

30. **Calculate** the overall mass of the part. The overall mass = 1280.33 grams.

31. **Locate** the Center of mass relative to the part Origin:

- X: 26.81 millimeters

- Y: 25.80 millimeters

- Z: -56.06 millimeters

32. **Save** the part. Name it Advanced Part 4-7.

33. **Close** the model.

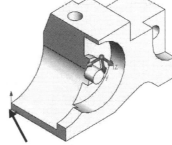

Mass = 1280.33 grams

Volume = 143857.58 cubic millimeters

Surface area = 26112.48 square millimeters

Center of mass: (millimeters)
    X = 26.81
    Y = 25.80
    Z = -56.06

This example was taken from the SOLIDWORKS website as an example of an Advanced Part on the CSWA exam. This model has thirteen features and twelve sketches. As stated throughout the text, there are numerous ways to create the models in these chapters.

A goal in this text is to display different design intents and techniques, and to provide you with the ability to successfully address the models in the given time frame of the CSWA exam.

🔆 Click on the different provided views to obtain additional model information.

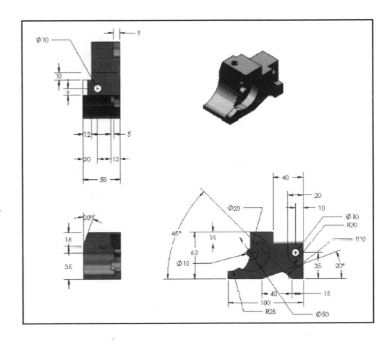

Summary

Advanced Part Creation and Modification is one of the five categories on the CSWA exam. The main difference between the Advanced Part Creation and Modification and the Basic Part Creation and Modification or the Intermediate Part Creation and Modification category is the complexity of the sketches and the number of dimensions and geometric relations along with an increased number of features.

There are three questions on the CSWA exam in this category. One question is in a multiple-choice single answer format and the other two questions (Modification of the model) are in the fill in the blank format.

Each question is worth fifteen (15) points for a total of forty-five (45) points. You are required to build a model with six or more features and to answer a question either on the overall mass, volume, or the location of the Center of mass for the created model relative to the default part Origin location. You are then requested to modify the model and answer a fill in the blank format question.

Assembly Creation and Modification (Bottom-up) is the next chapter in this book. Up to this point, a Basic Part, Intermediate Part or an Advanced part was the focus. The *Assembly Creation and Modification* category addresses an assembly with numerous sub-components. All sub-components are provided to you in the exam.

The next chapter covers the general concepts and terminology used in the *Assembly Creation and Modification* category and then addresses the core elements. Knowledge of Standard mates is required in this category.

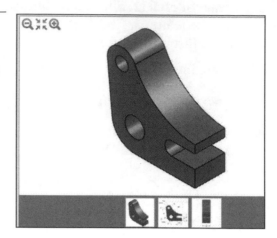

Create the part

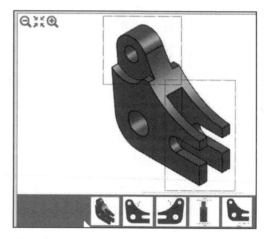

Use the created part in the previous question and modify it by removing material in the indicated areas.

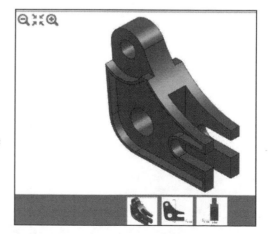

Use the part created in the previous question and modify it by adding a drafted pocket.

There are four questions. Multiple choice/single answer in the category. Each question is worth 30 points.

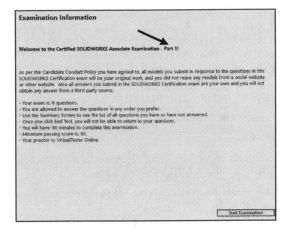

The CSWA Academic exam is provided either in a single 3-hour segment, or 2 - 90-minute segments. The CSWA exam for industry is only provided in a single 3-hour segment. All exams cover the same material.

This book addresses the CSWA Academic exam provided in 2 - 90-minute segments.

Part 1 of the CSWA Academic exam is 90 minutes, minimum passing score is 80, with 6 questions.

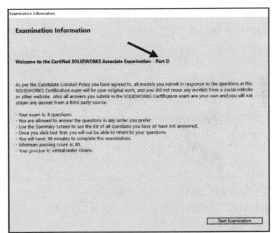

Part 2 of the CSWA Academic exam is 90 minutes, minimum passing score is 80 with 8 questions.

If your school is an academic certification provider, your instructor can allocate free exam credits for the CSWA (Segment 1 & 2), CSWA-SD, CSWA-S, and CSWA-AM certifications. The instructor will require your .edu email address.

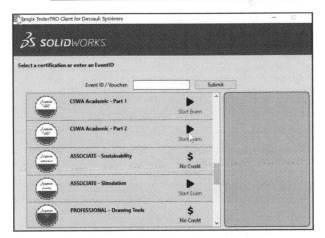

Homework Questions

Question 1:

In Tutorial, Advanced Part 4-1 you created the illustrated part. Modify the Base flange thickness from .40in to .50in. Modify the Chamfer feature angle from 45deg to 33deg. Modify the Fillet feature radius from .10in to .125in. Modify the material from 1060 Alloy to Nickel.

Calculate the overall mass of the part, volume, and locate the Center of mass with the provided information. Save the part. Name it Homework 4-1. **Note:** The original model is located in the SOLIDWORKS CSWA Model 20XX Folder\Chapter 4 folder.

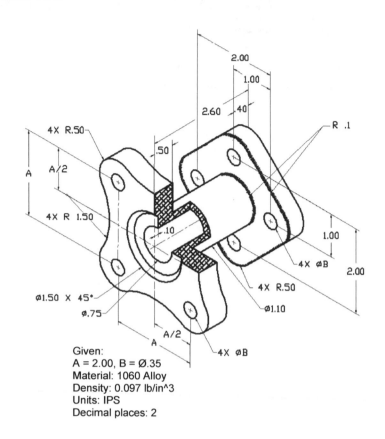

Given:
A = 2.00, B = Ø.35
Material: 1060 Alloy
Density: 0.097 lb/in^3
Units: IPS
Decimal places: 2

Question 2:

In Tutorial, Advanced Part 4-2 you created the illustrated part. Modify the CirPattern1 feature. Modify the number of instances from 6 to 8. Modify the seed feature from an 8mm diameter to a 6mm diameter.

Calculate the overall mass, volume, and the location of the Center of mass relative to the part Origin. Save the part. Name it Homework 4-2. **Note:** The original model is located in the SOLIDWORKS CSWA Model 20XX Folder\Chapter 4 folder.

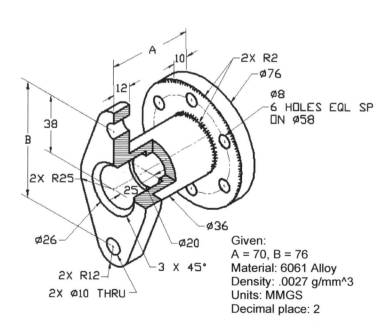

Given:
A = 70, B = 76
Material: 6061 Alloy
Density: .0027 g/mm^3
Units: MMGS
Decimal place: 2

Question 3:

In Tutorial, Advanced Part 4-3 you created the illustrated part. Modify the material from 6061 Alloy to Copper. Modify the B dimension from 4.000in. to 3.500in.

Modify the Fillet radius from .12in to .14in. Modify the unit system from IPS to MMGS.

Calculate the volume of the part and the location of the Center of mass.

Save the part. Name it Homework 4-3. **Note:** The original model is located in the SOLIDWORKS CSWA Model 20XX Folder\Chapter 4 folder.

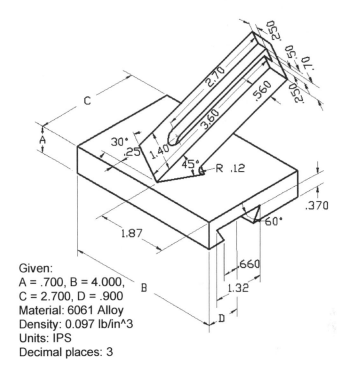

Given:
A = .700, B = 4.000,
C = 2.700, D = .900
Material: 6061 Alloy
Density: 0.097 lb/in^3
Units: IPS
Decimal places: 3

Question 4:

Build the illustrated model. Set document properties, identify the correct Sketch planes, apply the correct Sketch and Feature tools, and apply material. Calculate the overall mass of the part, volume and locate the Center of mass with the provided information. Save the model. Name it Homework 4-4.

- Material: 6061 Alloy
- Units: MMGS
- Decimal place: 2

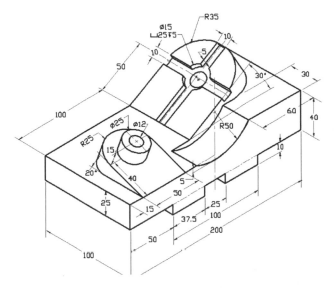

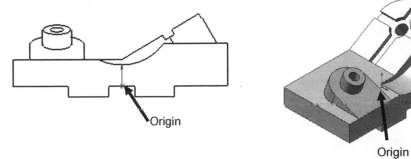

Origin

Origin

Question 5:

The model is located in the SOLIDWORKS CSWA Model 20XX Folder\Chapter 4 folder.

Open Homework 4-5.

Calculate the Center of mass with the new coordinate system. Save the model.

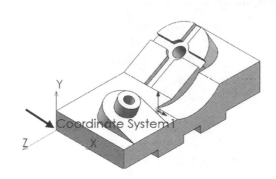

Question 6:

Build the illustrated model in SOLIDWORKS. Calculate the overall mass of the part, volume, and locate the Center of mass with the provided information. Save the model.

Name it Homework 4-6.

- Material: 6061 Alloy
- Units: MMGS
- Decimal place: 2

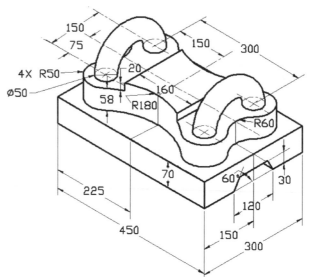

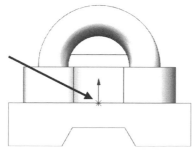

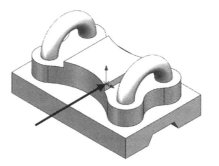

Right view

Question 7:

The model is located in the SOLIDWORKS CSWA Model 20XX Folder\Chapter 4 folder.

Open Homework 4-6.

Create Coordinate System1 to locate the Center of mass for the model. Calculate the Center of mass.

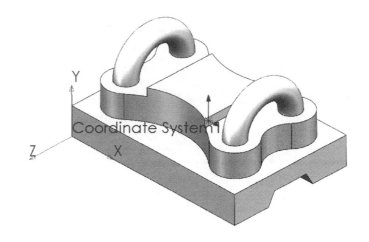

- Material: 6061 Alloy

- Units: MMGS

- Decimal place: 2

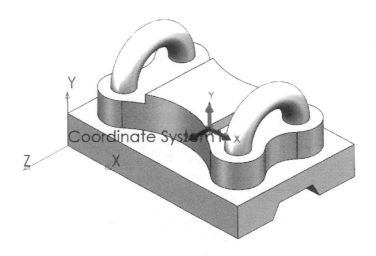

Question 8:

Build the illustrated model.

The model is located in the SOLIDWORKS CSWA Model 20XX Folder\Chapter 4 folder.

Open Homework 4-8.

Utilize the rollback bar to view the dimension, sketches and features. Note: There are many ways to create this model.

Calculate the overall mass and volume of part with the provided information.

- Decimal place: 2

- Material: Case Stainless Steel

- Units: MMGS

Note: As an exercise, use the rollback bar to view the dimensions, sketches and features.

Create the part using different sketches and features.

Save the part. Name it Homework 4-8Modify.

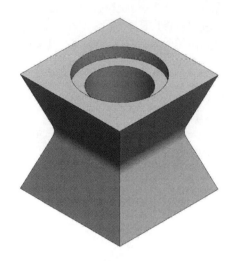

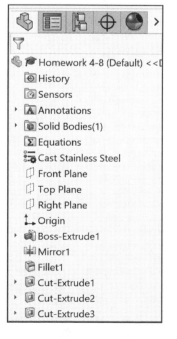

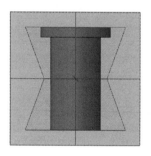

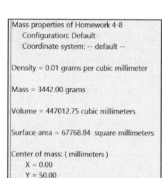

Mass properties of Homework 4-8
    Configuration: Default
    Coordinate system: -- default --

Density = 0.01 grams per cubic millimeter

Mass = 3442.00 grams

Volume = 447012.75 cubic millimeters

Surface area = 67768.84 square millimeters

Center of mass: (millimeters)
    X = 0.00
    Y = 50.00
    Z = 0.00

Homework 4-8 (Default) <<[
- History
- Sensors
- Annotations
- Solid Bodies(1)
- Equations
- Cast Stainless Steel
- Front Plane
- Top Plane
- Right Plane
- Origin
- Boss-Extrude1
- Mirror1
- Fillet1
- Cut-Extrude1
- Cut-Extrude2
- Cut-Extrude3

Question 9:

Build the illustrated model.

The model is located in the SOLIDWORKS CSWA Model 20XX Folder\Chapter 4 folder.

Open Homework 4-9.

Utilize the rollback bar to view the dimensions, sketches and features.

There are many ways to create this model.

Calculate the overall mass and volume of part with the provided information.

- Decimal place: 2

- Material: 1060 Alloy

- Units: MMGS

Note: As an exercise, use the rollback bar to view the dimensions, sketches and features. Create the part using different sketches and features.

Save the part. Name it Homework 4.9Modify.

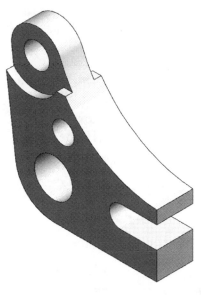

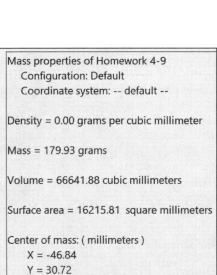

Mass properties of Homework 4-9
    Configuration: Default
    Coordinate system: -- default --

Density = 0.00 grams per cubic millimeter

Mass = 179.93 grams

Volume = 66641.88 cubic millimeters

Surface area = 16215.81 square millimeters

Center of mass: (millimeters)
    X = -46.84
    Y = 30.72
    Z = 0.00

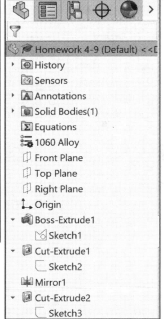

Question 9A:

The model is provided in the SOLIDWORKS CSWA Model 20XX Folder\Chapter 4 folder.

Open model 4-9A.

Create a new coordinate system as illustrated. The new coordinate system location is the front right bottom vertex of the model. Save the model. Name it Homework 4-9A.

Enter the Center of Mass in millimeters:

X = **-46.84 millimeters**

Y = **30.72 millimeters**

Z = **-10.00 millimeters**

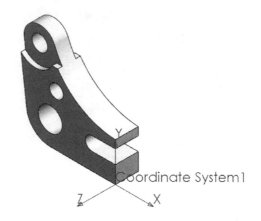

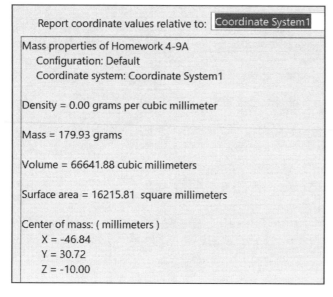

Report coordinate values relative to: Coordinate System1

Mass properties of Homework 4-9A
    Configuration: Default
    Coordinate system: Coordinate System1

Density = 0.00 grams per cubic millimeter

Mass = 179.93 grams

Volume = 66641.88 cubic millimeters

Surface area = 16215.81 square millimeters

Center of mass: (millimeters)
    X = -46.84
    Y = 30.72
    Z = -10.00

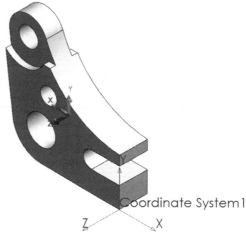

Question 10:

Build the illustrated model.

Calculate the overall mass and volume of the part with the provided information. Save the model. Name it Homework 4-10.

- Decimal place: 2

- Material: AISI 304

- Units: MMGS

- All Holes ⊽ 25mm

- All Rounds 5mm

- All Holes Ø4mm

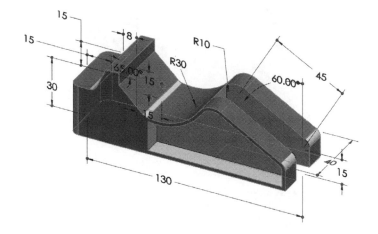

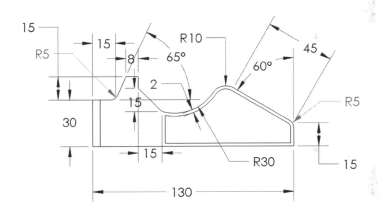

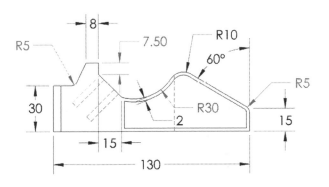

Front views

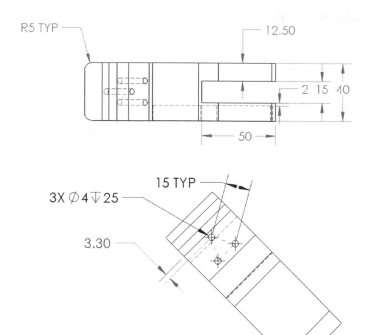

Top and Auxiliary view

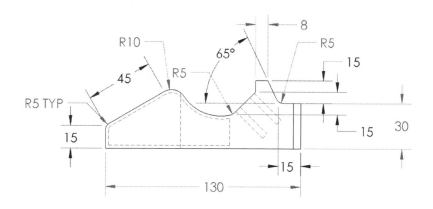

Calculate the mass in grams:

A = 888.48grams

B = 990.50grams

Back view

C = 788.48grams

D = 820.57grams

💡 If you don't find your answer (within 1%) in the multiple-choice single answer format section, recheck your solid model for precision and accuracy. It could be as simple as missing a few fillets.

Calculate the volume in cubic millimeters:

A = 102259.43 cubic millimeters

B = 133359.47 cubic millimeters

C = 111059.43 cubic millimeters

D = 125059.49 cubic millimeters

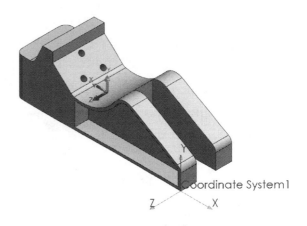

Question 10A:

The model is located in the SOLIDWORKS CSWA Model 20XX Folder\Chapter 4 folder.

Open Homework 4-10. Create a new coordinate system. The new coordinate system location is at the front right bottom vertex of the model. Save the model. Name it Homework 4.10.A Enter the Center of Mass in millimeters:

X = **-80.39 millimeters**

Y = **-15.93 millimeters**

Z = **-22.65 millimeters**

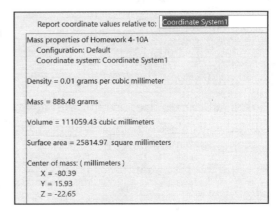

Question 10B:

Modify the model.

Calculate the overall mass and volume.

- Modify all fillets (rounds) to 7mm.

- Modify the overall length to 140mm.

- Modify material to 1060 alloy.

Save the model. Name it Homework 4.10B.

Enter the mass in grams:

309.75 grams

Enter the volume in cubic millimeters:

114721.22 cubic millimeters

If you don't find your answer (within 1%) in the multiple-choice single answer format section, recheck your solid model for precision and accuracy. It could be as simple as missing a few fillets.

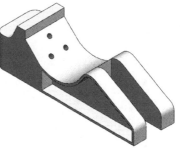

Question 11:

Build the illustrated model.

The model is located in the SOLIDWORKS CSWA Model 20XX Folder\Chapter 4 folder.

Open Homework 4-11.

Utilize the rollback bar to view the dimensions, sketches and features.

There are many ways to create this model. The Mirror feature is used in this procedure.

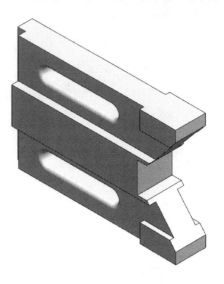

Calculate the overall mass and volume of part with the provided information.

- Decimal place: 2

- Material: Plain Carbon Steel

- Units: MMGS

Note: As an exercise, use the rollback bar to view the dimensions, sketches and features. Create the part using different sketches and features.

Save the part. Name it Homework 4-11Modify.

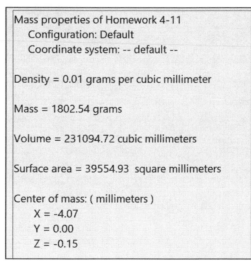

Mass properties of Homework 4-11
    Configuration: Default
    Coordinate system: -- default --

Density = 0.01 grams per cubic millimeter

Mass = 1802.54 grams

Volume = 231094.72 cubic millimeters

Surface area = 39554.93 square millimeters

Center of mass: (millimeters)
    X = -4.07
    Y = 0.00
    Z = -0.15

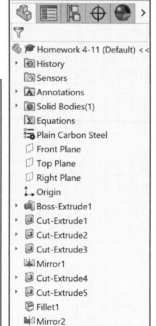

Question 11A:

The model is located in the SOLIDWORKS CSWA Model 20XX Folder\Chapter 4 folder.

Open Homework 4-11.

Create a new coordinate system. The new coordinate system location is the front right bottom vertex of the model as illustrated. Save the model. Name it Homework 4-11A

Enter the Center of Mass in millimeters:

X = -74.07millimeters

Y = 48.00 millimeters

Z = -12.65 millimeters

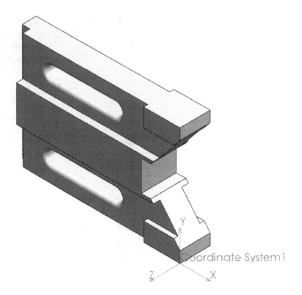

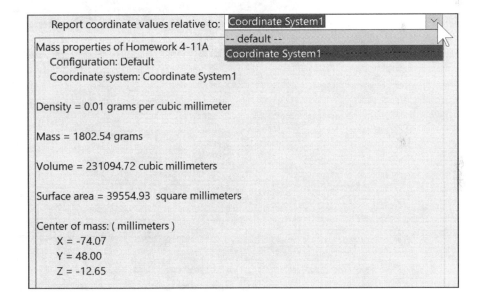

Report coordinate values relative to: Coordinate System1

-- default --
Coordinate System1

Mass properties of Homework 4-11A
    Configuration: Default
    Coordinate system: Coordinate System1

Density = 0.01 grams per cubic millimeter

Mass = 1802.54 grams

Volume = 231094.72 cubic millimeters

Surface area = 39554.93 square millimeters

Center of mass: (millimeters)
    X = -74.07
    Y = 48.00
    Z = -12.65

Zoom in on the
part if needed.

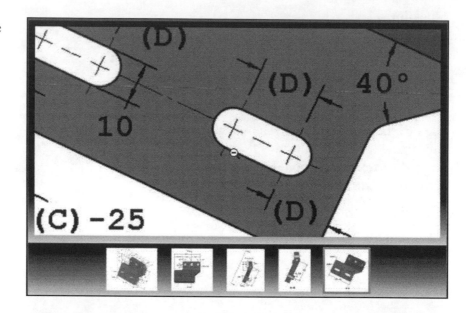

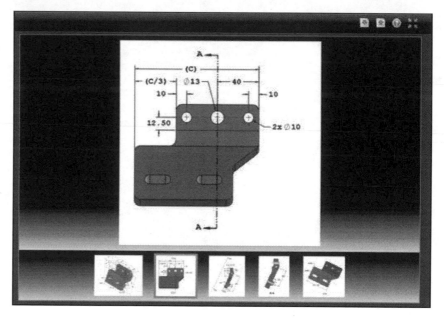

Click on the additional views to understand the part and provided information.
Read each question carefully.
Understand the dimensions, center of mass and units.

Apply needed material.

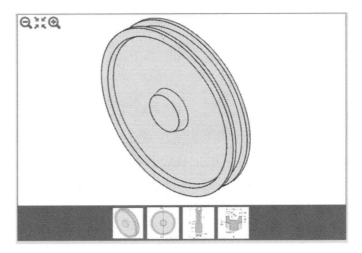

Notes:

CHAPTER 5 - ASSEMBLY CREATION AND MODIFICATION

Objectives

Provide in-depth coverage of the *Assembly Creation and Modification* category on the CSWA exam. Knowledge to create and modify models in this category from detailed dimensioned illustrations.

Introduction

The chapter covers the general concepts and terminology used in assembly modeling and then addresses the core elements that are aligned to the CSWA exam. Knowledge to insert mates and to create a new Coordinate system location is required in this category.

There are four questions on the CSWA Academic exam (2 questions in part 1, 2 questions in part 2) in the Assembly Creation and Modification category: (2) different assemblies - (4) questions - (2) multiple choice/(2) single answer - 30 points each.

The first question is in a multiple-choice single answer format. You should be within 1% of the multiple-choice answer before you move on to the modification single answer section (fill in the blank format).

Download the needed components (6 - 7) from the provided zip file to create the assembly.

🔆 Download all needed model files for this book from the SDC Publication website www.SDCpublications.com/downloads/978-1-63057-567-0.

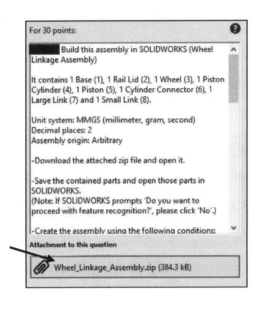

For 30 points:

_____ Build this assembly in SOLIDWORKS (Wheel Linkage Assembly)

It contains 1 Base (1), 1 Rail Lid (2), 1 Wheel (3), 1 Piston Cylinder (4), 1 Piston (5), 1 Cylinder Connector (6), 1 Large Link (7) and 1 Small Link (8).

Unit system: MMGS (millimeter, gram, second)
Decimal places: 2
Assembly origin: Arbitrary

-Download the attached zip file and open it.

-Save the contained parts and open those parts in SOLIDWORKS.
(Note: If SOLIDWORKS prompts 'Do you want to proceed with feature recognition?', please click 'No'.)

-Create the assembly using the following conditions:

Attachment to this question

Wheel_Linkage_Assembly.zip (384.3 kB)

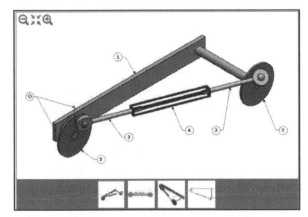

On the completion of the chapter, you will be able to:

- Specify Document Properties.

- Identify the first fixed component in an assembly.

- Create a Bottom-up assembly with the following mates:

  - Coincident, Concentric, Perpendicular, Parallel, Tangent, Distance, Angle, Advanced Distance, Advanced Angle, and Aligned, Anti-Aligned options.

- Apply the Measure tool.

- Apply the Mirror Component tool.

- Locate the Center of mass relative to the assembly Origin.

- Locate the Center of mass relative to a created Coordinate system.

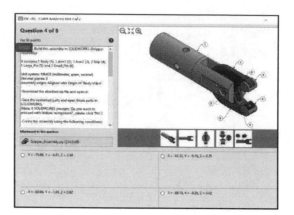

Assembly Modeling

There are two key Assembly Modeling techniques:

- Top-down, "In-Context" assembly modeling.

- Bottom-up assembly modeling.

In Top-down assembly modeling, one or more features of a part are defined by something in an assembly, such as a sketch or the geometry of another component. The design intent comes from the top, and moves down into the individual components, hence the name Top-down assembly modeling.

💡 Mate the first component with respect to the assembly reference planes.

Bottom-up assembly modeling is a traditional method that combines individual components. Based on design criteria, the components are developed independently. The three major steps in a Bottom-up design approach are:

1. Create each part independent of any other component in the assembly.

2. Insert the parts into the assembly.

3. Mate the components in the assembly as they relate to the physical constraints of your design.

Build an Assembly from a Detailed Dimensioned illustration

An exam question in this category could read: Build this assembly in SOLIDWORKS. Locate the Center of mass of the model with respect to the illustrated coordinate system. Decimal place: 2. Unit system: IPS.

The assembly contains the following: (1) Clevis component, (3) Axle components, (2) 5 Hole Link components, (2) 3 Hole Link components, and (6) Collar components. All holes Ø.190 THRU unless otherwise noted. Angle A = 150deg. Angle B = 120deg.

Note: The location of the illustrated coordinate system.

In the exam, download the zip file of the components. Unzip the components. Do not use feature recognition when you open the downloaded components. This is a timed exam.

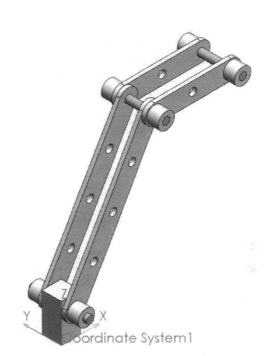

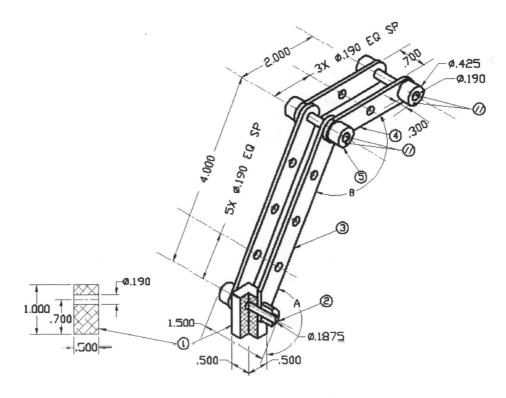

- Clevis, (Item 1): Material: 6061 Alloy. The two (5) Hole Link components are positioned with equal Angle mates, (150deg) to the Clevis component.

- Axle, (Item 2): Material: AISI 304. The first Axle component is mated Concentric and Coincident to the Clevis. The second and third Axle components are mated Concentric and Coincident to the 5 Hole Link and the 3 Hole Link components respectively.

- 5 Hole Link, (Item 3): Material: 6061 Alloy. Material thickness = .100in. Radius = .250in. Five holes located 1in. on center. The 5 Hole Link components are positioned with equal Angle mates, (120deg) to the 3 Hole Link components.

- 3 Hole Link, (Item 4): Material: 6061 Alloy. Material thickness = .100in. Radius = .250in. Three holes located 1in. on center. The 3 Hole Link components are positioned with equal Angle mates, (120deg) to the 5 Hole Link components.

- Collar, (Item 5): Material: 6061 Alloy. The Collar components are mated Concentric and Coincident to the Axle and the 5 Hole Link and 3 Hole Link components respectively.

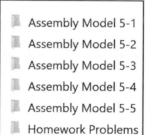

- Assembly Model 5-1
- Assembly Model 5-2
- Assembly Model 5-3
- Assembly Model 5-4
- Assembly Model 5-5
- Homework Problems

Think about the steps that you would take to build the illustrated assembly. Identify the first fixed component. Insert the required mates.

Locate the Center of mass of the model with respect to the illustrated coordinate system. In this example, start with the Clevis component.

Illustrations will vary depending on your version and system setup.

⚡ When you create a new part or assembly, the three default Planes (Front, Right and Top) are aligned with specific views. The Plane you select for the Base sketch determines the orientation of the part.

Tutorial: Assembly Model 5-1

Build the assembly in SOLIDWORKS.

1. **Download** the needed components from the SOLIDWORKS CSWA Model folder XXX\Chapter 5\Assembly Model 5-1 folder.

2. **Create** a new IPS assembly in SOLIDWORKS.

3. Click **Cancel** in the Open dialog box.

4. Click **Cancel** ✖ from the Begin Assembly PropertyManager. Assem1 is the default document name. Assembly documents end with the extension.sldasm.

5. **Set** document properties for the model.

6. **Insert** the Clevis part.

7. **Fix** the component to the assembly Origin. Click OK from the Insert Component PropertyManager. The Clevis is displayed in the Assembly FeatureManager and in the Graphics window.

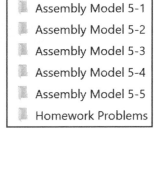

The first component or sub-assembly should be fixed **(f)** to the origin, fully defined to the assembly document or mated to an axis about the assembly origin.

⚡ Only insert the required mates (timed exam) to obtain the needed Mass properties information.

8. **Insert** the Axle part above the Clevis component as illustrated.

9. **Insert** a Concentric mate between the inside cylindrical face of the Clevis and the outside cylindrical face of the Axle.

10. **Insert** a Coincident mate between the Right Plane of the Clevis and the Right Plane of the Axle.

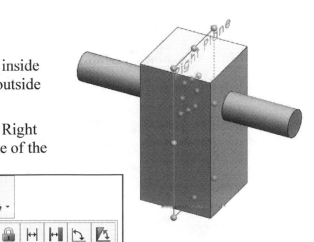

11. **Insert** the 5 Hole Link part.

12. **Rotate** the component as illustrated.

13. **Insert** a Concentric mate between the outside cylindrical face of the Axle and the inside cylindrical face of the 5 Hole Link. Concentric2 is created.

14. **Insert** a Coincident mate between the right face of the Clevis and the left face of the 5 Hole Link. Coincident2 is created.

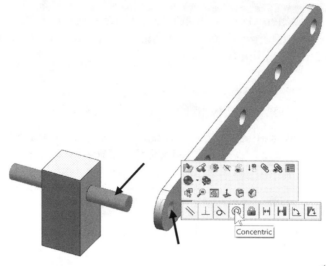

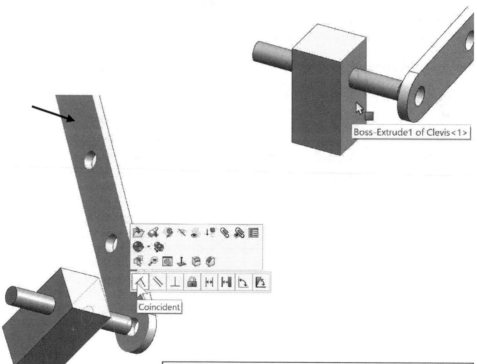

Mates
- ◎ Concentric1 (Clevis<1>,Axle<1>)
- 人 Coincident1 (Clevis<1>,Axle<1>)
- ◎ Concentric2 (Axle<1>,5 Hole Link<1>)
- 人 Coincident2 (Clevis<1>,5 Hole Link<1>)

15. **Insert** an Angle mate between the bottom face of the 5 Hole Link and the back face of the Clevis. Angle = 30deg. If needed, click the Flip direction box and or the Aligned box.

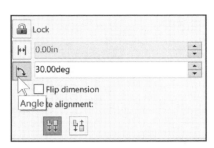

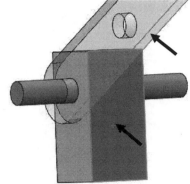

🔅 Depending on the component orientation, select the Flip direction option and/or enter the supplement of the angle.

16. **Insert** the second Axle part. Locate the second Axle component near the end of the 5 Hole Link as illustrated.

17. **Insert** a Concentric mate between the inside cylindrical face of the 5 Hole Link and the outside cylindrical face of the Axle. Concentric3 is created.

18. **Insert** a Coincident mate between the Right Plane of the assembly and the Right Plane of the Axle. Coincident3 is created.

19. **Insert** the 3 Hole Link part.

20. **Rotate** the component as illustrated.

21. **Insert** a Concentric mate between the outside cylindrical face of the Axle and the inside cylindrical face of the 3 Hole Link. Concentric4 is created.

22. **Insert** a Coincident mate between the right face of the 5 Hole Link and the left face of the 3 Hole Link.

23. **Insert** an Angle mate between the bottom face of the 5 Hole Link and the bottom face of the 3 Hole Link. Angle = 60deg. Angle2 is created.

🔅 Depending on the component orientation, select the Flip direction option and/or enter the supplement of the angle when needed.

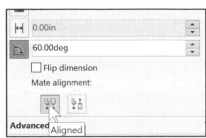

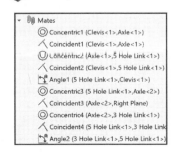

Apply the Measure tool to check the angle.

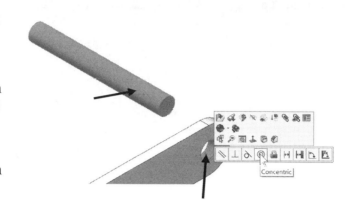

24. **Insert** the third Axle part.

25. **Insert** a Concentric mate between the inside cylindrical face of the 3 Hole Link and the outside cylindrical face of the Axle.

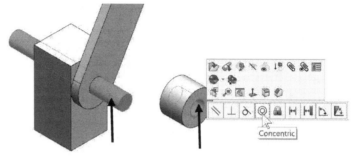

26. **Insert** a Coincident mate between the Right Plane of the assembly and the Right Plane of the Axle.

27. **Insert** the Collar part. Locate the Collar near the first Axle component.

28. **Insert** a Concentric mate between the inside cylindrical face of the Collar and the outside cylindrical face of the first Axle.

29. **Insert** a Coincident mate between the right face of the 5 Hole Link and the left face of the Collar.

30. **Insert** the second Collar part. Locate the Collar near the second Axle component.

31. **Insert** a Concentric mate between the inside circular face of the second Collar and the outside circular face of the second Axle.

32. **Insert** a Coincident mate between the right face of the 3 Hole Link and the left face of the second Collar.

33. **Insert** the third Collar part. Locate the Collar near the third Axle component.

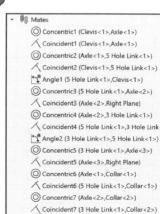

34. **Insert** a Concentric mate between the inside cylindrical face of the Collar and the outside cylindrical face of the third Axle.

35. **Insert** a Coincident mate between the right face of the 3 Hole Link and the left face of the third Collar.

36. **Mirror** the components. Mirror the three Collars, 5 Hole Link and 3 Hole Link about the Right Plane. If using an older version of SOLIDWORKS, check the Recreate mates to new components box. Click Next in the Mirror Components PropertyManager. View your options.

Click **Insert, Mirror Components** from the Main menu or click the **Mirror Components** tool from the Linear Component Pattern Consolidated toolbar.

If using an older release of SOLIDWORKS, no check mark in the Components to Mirror box indicates that the components are copied. The geometry of a copied component is unchanged from the original, only the orientation of the component is different.

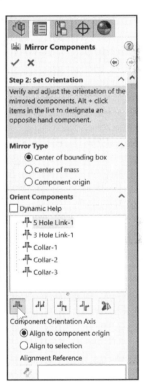

Create the coordinate system location for the assembly.

37. Select the front right **vertex** of the Clevis component as illustrated.

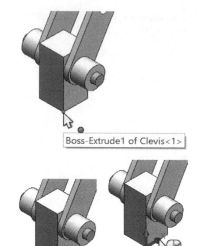

Boss-Extrude1 of Clevis<1>

38. Click the **Coordinate System** tool from the Reference Geometry Consolidated toolbar. The Coordinate System PropertyManager is displayed.

39. Click the **right bottom edge** of the Clevis component.

40. Click the **front bottom edge** of the Clevis component as illustrated.

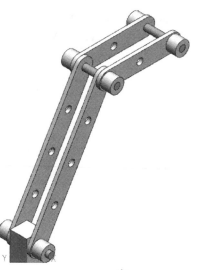

41. Address the **direction** for X, Y, Z as illustrated.

42. Click **OK** from the Coordinate System PropertyManager. Coordinate System1 is displayed.

43. **Locate** the Center of mass based on the location of the illustrated coordinate system. Select Coordinate System1.

- X: 1.79 inches

- Y: 0.25 inches

- Z: 2.61 inches

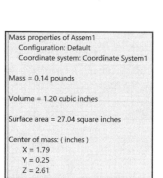

Mass properties of Assem1
   Configuration: Default
   Coordinate system: Coordinate System1

Mass = 0.14 pounds

Volume = 1.20 cubic inches

Surface area = 27.04 square inches

Center of mass: (inches)
   X = 1.79
   Y = 0.25
   Z = 2.61

44. **Save** the part. Name it Assembly Modeling 5-1.

45. **Close** the model.

There are numerous ways to create the models in this chapter. A goal in this text is to display different design intents and techniques.

☀ If you don't find an option within 1% of your answer on the exam re-check your assembly.

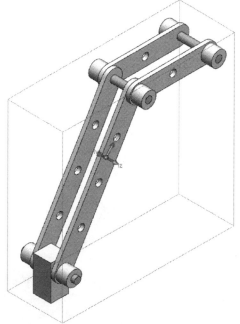

Tutorial: Assembly Model 5-2

An exam question in this category could read: Build this assembly in SOLIDWORKS. Locate the Center of mass of the model with the illustrated coordinate system. Decimal: 2. Unit system: MMGS.

The assembly contains the following: (2) U-Bracket components, (4) Pin components and (1) Square block component.

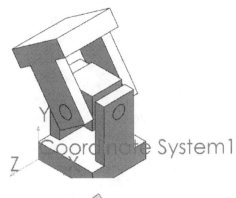

- U-Bracket, (Item 1): Material: AISI 304. Two U-Bracket components are combined together Concentric to opposite holes of the Square block component. The second U-Bracket component is positioned with an Angle mate to the right face of the first U-Bracket and a Parallel mate between the top face of the first U-Bracket and the top face of the Square block component. Angle A = 125deg.

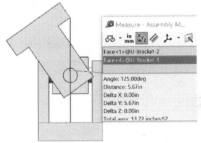

- Square block, (Item 2): Material: AISI 304. The Pin components are mated Concentric and Coincident to the 4 holes in the Square block (no clearance). The depth of each hole = 10mm.

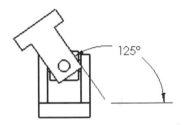

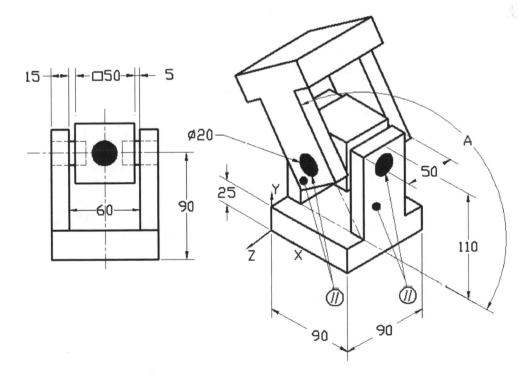

- Pin, (Item 3): Material: AISI 304. The Pin components are mated Concentric to the hole (no clearance). The end face of the Pin components are Coincident to the outer face of the U-Bracket components. The Pin component has a 5mm spacing between the Square block component and the two U-Bracket components.

Think about the steps that you would take to build the illustrated assembly. Identify the first fixed component. This is the Base component of the assembly. Insert the required Standard mates. Locate the Center of mass of the model with respect to the illustrated coordinate system. In this example, start with the U-Bracket part.

Create the assembly.

1. **Download** the needed components from the SOLIDWORKS CSWA Model folder XXX\Chapter 5\Assembly Model 5-2 folder.

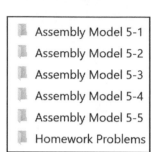

2. **Create** a new assembly in SOLIDWORKS. The created models are displayed in the Open documents box.

3. Click **Cancel** in the Open dialog box.

4. Click **Cancel** ✖ from the Begin Assembly PropertyManager.

5. **Set** document properties for the model.

6. **Insert** the first U-Bracket component into the assembly document.

7. **Fix** the component to the assembly Origin. Click OK from the PropertyManager. The U-Bracket is displayed in the Assembly FeatureManager and in the Graphics window.

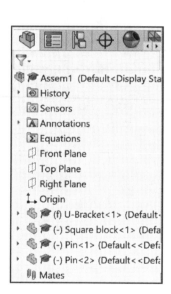

8. **Insert** the Square block above the U-Bracket component as illustrated.

9. **Insert** the first Pin part. Locate the first Pin to the front of the Square block.

10. **Insert** the second Pin part. Locate the second Pin to the back of the Square block.

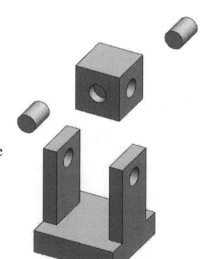

11. **Insert** the third Pin part. Locate the third Pin to the left side of the Square block.

12. **Rotate** the Pin as illustrated.

13. **Insert** the fourth Pin part. Locate the fourth Pin to the right side of the Square block.

14. **Rotate** the Pin as illustrated.

15. **Insert** a Concentric mate between the inside cylindrical face of the Square block and the outside cylindrical face of the first Pin. Concentric1 is created.

16. **Insert** a Coincident mate between the inside back circular face of the Square block and the flat back face of the first Pin. Coincident1 mate is created.

17. **Insert** a Concentric mate between the inside cylindrical face of the Square block and the outside cylindrical face of the second Pin. Concentric2 is created.

18. **Insert** a Coincident mate between the inside back circular face of the Square block and the front flat face of the second Pin. Coincident2 mate is created.

19. **Insert** a Concentric mate between the inside cylindrical face of the Square block and the outside cylindrical face of the third Pin. Concentric3 is created.

20. **Insert** a Coincident mate between the inside back circular face of the Square block and the right flat face of the third Pin. Coincident3 mate is created.

21. **Insert** a Concentric mate between the inside circular face of the Square block and the outside cylindrical face of the fourth Pin. Concentric4 is created.

22. **Insert** a Coincident mate between the inside back circular face of the Square block and the left flat face of the fourth Pin. Coincident4 mate is created.

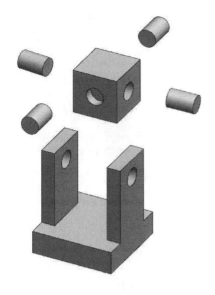

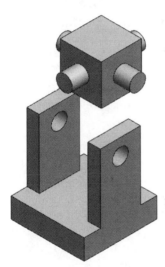

- Origin
- ⊳ 🖤 🚩 (f) U-Bracket<1> (Default<<Default>_
- ⊳ 🖤 🚩 (-) Square block<1> (Default<<Defau
- ⊳ 🖤 🚩 (-) Pin<1> (Default<<Default>_Appea
- ⊳ 🖤 🚩 (-) Pin<2> (Default<<Default>_Appea
- ⊳ 🖤 🚩 (-) Pin<3> (Default<<Default>_Appea
- ⊳ 🖤 🚩 (-) Pin<4> (Default<<Default>_Appea
- ▾ 🕅 Mates
  - ◎ Concentric1 (Square block<1>,Pin<1>
  - ⋀ Coincident1 (Square block<1>,Pin<1>
  - ◎ Concentric2 (Square block<1>,Pin<2>
  - ⋀ Coincident2 (Square block<1>,Pin<2>
  - ◎ Concentric3 (Square block<1>,Pin<3>
  - ⋀ Coincident3 (Square block<1>,Pin<3>
  - ◎ Concentric4 (Square block<1>,Pin<4>
  - ⋀ Coincident4 (Square block<1>,Pin<4>

23. **Insert** a Concentric mate between the inside right cylindrical face of the Cut-Extrude feature on the U-Bracket and the outside cylindrical face of the right Pin. Concentric5 is created.

24. **Insert** a Coincident mate between the Right Plane of the Square block and the Right Plane of the assembly. Coincident5 is created.

25. **Insert** the second U-Bracket part above the assembly. Position the U-Bracket as illustrated.

26. **Insert** a Concentric mate between the inside cylindrical face of the second U-Bracket component and the outside cylindrical face of the second Pin. The mate is created.

27. **Insert** a Coincident mate between the outside circular edge of the second U-Bracket and the back flat face of the second Pin. The mate is created.

There are numerous ways to mate the models in this chapter. A goal is to display different design intents and techniques.

28. **Insert** an Angle mate between the top flat face of the first U-Bracket component and the front narrow face of the second U-Bracket component as illustrated. Angle1 is created. An Angle mate is required to obtain the correct Center of mass.

29. **Insert** a Parallel mate between the top flat face of the first U-Bracket and the top flat face of the Square block component.

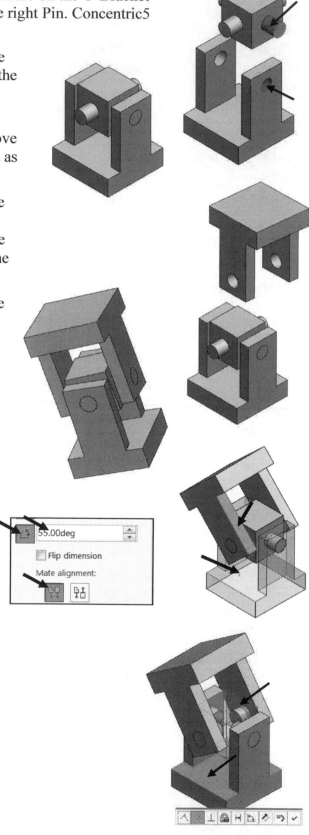

30. **Expand** the Mates folder and the components from the FeatureManager. View the created mates.

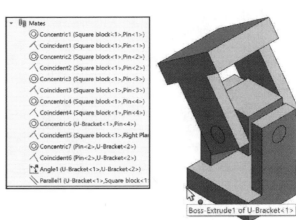

Create the coordinate location for the assembly.

31. Select the front **bottom left vertex** of the first U-Bracket component as illustrated.

32. Click the **Coordinate System** tool from the Reference Geometry Consolidated toolbar. The Coordinate System PropertyManager is displayed.

33. Click **OK** from the Coordinate System PropertyManager. Coordinate System1 is displayed.

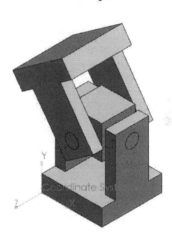

34. **Locate** the Center of mass based on the location of the illustrated coordinate system. Select Coordinate System1.

- X: 31.54 millimeters

- Y: 85.76 millimeters

- Z: -45.00 millimeters

35. **Save** the part. Name it Assembly Modeling 5-2.

36. **Close** the model.

🔆 If you don't find an option within 1% of your answer on the exam re-check your assembly.

🔆 Click on the additional views during the CSWA exam to better understand the assembly/component. Read each question carefully. Identify the dimensions, center of mass and units. Apply needed material.

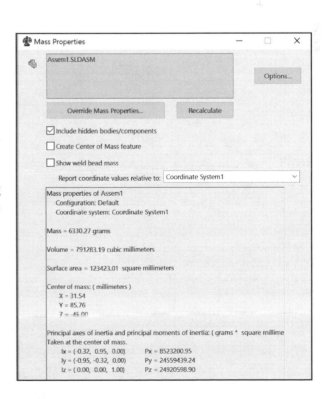

Tutorial: Assembly Model 5-3

An exam question in this category could read: Build this assembly in SOLIDWORKS. Locate the Center of mass using the illustrated coordinate system. Decimal place: 2. Unit system: MMGS.

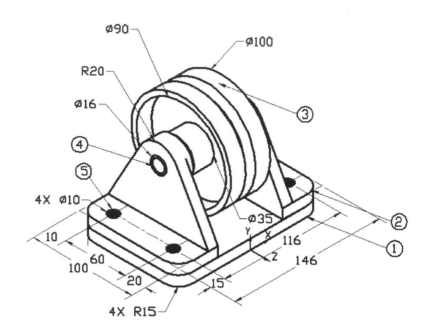

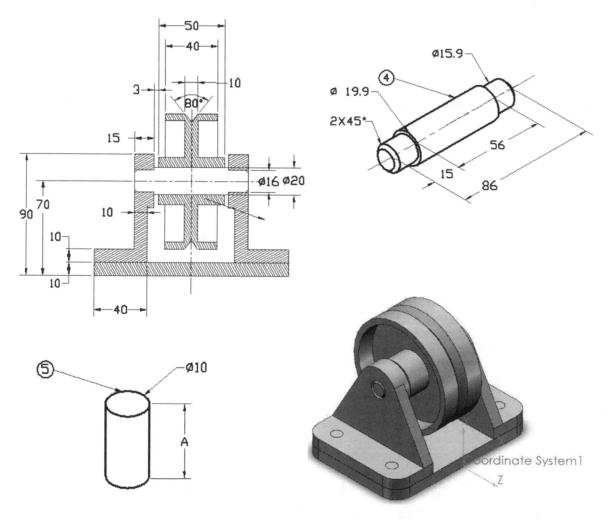

The assembly contains the following: (1) WheelPlate component, (2) Bracket100 components, (1) Axle40 component, (1) Wheel1 component and (4) Pin-4 components.

- WheelPlate, (Item 1): Material: AISI 304. The WheelPlate contains 4-Ø10 holes. The holes are aligned to the left Bracket100 and the right Bracket100 components. All holes are THRU ALL. The thickness of the WheelPlate = 10 mm.

- Bracket100, (Item 2): Material: AISI 304. The Bracket100 component contains 2-Ø10 holes and 1- Ø16 hole. All holes are through-all.

- Wheel1, (Item 3): Material AISI 304: The center hole of the Wheel1 component is Concentric with the Axle40 component. There is a 3mm gap between the inside faces of the Bracket100 components and the end faces of the Wheel hub.

- Axle40, (Item 4): Material AISI 304: The end faces of the Axle40 are Coincident with the outside faces of the Bracket100 components.

- Pin-4, (Item 5): Material AISI 304: The Pin-4 components are mated Concentric to the holes of the Bracket100 components (no clearance). The end faces are Coincident to the WheelPlate bottom face and the Bracket100 top face.

Think about the steps that you would take to build the illustrated assembly. Identify the first fixed component. This is the Base component of the assembly. Insert the required mates.

Locate the Center of mass of the illustrated model with respect to the referenced coordinate system.

The referenced coordinate system is located at the bottom, right, midpoint of the Wheelplate. In this example, start with the WheelPlate part.

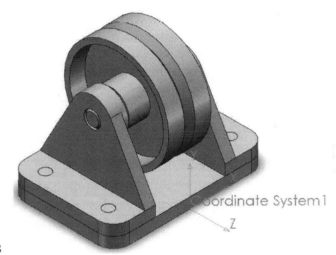

1. **Download** the needed components from the SOLIDWORKS CSWA Model folder XXX\Chapter 5\Assembly Model 5-3 folder.

2. **Create** a new assembly in SOLIDWORKS.

3. Click **Cancel** in the Open dialog box.

4. Click **Cancel** ✖ from the Begin Assembly PropertyManager.

5. **Set** document properties for the assembly.

- Assembly Model 5-1
- Assembly Model 5-2
- Assembly Model 5-3
- Assembly Model 5-4
- Assembly Model 5-5
- Homework Problems

6. **Insert** the first component. Insert the WheelPlate. Fix the component to the assembly Origin. The WheelPlate is displayed in the Assembly FeatureManager and in the Graphics window. The WheelPlate component is fixed.

7. **Insert** the first Bracket100 part above the WheelPlate component as illustrated.

8. **Insert** a Concentric mate between the inside front left cylindrical face of the Bracket100 component and the inside front left cylindrical face of the WheelPlate. Concentric1 is created.

9. **Insert** a Concentric mate between the inside front right cylindrical face of the Bracket100 component and the inside front right cylindrical face of the WheelPlate. Concentric2 is created.

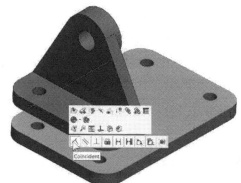

10. **Insert** a Coincident mate between the bottom flat face of the Bracket100 component and the top flat face of the WheelPlate component. Coincident1 is created.

11. **Insert** the Axle40 part above the first Bracket100 component as illustrated.

12. **Insert** a Concentric mate between the outside cylindrical face of the Axle40 component and the inside cylindrical face of the Bracket100 component. Concentric3 is created.

13. **Insert** a Coincident mate between the flat face of the Axle40 component and the front outside edge of the first Bracket100 component. Coincident2 is created.

☀ To verify that the distance between holes of mating components is equal, utilize Concentric mates between pairs of cylindrical hole faces.

14. **Insert** the first Pin-4 part above the Bracket100 component.

15. **Insert** the second Pin-4 part above the Bracket100 component.

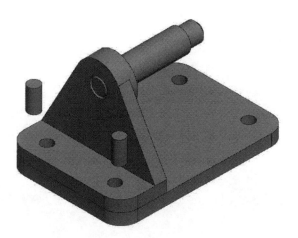

16. **Insert** a Concentric mate between the outside cylindrical face of the first Pin-4 component and the inside front left cylindrical face of the Bracket100 component. Concentric4 is created.

17. **Insert** a Coincident mate between the flat top face of the first Pin-4 component and the top face of the first Bracket100 component. Coincident3 is created.

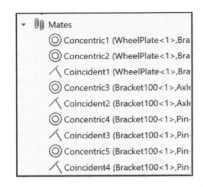

18. **Insert** a Concentric mate between the outside cylindrical face of the second Pin-4 component and the inside front right cylindrical face of the Bracket100 component. Concentric5 is created.

19. **Insert** a Coincident mate between the flat top face of the second Pin-4 component and the top face of the first Bracket100 component. Coincident4 is created.

20. **Insert** the Wheel1 part as illustrated.

21. **Insert** a Concentric mate between the outside cylindrical face of Axle40 and the inside front cylindrical face of the Wheel1 component. Concentric6 is created.

22. **Insert** a Coincident mate between the Front Plane of Axle40 and the Front Plane of Wheel1. Coincident5 is created.

23. **Mirror** the components. Mirror the Bracket100 and the two Pin-4 components about the Front Plane.

🔆 Click **Insert, Mirror Components** from the Main menu or click the **Mirror Components** tool from the Linear Component Pattern Consolidated toolbar.

Create the coordinate location for the assembly.

24. Click the **Coordinate System** tool from the Reference Geometry Consolidated toolbar. The Coordinate System PropertyManager is displayed.

25. **Select** the right bottom midpoint as the Origin location.

26. **Select** the bottom right edge as the X axis direction reference as illustrated.

27. Click **OK** from the Coordinate System PropertyManager. Coordinate System1 is displayed.

28. **Locate** the Center of mass based on the location of the illustrated coordinate system. Select Coordinate System1.

- X: = 0.00 millimeters

- Y: = 37.14 millimeters

- Z: = -50.00 millimeters

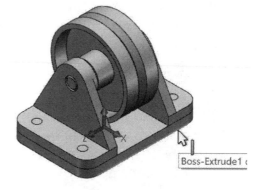

If you don't find an option within 1% of your answer on the exam re-check your assembly.

29. **Save** the part. Name it Assembly Modeling 5-3.

30. **Close** the model.

Mass = 3797.32 grams

Volume = 474665.19 cubic millimeters

Surface area = 130119.83 square millimeters

Center of mass: (millimeters)
    X = 0.00
    Y = 37.14
    Z = -50.00

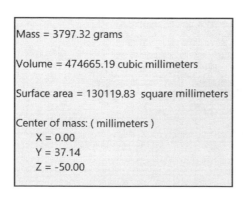

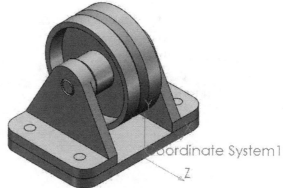

Mate the First Component with Respect to the Assembly Reference Planes

You can fix the position of a component so that it cannot move with respect to the assembly Origin. By default, the first part in an assembly is fixed - however, you can float it at any time.

It is recommended that at least one assembly component is either fixed, or mated to the assembly planes or Origin. This provides a frame of reference for all other mates, and prevents unexpected movement of components when mates are added.

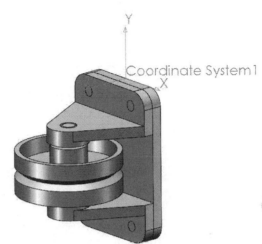

Up to this point, you identified the first fixed component, and built the required Base component of the assembly. The component features were orientated correctly to the illustrated assembly. In the exam, what if you created the Base component where the component features were not orientated correctly to the illustrated assembly.

In the next tutorial, build the illustrated assembly. Insert the Base component, float the component, then mate the first component with respect to the assembly reference planes.

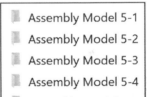
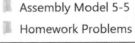

Complete the assembly with the components from the Tutorial: Assembly model 5-3. Decimal place: 2.

Tutorial: Assembly Model 5-4

1. **Create** a new assembly in SOLIDWORKS.

2. **Set** document settings. Decimal place: 2. Unit system: MMGS.

3. **Insert** the illustrated WheelPlate part from the book.

4. **Float** the WheelPlate component from the FeatureManager.

5. **Insert** a Coincident mate between the Front Plane of the assembly and the bottom flat face of the WheelPlate. Coincident1 is created.

6. **Insert** a Coincident mate between the Right Plane of the assembly and the Right Plane of the WheelPlate. Coincident2 is created.

7. **Insert** a Coincident mate between the Top Plane of the assembly and the Front Plane of the WheelPlate. Coincident3 is created.

☀ When the Base component is mated to three assembly reference planes, no component status symbol is displayed in the Assembly FeatureManager.

8. **Insert** the first Bracket100 part as illustrated. Rotate the component if required.

9. **Insert** a Concentric mate between the inside back left circular face of the Bracket100 component and the inside top left circular face of the WheelPlate. Concentric1 is created.

10. **Insert** a Concentric mate between the inside back right cylindrical face of the Bracket100 component and the inside top right cylindrical face of the WheelPlate. Concentric2 is created.

11. **Insert** a Coincident mate between the flat back face of the Bracket100 component and the front flat face of the WheelPlate component. Coincident4 is created.

12. **Insert** the Axle40 part as illustrated. Rotate the component if required.

13. **Insert** a Concentric mate between the outside cylindrical face of the Axle40 component and the inside cylindrical face of the Bracket100 component. Concentric3 is created.

14. **Insert** a Coincident mate between the top flat face of the Axle40 component and the top outside circular edge of the Bracket100 component. Coincident5 is created.

15. **Insert** the first Pin-4 part above the Bracket100 component. Rotate the component.

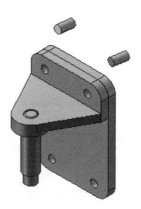

16. **Insert** the second Pin-4 part above the Bracket100 component. Rotate the component.

17. **Insert** a Concentric mate between the outside cylindrical face of the first Pin-4 component and the inside front left cylindrical face of the Bracket100 component. Concentric4 is created.

18. **Insert** a Coincident mate between the flat front face of the first Pin-4 component and the top flat front face of the Bracket100 component. Coincident6 is created.

19. **Insert** a Concentric mate between the outside cylindrical face of the second Pin-4 component and the inside front right cylindrical face of the Bracket100 component. Concentric5 is created.

20. **Insert** a Coincident mate between the flat front face of the second Pin-4 component and the top flat front face of the Bracket100 component. Coincident7 is created.

21. **Insert** the Wheel1 part as illustrated.

22. **Insert** a Concentric mate between the outside cylindrical face of Axle40 and the inside top cylindrical face of the Wheel1 component. Concentric6 is created.

23. **Insert** a Coincident mate between the Right Plane of Axle40 and the Right Plane of the Wheel1 component. Coincident8 is created.

24. **Insert** a Coincident mate between the Front Plane of Axle40 and the Front Plane of the Wheel1 component. Coincident9 is created.

25. **Mirror** the components. Mirror the Bracket100, and the two Pin-4 components about the Top Plane. Do not check any components in the Components to Mirror box. Check the Recreate mates to new components box. Click Next in the PropertyManager. Check the Preview instanced components box.

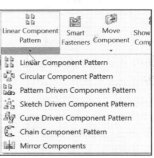

Create the coordinate location for the assembly.

26. Click the **Coordinate System** tool from the Reference Geometry Consolidated toolbar. The Coordinate System PropertyManager is displayed.

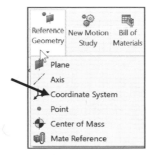

27. **Select** the top back midpoint for the Origin location as illustrated.

28. Click **OK** from the Coordinate System PropertyManager. Coordinate System1 is displayed.

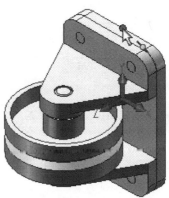

29. **Locate** the Center of mass based on the location of the illustrated coordinate system. Select Coordinate System1.

- X: = 0.00 millimeters

- Y: = -73.00 millimeters

- Z: = 37.14 millimeters

30. **Save** the part. Name it Assembly Modeling 5-4.

31. **Close** the model.

In a multi choice format - the question is displayed as:

What is the center of mass of the assembly (millimeters)?

A) X = 0.00, Y = -73.00, Z = 37.14

B) X = 308.53, Y = -109.89, Z = -61.40

C) X = 298.66, Y = -17.48, Z = -89.22

D) X = 448.66, Y = -208.48, Z = -34.64

The answer would be A.

If you don't find an option within 1% of your answer on the exam re-check your assembly.

In the exam you are allowed to answer the questions in any order. Use the Summary Screen during the exam to view the list of all questions you have or have not answered.

Tutorial: Assembly Model 5-5

An exam question in this category could read:

Build this assembly in SOLIDWORKS (Chain Link Assembly). It contains 2 long_pins (1), 3 short_pins (2), and 4 chain_links (3).

- Unit system: MMGS

- Decimal places: 2

- Assembly origin: Arbitrary

Mass properties of Assembly Modeling 5-4
    Configuration: Default
    Coordinate system: Coordinate System1

Mass = 3797.32 grams

Volume = 474665.19 cubic millimeters

Surface area = 130119.83 square millimeters

Center of mass: (millimeters)
    X = 0.00
    Y = -73.00
    Z = 37.14

Assembly Model 5-1
Assembly Model 5-2
Assembly Model 5-3
Assembly Model 5-4
Assembly Model 5-5
Homework Problems

IMPORTANT: Create the assembly with respect to the Origin as shown in the Isometric view. (This is important for calculating the proper Center of Mass). Create the assembly using the following conditions:

1. Pins are mated concentric to chain link holes (no clearance).

2. Pin end faces are coincident to chain link side faces.

A = 25 degrees, B = 125 degrees, C = 130 degrees

What is the center of mass of the assembly (millimeters)?

Hint: If you don't find an option within 1% of your answer please re-check your assembly.

A) X = 348.66, Y = -88.48, Z = -91.40

B) X = 308.53, Y = -109.89, Z = -61.40

C) X = 298.66, Y = -17.48, Z = -89.22

D) X = 448.66, Y = -208.48, Z = -34.64

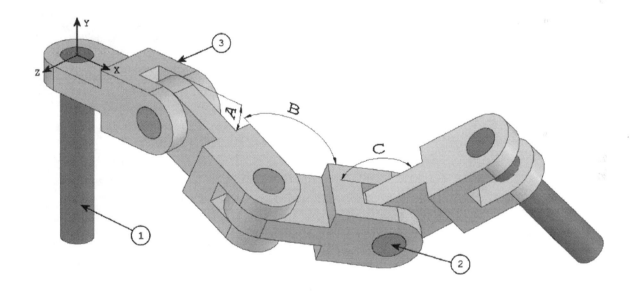

Assembly Creation and Modification

There are no step-by-step procedures in this section. Below are various Assembly FeatureManagers that created the above assembly and obtained the correct answer.

The correct answer is:

A) X = 348.66, Y = -88.48, Z = -91.40

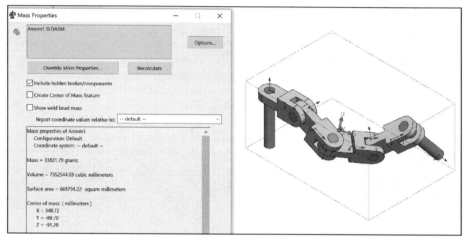

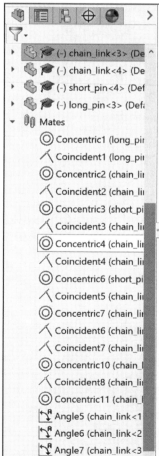

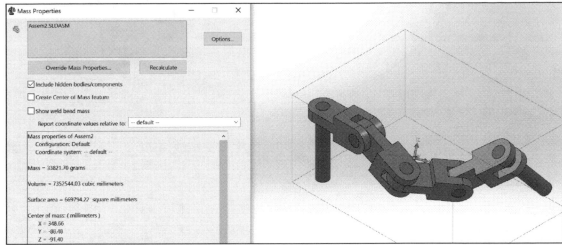

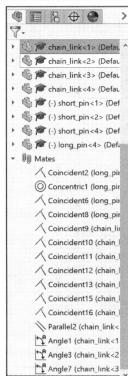

Summary

Assembly Creation and Modification is one of the five categories on the CSWA exam. In the last two chapters, a Basic, Intermediate or Advanced model was the focus. The Assembly Creation and Modification (Bottom-up) category addresses an assembly with numerous components.

There are four questions on the CSWA exam in the Assembly Creation and Modification category: (2) different assemblies - (4) questions - (2) multiple choice\(2) single answers - 30 points each.

You are required to download the needed components from a provided zip file and insert them correctly to create the assembly as illustrated. You are then requested to modify the assembly and answer fill in the blank format questions.

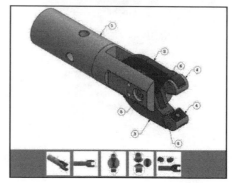

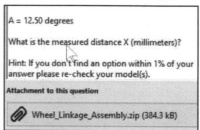

Use the Measure tool in the assembly between components.

Take the free sample CSWA online exam, visit https://3dexperience.virtualtester.com/#home

View the provided SOLIDWORKS CSWA Sample Exam folder included in the downloaded materials. The CSWA Sample Exam folder contains a pdf with information on the following: exam details, how to prepare for the exam, how to take the practice exam, taking the exam, sample test quesitons, test answers, and helpful sites.

If your school is an academic certification provider, your instructor can allocate free exam credits for Segment 1 & 2. The instructor will require your .edu email address.

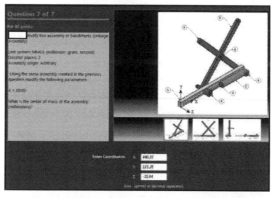

Screen shots from a CSWA exam

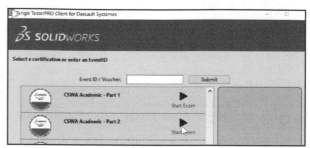

Questions

1: Build this assembly from the provided information in SOLIDWORKS. Unit System: MMGS. Decimal precision: 2.

Calculate the overall mass and volume of the assembly.

Locate the Center of mass using the illustrated coordinate system.

The assembly contains the following: one Base100 component, one Yoke component, and one AdjustingPin component.

- Base100, (Item 1): Material 1060 Alloy. The distance between the front face of the Base100 component and the front face of the Yoke = 60mm.
- Yoke, (Item 2): Material 1060 Alloy. The Yoke fits inside the left and right square channels of the Base100 component (no clearance). The top face of the Yoke contains a Ø12mm through all hole.
- AdjustingPin, (Item 3): Material 1060 alloy. The bottom face of the AdjustingPin head is located 40mm from the top face of the Yoke component. The AdjustingPin component contains an Ø5mm Through All hole.

The Coordinate system is located in the lower left corner of the Base100 component. The X axis points to the right. The Y axis points upwards.

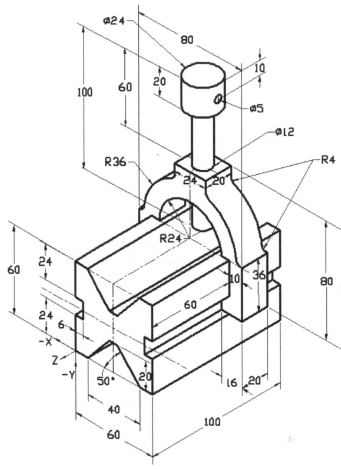

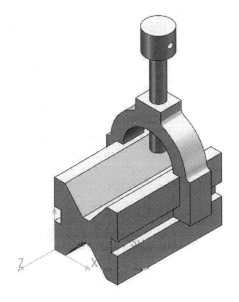

2. Build the assembly from the provided information in SOLIDWORKS. Calculate the overall mass and volume of the assembly. Locate the Center of mass using the illustrated coordinate system. The assembly contains the following: three MachinedBracket components and two Pin-5 components. Unit system: MMGS. Decimal place: 2.

Insert the Base component, float the component, then mate the first component with respect to the assembly reference planes.

- MachinedBracket, (Item 1): Material 6061 Alloy. The MachineBracket component contains two Ø10mm through all holes. Each MachinedBracket component is mated with two Angle mates. The Angle mate = 45deg. The top edge of the notch is located 20mm from the top edge of the MachinedBracket.
- Pin-5, (Item 2): Material Titanium. The Pin-5 component is 5mms in length and equal in diameter. The Pin-5 component is mated Concentric to the MachinedBracket (no clearance). The end faces of the Pin-5 component are Coincident with the outer faces of the MachinedBracket. There is a 1mm gap between the Machined Bracket components.

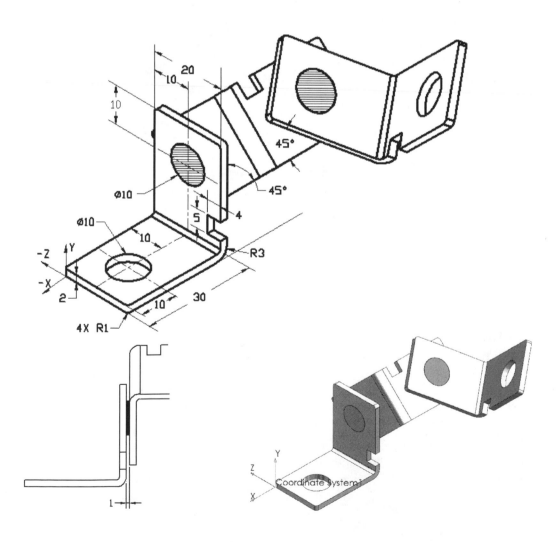

3. Build the assembly from the provided information in SOLIDWORKS. Calculate the overall mass and volume of the assembly. Locate the Center of mass using the illustrated coordinate system. The illustrated assembly contains the following components: three Machined-Bracket components and two Pin-6 components. Unit System: MMGS. Decimal place: 2.

☼ Insert the Base component, float the component, then mate the first component with respect to the assembly reference planes.

- Machined-Bracket, (Item 1): Material 6061 Alloy. The Machine-Bracket component contains two Ø10mm through all holes. Each Machined-Bracket component is mated with two Angle mates. The Angle mate = 45deg. The top edge of the notch is located 20mm from the top edge of the MachinedBracket.
- Pin-6, (Item 2): Material Titanium. The Pin-6 component is 5mms in length and equal in diameter. The Pin-5 component is mated Concentric to the Machined-Bracket (no clearance). The end faces of the Pin-6 component are Coincident with the outer faces of the Machined-Bracket. There is a 1mm gap between the Machined-Bracket components.

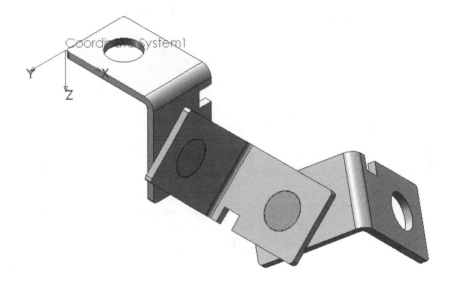

Sample screen shots from an older CSWA exam for an assembly. Click on the additional views to understand the assembly and provided information. Read each question carefully. Understand the dimensions, center of mass and units. Apply needed materials.

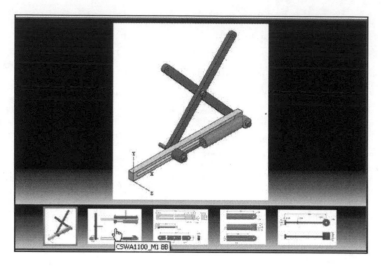

Zoom in on the part if needed.

Use the Mass Properties tool to locate the new coordinate system.

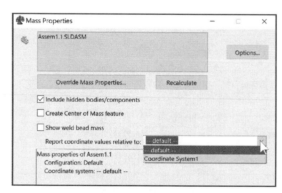

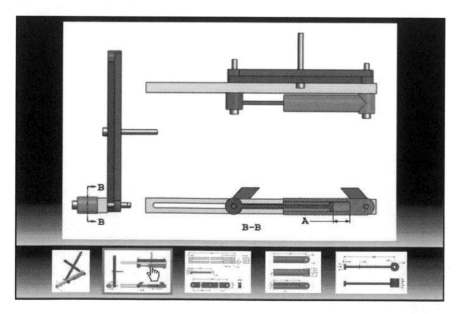

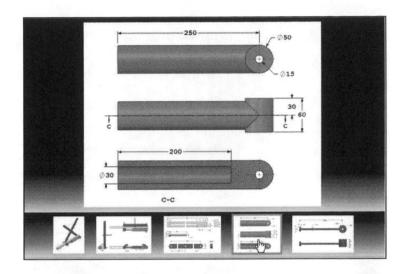

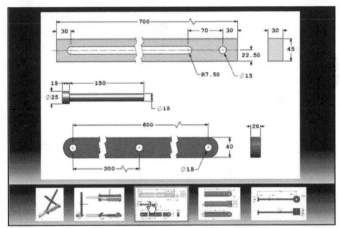

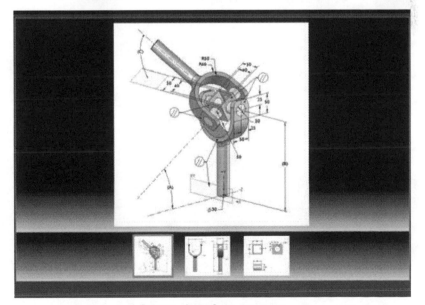

Screen shots from an exam

💡 Zoom in on the part if needed.

💡 If needed, use the Measure tool in the assembly on the components for distance or angle.

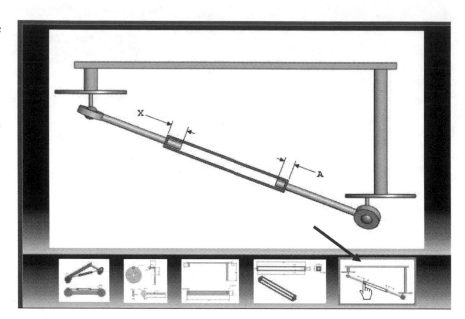

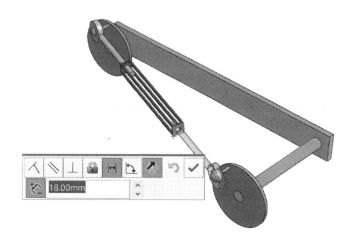

Screen shots from an exam

If needed, use the Measure tool in the assembly on the components for distance or angle.

Screen shots from an exam

Notes:

CHAPTER 6 - CERTIFIED SOLIDWORKS ASSOCIATE SUSTAINABLE DESIGN (CSWA-SD)

Introduction

The Certified SOLIDWORKS Associate Sustainable Design (CSWA-SD) certification indicates a foundation in and apprentice knowledge of demonstrating an understanding in the principles of environmental assessment and sustainable design. Formally called the Certified Sustainable Design Associate, old acronym (CSDA).

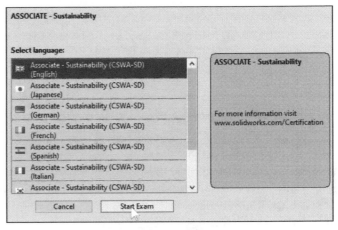

The main requirement for obtaining the Certified SOLIDWORKS Associate Sustainable Design (CSWA-SD) certification is to take and pass the on-line 60 minute exam (minimum of 24 out of 30 points). There are 30 questions, each question is worth 1 point.

The 30 questions address various key categories: Environmental Assessment, Introduction to sustainability and Sustainable design.

All questions are in a multiple choice/multi answer format. SOLIDWORKS does not require that you have a copy of SOLIDWORKS Sustainability, or even SOLIDWORKS. No SOLIDWORKS models need to be created for the exam.

You are allowed to answer the questions in any order you prefer. Use the Summary Screen during the CSWA-SD exam to view the list of all questions you have or have not answered.

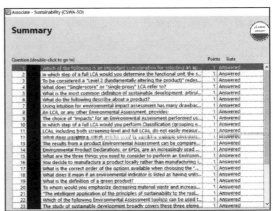

If you fail, there is a 14-day waiting period before retaking the (CSWA-SD) certification.

With a Certified SOLIDWORKS Associate Sustainable Design (CSWA-SD) certification, you will be prepared for today's competitive job market by demonstrating an understanding in the principles of environmental assessment and sustainable design.

Visit the following link http://www.SOLIDWORKS.com/ sustainability/sustainable-design-guide.htm to download the Guide zip folder. The Guide is provided in a MS word document. Read, learn to better understand the principles and industry standards of sustainability.

SOLIDWORKS Sustainability Guide

Sustainable Design Guide

The ideas of sustainability and sustainable design are a growing part of today's product design conversations. But exactly what is sustainable design, and how do you create a greener product? We'll answer these questions through this Guide to Sustainable Design with interactive content and detailed examples.

Chapter 1: Introduction and Terminology

The idea of "Sustainable Design" is cropping up more and more in today's product design conversations. But what is sustainable design, and how do I do it? We hope to answer this question throughout this Guide.

Why should you read this guide?

There are probably as many reasons to read this guide as there are people reading it. That said, design engineers will want to incorporate sustainability principles into their work for at least one of four general reasons.

Personal interest

Many people are drawn to sustainable design because they want to use their talents and expertise to make the world a better place. As naïve as that sentiment might seem, it is a powerful driver behind a great deal of innovation and creative engineering. And, given the state of the world today, we could use all the help we can get.

Company intent

Many readers may be here not because of their own interest, but because sustainable design is part of a company initiative. Whether driven by stockholders, customers, or senior leadership, "sustainability" is increasingly on corporate agendas. While social and environmental responsibility is often at the root of such efforts, many companies are also finding that sustainable design is just "good business." Through it, companies find new ways to decrease material and energy costs, and increase revenue through resulting new product innovations.

solidworks_sustainability_guide

On the completion of the chapter and reading the guide, you will be able to:

- Understand basic concepts of sustainability.

- Comprehend sustainable design concepts, such as Design for Disassembly (DfD), Extended Producer Responsibility (EPR) and Biomimicry.

- Recognize stages of a product life cycle.

- Understand Sustainability and Sustainable Business.

- Identify the many faces of Sustainable Design.

- Set up an Environmental Assessment study, such as environmental indicators, scope of the assessment.

- Appreciate goal and scope variables for an Environmental Assessment study, such as system boundary and functional unit.

- Basic steps for performing a Life Cycle Assessment (LCA) study.

- Value a product Environmental Assessment.

- Making Theory Matter - Initial Analysis Decisions.

It is recommended to review the SOLIDWORKS SustainabilityXpress and SOLIDWORKS Sustainability tutorials.

SOLIDWORKS SustainabilityXpress for parts is supplied in all SOLIDWORKS product configurations.

SOLIDWORKS Sustainability for assemblies is an Add-in on the commercial versions of SOLIDWORKS and is contained in the SOLIDWORKS Education Edition and SOLIDWORKS Student Edition.

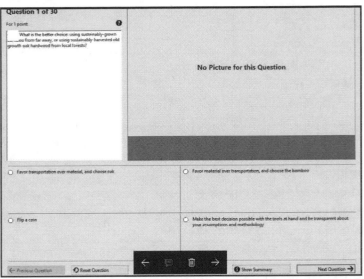

You are allowed to answer the questions in any order you prefer. Use the Summary Screen during the CSWA-SD exam to view the list of all questions you have or have not answered.

During the exam, use the control keys at the bottom of the screen to:

- *Show the Previous the Question.*

- *Reset the Question.*

- *Show the Summary Screen.*

- *Move to the Next Question.*

Goals

The primary goal is not only to help you pass the Certified SOLIDWORKS Associate Sustainable Design (CSWA-SD) exam, but also to ensure that you understand and comprehend the concepts and implementation details of the CSWA-SD process.

The second goal is to provide the most comprehensive coverage of CSWA-SD exam related topics available, without too much coverage of topics not on the exam.

The third and ultimate goal is to get you from where you are today to the point that you can confidently pass the CSWA-SD exam.

ASSOCIATE Sustainable Design

DS SOLIDWORKS

Background

Sustainable design, like quality, time to market, and cost, will soon dictate how engineers approach most every product they develop. Choosing products based on their carbon footprint will be equally as important as design validation. To stay ahead of the curve, you and your company need to understand sustainable design and how to implement it now.

Sustainable engineering is the integration of social, environmental, and economic conditions into a product or process. Soon all design will be Sustainable Design.

SOLIDWORKS Sustainability allows students and designers to be environmentally conscious about their designs.

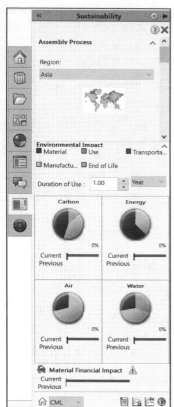

Every license of SOLIDWORKS obtains a copy of SustainabilityXpress. SustainabilityXpress which calculates environmental impact on a part in four key areas:

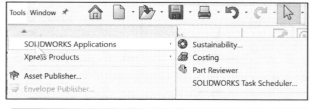

- *Carbon Footprint*

- *Energy Consumption*

- *Air Acidification*

- *Water Eutrophication*

Material and Manufacturing process region and Transportation Usage region are used as input variables. Two SOLIDWORKS Sustainability products are available: *SOLIDWORKS SustainabilityXpress* and *SOLIDWORKS Sustainability*.

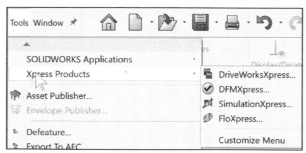

SOLIDWORKS SustainabilityXpress: Handles part documents and is included in the core software.

SOLIDWORKS Sustainability: Provides the following functions:

- Same functions as SustainabilityXpress

- Life Cycle Assessment (LCA) of assemblies

- Configuration support:

  o Save inputs and results per configuration

- Expanded reporting capabilities for assemblies

- Specify amount & type of energy consumed during use

- Specify method of transportation

- Support for Assembly Visualization

SOLIDWORKS Sustainability provides real-time feedback on key impact factors in the Environmental Impact Dashboard, which updates dynamically with any changes by the user. You can generate customized reports to share the results.

SOLIDWORKS Sustainability for assemblies, is an Add-in on the commercial versions of SOLIDWORKS and is contained in the SOLIDWORKS Education Edition and SOLIDWORKS Student Edition.

Life Cycle Assessment

Life Cycle Assessment is a method to quantitatively assess the environmental impact of a product throughout its entire lifecycle, from the procurement of the raw materials, through the production, distribution, use, disposal and recycling of that product.

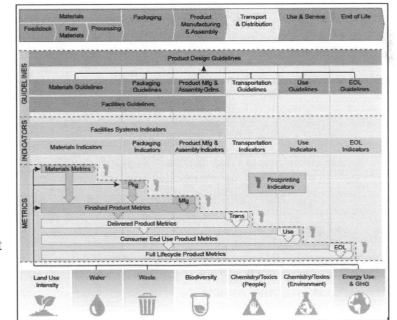

- **Raw Material Extraction:**

  o Planting, growing, and harvesting of trees

  o Mining of raw ore (example: bauxite)

  o Drilling and pumping of oil

- **Material Processing** - The processing of raw materials into engineered materials:
  - o Oil into Plastic
  - o Iron into Steel
  - o Bauxite into Aluminum
- **Part Manufacturing** - Processing of material into finished parts:
  - o Injection molding
  - o Milling and Turning
  - o Casting
  - o Stamping
- **Assembly** - Assemble all of the finished parts to create the final product
- **Product Use** - End consumer uses product for intended lifespan of product
- **End of Life** - Once the product reaches the end of its useful life, how is it disposed of:
  - o Landfill
  - o Recycled
  - o Incinerated

Life Cycle Assessment Key Elements

SOLIDWORKS Sustainability provides the ability to assess the following key elements on the life cycle:

- **Identify and quantify the environmental loads involved:**
  - o Energy and raw materials consumed
  - o Emissions and wastes generated
- **Evaluate the potential environmental impacts of these loads**
- **Assess the options available for reducing these environmental impacts**

Design Categories

Materials: Provides the ability to select various material classes and Names for the part from the drop-down menu.

Manufacturing: Provides the ability to select both the process and usage region for the part. At this time, there are four different areas to choose from: North America, Europe, Asia and Japan.

Process: Provides the ability to select between multiple different production techniques to manufacture the part.

Environmental Impact: This area included four quantities: *Carbon Footprint*, *Total Energy*, *Air Acidification*, and *Water Eutrophication*. Each graph presents a graphical breakdown in: Material Impact, Transportation and Use, Manufacturing and End of Life.

- **Carbon Footprint**: A measure of carbon-dioxide and other greenhouse gas emissions such as methane (in CO_2 equivalent units, CO_{2e}) which contributes to emissions, predominantly caused by burning fossil fuels. Global warming Potential (GWP) is also commonly referred to as a carbon footprint.

- **Energy Consumption**: A measure of the non-renewable energy sources associated with the part's lifecycle in units of Mega Joules (MJ). This impact includes not only the electricity of fuels used during the product's lifecycle, but also the upstream energy required to obtain and process these fuels, and the embodied energy of materials which would be released if burned. Energy Consumed is expressed as the net calorific value of energy demand from non-renewable resources (petroleum, natural gas, etc.).

☀ Efficiencies in energy conversion (power, heat, steam, etc.) are taken into account.

- **Air Acidification**: Sulfur dioxide, nitrous oxides other acidic emissions to air cause an increase in the acidity of rain water, which in turn acidifies lakes and soil. These acids can make the land and water toxic for plants and aquatic life. Acid rain can also slowly dissolve man-made building materials such as concrete. This impact is typically measured in units of either kg sulfur dioxide equivalent SO_{2e} or moles H+ equivalent.

- **Water Eutrophication**: When an overabundance of nutrients are added to a water ecosystem, Eutrophication occurs, nitrogen and phosphorous from waste water and agricultural fertilizers causes an overabundance of algae to bloom, which then depletes the water of oxygen and results in the death of both plant and animal life. This impact is typically measured in either kg phosphate equivalent (PO_{4e}) or kg nitrogen (N) equivalent.

Generate a Report: Provides the ability to generate a customer report that captures your baseline design and comparison between the various input parameters.

The CSWA-SD exam is timed. Work efficiently. Use the information presented in the MS word document from the following link: http://www.SOLIDWORKS.com/sustainability/sustainable-design-guide.htm during the exam. Note: The MS word document is provided in the Download model files in this chapter.

Download all needed model files from the SDC Publication website www.SDCpublications.com/downloads/978-1-63057-567-0.

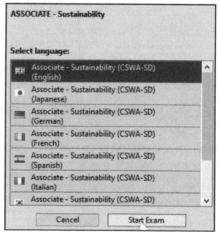

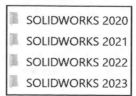

References

It is very important to understand definitions and how SOLIDWORKS information is obtained on the above areas for baseline calculations and comparative calculations. The following standards and agencies were used in the development cycles of this tool:

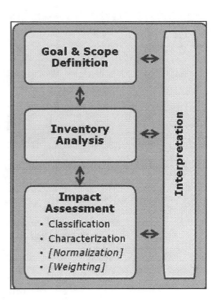

- **Underlying LCA Technology: PE International**:
  - 20 years of LCA experience
  - LCA international database
  - GaBi 4 - leading software application for product sustainability
  - www.pe-international.com

- **International LCA Standards:**
  - Environmental Management Life Cycle Assessment Principles and Framework ISO 14040/44 www.iso.org

- **US EPA LCA Resources:**
  - http://www.epa.gov/nrmrl/lcaccess/

SOLIDWORKS Sustainability Methodology

The following chart was created to provide a visualization of the Methodology used in SOLIDWORKS Sustainability. Note the Input variables and the Output areas along with the ability to create and send a customer report are based on your selected decisions.

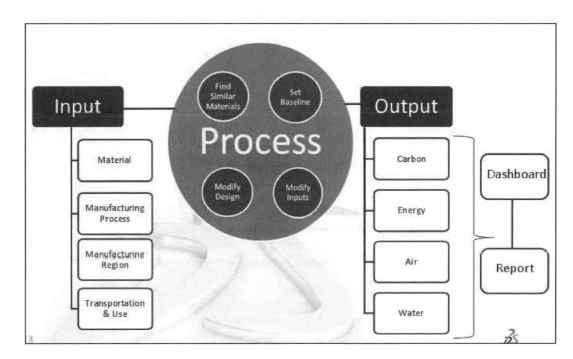

Sustainable Design Guide

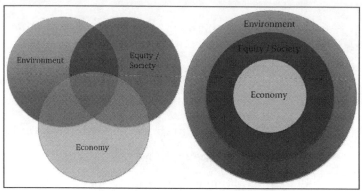

The idea of "Sustainable Design" is cropping up more and more in today's product design conversations. But what is sustainable design, and how do I do it? Visit the following link http://www.SOLIDWORKS.com/sustainability/sustainable-design-guide.htm to download the MS word document. Read, learn to better understand the principles and industry standards of sustainability.

The Guide is divided into ten chapters. First, feel free to jump around. This guide was made to be consumed in sections, and in no particular order. The Guide addresses terminology early on, so if you find concepts that you're unfamiliar with, navigate to the appropriate section to learn more. Going through the entire Guide and playing with some of the examples should take approximately 5-7 hours, so it's a good idea to pace yourself.

SOLIDWORKS does not require that you have a copy of SOLIDWORKS Sustainability, or even SOLIDWORKS to take the CSWA-SD certification. The Guide is designed to be interesting and informative without having access to the design software.

However, there are examples (in the Guide) that you can download in SOLIDWORKS to make the theory come alive.

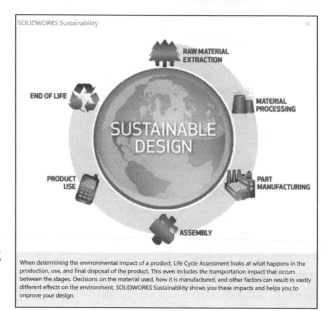

When determining the environmental impact of a product, Life Cycle Assessment looks at what happens in the production, use, and final disposal of the product. This even includes the transportation impact that occurs between the stages. Decisions on the material used, how it is manufactured, and other factors can result in vastly different effects on the environment. SOLIDWORKS Sustainability shows you these impacts and helps you to improve your design.

Activity: Run Sustainability - Analyze a simple part

Close all parts, assemblies and drawings. Run Sustainability. Perform a simple analysis on a part. The below procedure uses SOLIDWORKS 2021. There are a few minor variations between years.

1. Open the **CLAMP2** part from the CSWA-SD\Sustainability folder.

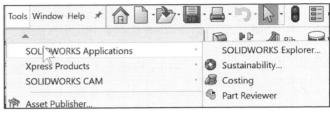

Activate Sustainability.

2. Click **Tools**, **SOLIDWORKS Applications**, **Sustainability** from the Main menu. The Sustainability Wizard is displayed in the Task Pane area.

View the SOLIDWORKS Sustainability dialog box.

3. Click **Continue**.

Select the Material Class.

4. Click the **Drop-down arrow** under Class. View your options. Select **Steel** from the drop-down menu as illustrated.

Select the Material Name. Material Name is Class dependent.

5. Click the **Drop-down arrow** under Name. View your options. Select **Stainless Steel (ferritic)** from the drop-down menu.

Select the Manufacturing Region. Each region produces energy by different method combinations. Impact of a kWh is different for each region. Example methods include: Fossil Fuels, Nuclear and Hydro-electric.

6. Position the **mouse pointer** over the map. View your available options. Click **Asia** as illustrated.

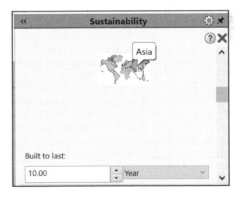

Select Build to last Period.

7. Select **10 years**.

Select Manufacturing Process. Manufacturing Process is Material Name dependent.

8. Select **Milled** from the drop-down menu. View the provided information on energy and scrap rate. Note: No paint is selected by default.

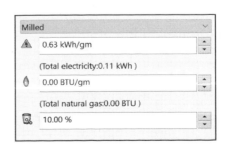

Select the Use region.

9. Position the **mouse pointer** over the map. View your available options.

10. Click **North America**.

View the provided Transportation modes.

11. Accept the default: **Boat**. This option estimates the environmental impact associated with transporting the product from its manufacturing location to its use location.

View the provided information on End of Life.

12. **Accept** the default.

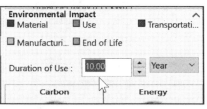

Set Duration of Use period.

13. Select **10 years**.

Set your design Baseline. A Baseline is required to comprehend the environmental impact of the original design.

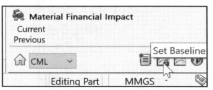

14. Click the **Set Baseline** 🔒 tool from the bottom of the Environmental Impact screen. The Environmental Impact of this part is displayed. The Environmental Impact is calculated in four key areas: *Carbon Footprint, Energy Consumption, Air Acidification* and *Water Eutrophication*.

15. Position the **mouse pointer** over the Carbon box. View Factor percentage. 75.05% represents the Material percentage of the total Carbon footprint of the part (0.871kg). Note: Your numbers will differ if you are using a different SOLIDWORKS version.

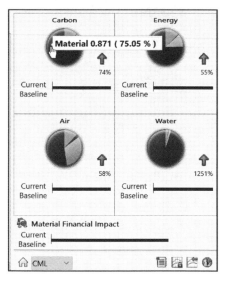

💡 75.05% represents the Material percentage of the Energy footprint of the part.

16. Click inside the **Carbon** box to display a Baseline bar chart of the Carbon Footprint. View the results.

17. Click the **Home** 🏠 icon to return to the Environmental Impact display.

18. Click inside the **Energy** Consumption impact screen to display a Baseline bar chart of Energy Consumption. View the results.

19. Click the **Home** 🏠 icon to return to the Environmental Impact display.

20. Click inside the **Air** Consumption impact screen to display a Baseline bar chart of Air Acidification. View the results.

21. Click the **Home** 🏠 icon to return to the Environmental Impact display.

22. Click inside the **Water** Consumption impact screen to display a Baseline bar chart of Water Eutrophication. View the results.

23. Click the **Home** 🏠 icon to return to the Environmental Impact display.

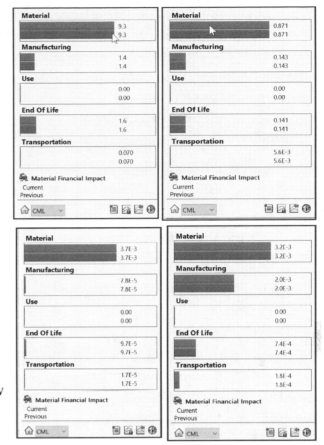

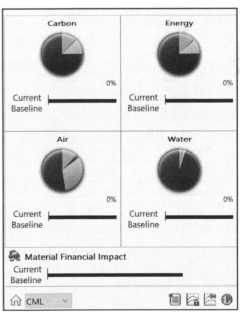

In the next section, compare the baseline design to a different Material Class, Name and Manufacturing Process. Let's compare the present material Stainless Steel (ferritic) to Nylon 6/10.

Select a new Material Class.
24. Click the **Drop-down arrow** under Class. View your options.

25. Select **Plastics** from the drop-down menu.

Select a new Material Name.
26. Click the **Drop-down arrow** under Name. View your options.

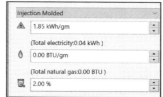

27. Select **Nylon 6/10** from the drop-down menu.

Select a new Manufacturing Process.
28. Select **Injection Molded** from the drop-down menu. Note: the energy and scrap changes. Asia is selected for Manufacturing Region and North America is selected for Use region.

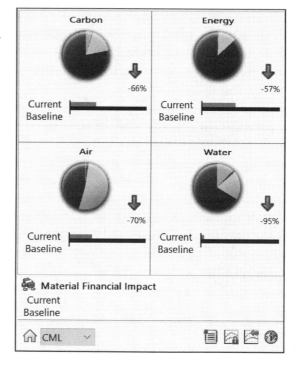

View the results.
29. Changing the material from **Stainless Steel (ferritic)** to **Nylon 6/10** and the manufacturing process from Milled to Injection Molded had a positive impact in all categories, but a further material change may provide a better result. Your numbers may vary depending on your release and version year. What is important is the trend.

Find a similar material and compare the Environmental Impact to Nylon 6/10. This function is a real time saver.

30. Click the **Find Similar** button. The Find Similar Material dialog box is displayed.

31. Click the **Value (-any-)** drop-down arrow. View your options.

32. Select **Plastics**. You can perform a general search or customize your search on physical properties of the material.

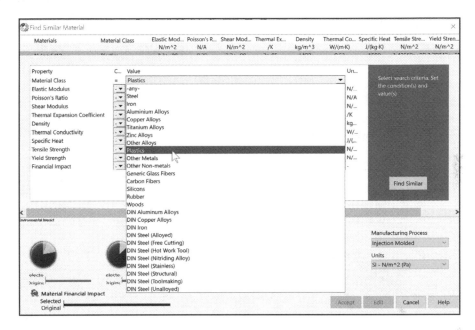

Select a Similar Material from the provided list. You can find similar materials based on the following Properties: *Material Class, Thermal Expansion Coefficient, Specific Heat, Density, Elastic Modulus, Shear Modulus, Thermal Conductivity, Poissons Ratio, Tensile Strength and Yield Strength.* The definitions are listed at the end of the chapter.

SOLIDWORKS Simulation provides the ability to search on various materials properties. Select either Greater than, Less than or approximately from the drop-down menu.

33. Click the **Find Similar** button. SOLIDWORKS provides a full list of comparable materials that you can further refine.

34. Check the **ABS** material box.

35. Check the **ABS PC** material box.

36. Click **inside the top left box (Show Selected Only)** as illustrated. The selected materials are displayed with their properties.

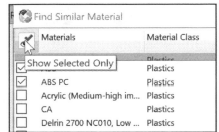

View the Environment Impact for the alternative materials.

37. Click the **ABS** material row as illustrated. View the results. The material is lower in Carbon Footprint, Energy Consumption, Air Acidification and Water Eutrophication than Nylon 6/10. Your numbers will vary depending on your release and version of SOLIDWORKS. What is important is the trend.

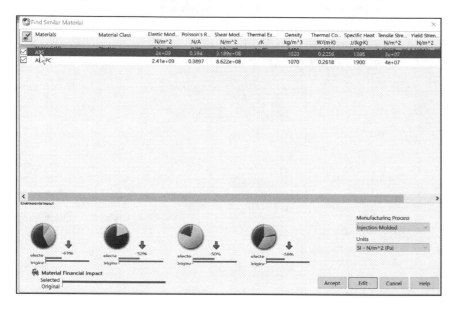

38. Click the **ABS PC** material row. View the results. You decide to stay with Nylon 6/10 due to cost and other design manufacturing issues. Your numbers may vary depending on your release and version year. What is important is the trend.

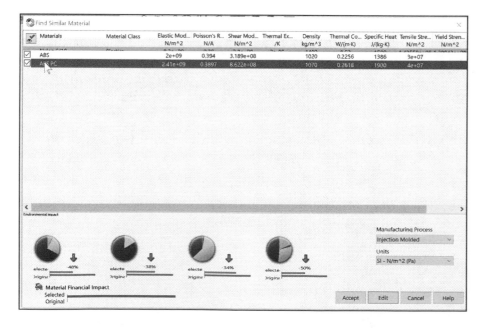

39. Click the **Cancel** button from the Find Similar Material dialog box.

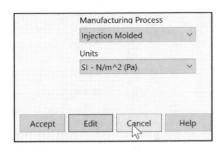

Run a Report.
40. Click the **Save As** or the **Generate Report** button as illustrated depending on your SOLIDWORKS version. SOLIDWORKS provides the ability to communicate this report information throughout your organization. Sustainability generates a report that will compare designs (material, regions) and explain each category of Environmental Impact and show how each design compares to the Base line.

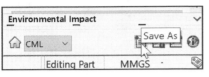

41. Click **OK**.

💡 You can't generate a report if Microsoft Word is running.

42. Review the Generated report. **Ctrl - Click here for alternative units such as Miles Driven In a Car.** View the results. Note: Internet access is required

43. **View the Glossary** in the report section and the other hyperlinks for additional information.

Close the Report and part. You are finished with this section.
44. **Close** the report.

45. **Close** the part.

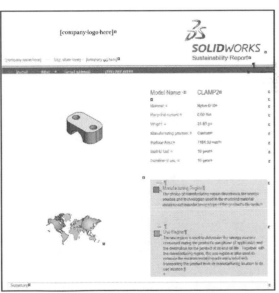

💡 The CSWA-SD exam consists of a total of 30 questions in various categories: *Environmental Assessment, Introduction to sustainability* and *Sustainable design* in a multiple choice/multi answer format. No SOLIDWORKS models need to be created for this exam. The above was a simple exercise to perform Sustainability.

Click here for alternative units such as 'Miles Driven in a Car'¤

.

Summary

The main requirement for obtaining the Certified SOLIDWORKS Associate Sustainable Design (CSWA-SD) certification is to take and pass the on-line 60 minute exam (minimum of 24 out of 30 points). There are 30 questions, each question is worth 1 point.

The 30 questions address various key categories: Environmental Assessment, Introduction to sustainability and Sustainable design.

All questions are in a multiple choice/multi answer format.

Visit the following link http://www.SOLIDWORKS.com/sustainability/sustainable-design-guide.htm to download the Guide zip folder. The Guide is provided in a MS word document. Read, learn to better understand the principles and industry standards of sustainability.

A copy of the Guide is included in this chapter.

Go to the 3DEXPERIENCE® Certification Center at https://3dexperience.virtualtester.com/#home.

The 3DEXPERIENCE® Certification Center is where you are able to log in to manage your certificates, take certifications, and make changes to your account settings.

If your school is an academic certification provider, your instructor can allocate a free exam credit.

The instructor will require your .edu email address.

If you fail, there is a 14-day waiting period before retaking the (CSWA-SD) certification.

solidworks_sustainability_guide

Sample Questions from the CSWA-SD exam

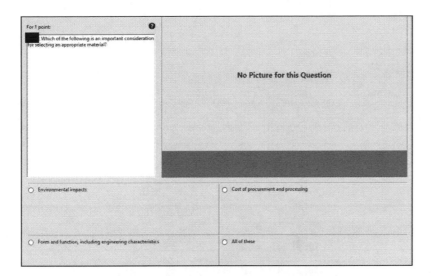

For 1 point:

Which of the following is an important consideration for selecting an appropriate material?

No Picture for this Question

○ Environmental impacts

○ Cost of procurement and processing

○ Form and function, including engineering characteristics

○ All of these

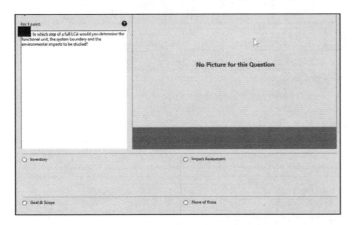

For 1 point:

In which step of a full LCA would you determine the functional unit, the system boundary and the environmental impacts to be studied?

No Picture for this Question

○ Inventory

○ Impact Assessment

○ Goal & Scope

○ None of these

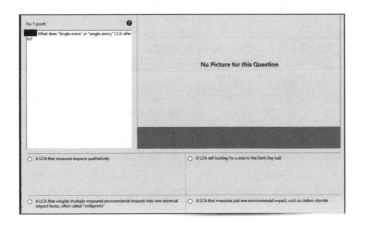

For 1 point:

What does "Single-score" or "single-proxy" LCA refer to?

No Picture for this Question

○ A LCA that measures impacts qualitatively

○ A LCA still looking for a date to the Earth Day ball

○ A LCA that weights multiple measured environmental impacts into one universal impact factor, often called "millipoints"

○ A LCA that measures just one environmental impact, such as carbon dioxide

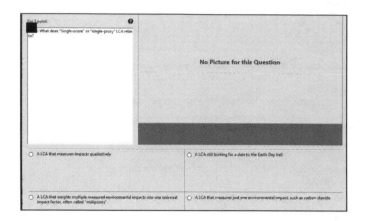

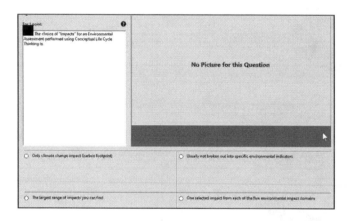

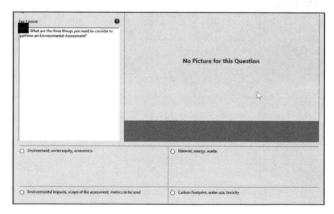

Sample Exam Questions

These questions are examples of what to expect on the certification exam. The multiple-choice questions should serve as a check for your knowledge of the exam materials.

1. Environmental Product Declarations, or EPDs, are an increasingly used method for communicating sustainability results with:

A. Suppliers and Customers.

B. Engineers.

C. Managers.

D. None of the above.

2. The commonly referenced definition of sustainable development put forth by the Brundtland Commission reads as follows:

A. "Sustainability requires closed material loops and energy independence."

B. "Sustainable development is development that meets the needs of the present without compromising the ability of future generations to meet their own needs."

C. "Sustainable development is the use of environmental claims in marketing."

D. None of the above.

3. The study of sustainable development broadly covers these three elements:

A. Land, air, and water.

B. Natural, man-made, hybrid.

C. Environment, social equity, economics.

D. Animal, vegetable, mineral.

4. This answer choice is NOT part of a long-term, working definition of a "sustainable company" ideal:

A. Generates wastes that are useful as inputs by industry or nature.

B. Sources recycled waste material and minimal virgin resources.

C. Follows all current environmental regulations.

D. Uses minimal energy that is ultimately from renewable sources.

5. "The intelligent application of the principles of sustainability to the realm of engineering and design" is a working definition for the following concept:

A. Sustainable design.

B. Sustainable business.

C. Life cycle assessment.

D. SOLIDWORKS Sustainability.

6. A focus on product design that ensures the ultimate recyclability of a product you're developing is a sustainable design technique most specifically called:

A. Design for Environment (DfE).

B. Design for Disassembly (DfD).

C. Life Cycle Assessment (LCA).

D. Design for Total Life Assessment (TLA).

7. The sustainable design technique that promotes systematically using natural inspiration and technologies found in nature to design products is known as:

A. Biomimicry.

B. Cradle to Cradle.

C. Environmental Management System (EMS).

D. Intelligent Design.

8. The sustainable design technique that can most simply be characterized by the concept that the waste from one entity equals the food of another is:

A. Cradle to Cradle.

B. Design for Disassembly (DfD).

C. Life Cycle Assessment (LCA).

D. Intelligent Design.

9. The sustainable design technique that focuses on re-formulating the raw materials we use to design out their toxicity and environmental impacts is known as:

A. Green chemistry.

B. Design for Environment (DfE).

C. Life cycle assessment (LCA).

D. Cradle to cradle.

10. The following is an example of green marketing:

A. A brochure of a product painted green, printed on 100% post-consumer recycled paper.

B. An ad touting the cost savings you can get from driving an efficient vehicle.

C. A label that indicates how many trees will be saved by purchasing this product.

D. None of the above.

11. A green product is defined as one that:

A. Is made of 100% recycled content and is itself recyclable.

B. Uses no energy or only renewable energy.

C. Has been designed using SOLIDWORKS Sustainability.

D. There is no such thing as a green product - the only "green" product is the one that's never made.

12. LCA stands for:

A. Life Cycle Analysis, because LCA is an exact science, similar to Finite Element Analysis (FEA).

B. Life Cycle Assessment, because LCA is an approximate and pragmatic method, like medicine.

C. Left Cymbal Assassination, because LCA practitioners rove the world destroying half of all percussion equipment.

D. None of the above.

13. Photochemical oxidation (smog) and ozone layer depletion are examples of environmental impacts that fall into the following domain:

A. Air impacts.

B. Terrestrial & aquatic impacts.

C. Natural resource depletion.

D. Climate effects.

14. The "global warming potential" (GWP) from greenhouse gases emitted throughout a product's lifecycle, such as carbon dioxide and methane, is a measure of the product's tendency to affect:

A. Human toxicity.

B. Climate change.

C. Ionizing radiation.

D. Air acidification.

15. The following: "(1) raw material extraction, (2) material processing, (3) part manufacturing, (4) assembly, (5) transportation, (6) product use, and (7) end of life" describes a product's:

A. Environmental indicators.

B. Metrics.

C. Lifecycle stages.

D. Good times.

CHAPTER 7 - CERTIFIED SOLIDWORKS ASSOCIATE - SIMULATION (CSWA-S) EXAM

Chapter Objective

Provide in-depth coverage of the material in the Certified SOLIDWORKS Associate - Simulation (CSWA-S) exam. Knowledge to create model studies, exam layout, question layouts, exam terminology, question types and grading procedure.

There are over 100 practice questions and examples in the chapter to help you pass the certification.

If your school is an academic certification provider, your instructor can allocate a free exam credit. The instructor will require your .edu email address.

Finite Element Modeling

SOLIDWORKS Simulation uses the displacement formulation of the finite element method to calculate component displacements, strains, and stresses under internal and external loads. The geometry under analysis is discretized using tetrahedral (3D), triangular (2D), and beam elements, and solved by either a direct sparse or iterative solver. SOLIDWORKS Simulation also offers the 2D simplification assumption for plane stress, plane strain, extruded, or axisymmetric options.

Introduction

The Certified SOLIDWORKS Associate - Simulation (CSWA-S) certification indicates a foundation in and apprentice knowledge of demonstrating an understanding in the principles of stress analysis and the Finite Element Method (FEM).

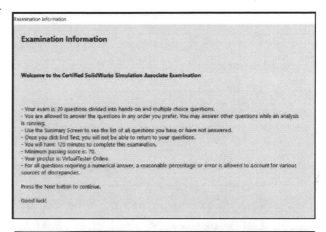

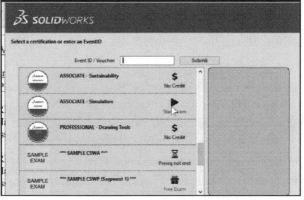

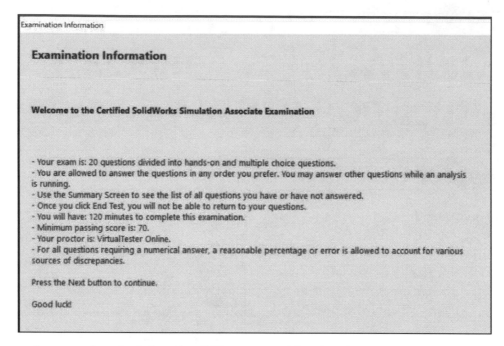

The main requirement for obtaining the CSWA-S certification is to take and pass the on-line **120-minute** exam which consists of **20** questions.

The questions consist of **3 hands-on problems (13 points, 7 points and 20 points)** single answer, and the rest are multiple choice questions for a total of 100 points.

The first hands-on problem is worth 13 points. Fill in the blank. Download the zip folder. Open the part and address the needed information for a Simulation (static) study. You are required to create and modify various elements (beam, solid, shell), apply and modify fixtures types, loads types, and results. You need to enter the answer in the correct units with the correct decimal place. There is no partial credit for any questions. Note the location of the triad in the illustration. Decimal place: 3.

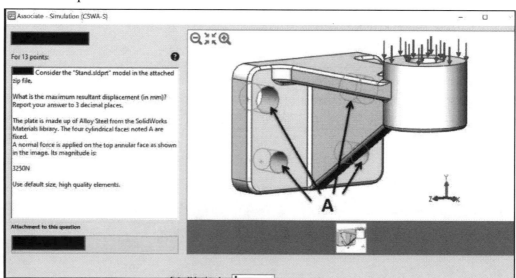

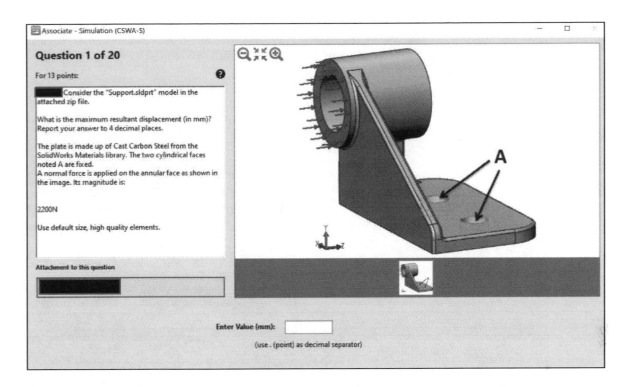

There are various orientations of the same part in the exam. Different fixed points. Different material. Decimal place: 4. You need to enter the answer in the correct units with the correct decimal place. There is no partial credit for any questions. View the location of the triad in the illustration.

Note: On the exam it will state, "Use default size, high quality elements". Set Jacobian points to 16. This could effect the results in the exam using an older SOLIDWORKS version.

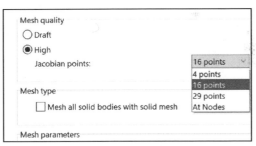

The second hands-on problem is worth 7 points. Fill in the blank. Download the zip folder. Open the part and address the needed information for a Simulation (static) study. You need to enter the answer in the correct units with the correct decimal place. Decimal place: 5. There is no partial credit for any questions. Note the location of the triad in the illustration.

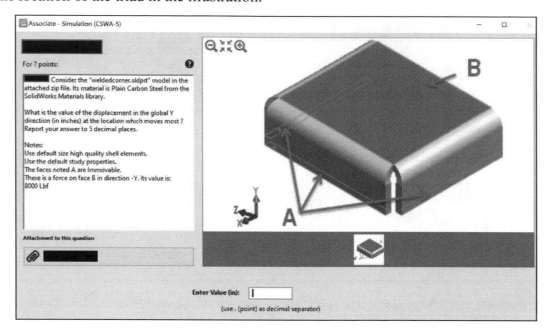

The third hands-on problem worth 20 points. Fill in the blank. Download the zip folder. Open the part and address the needed information for a Simulation (static) study. You need to enter the answer in the correct units with the correct decimal place. There is no partial credit for any questions. Note the location of the triad in the illustration. Decimal place: 2.

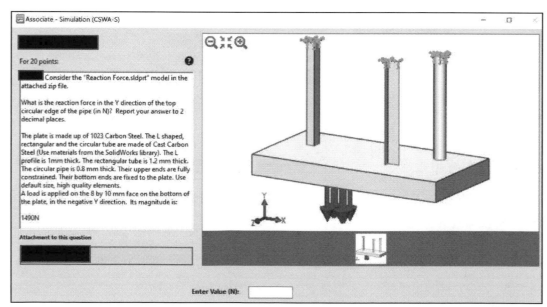

Point allocation ranges from **3 points** to **20 points**, depending on the difficulty. A passing grade is 70% (**70 out of 100**) or higher.

You are allowed to answer the questions in any order. Use the Summary Screen during the CSWA-S exam to view the list of all questions you have or have not answered.

Timing yourself is crucial. One exam strategy is to answer all the hands-on questions first within 90 minutes. Here you will gain the most points. Leave 30 minutes to answer the multiple choice yes/no, and multiple answer questions. Use SOLIDWORKS Simulation Help during the exam.

You are allowed to answer the questions in any order. Use the Summary Screen during the CSWA-S exam to view the list of all questions you have or have not answered.

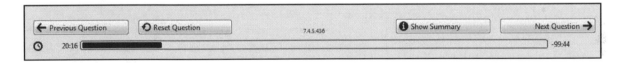

Associate - Simulation (CSWA-S)

Summary

Question (double-click to go to)	Points	State
1: Consider the "Stand.sldprt" model in the attached zip file	13	Answered
2: -1: Consider the "weldedcorner.sldprt" model in the attached zip file. Its m...	7	Answered
3: -1: In the same study, what is the bending stress in the X direction at the s...	5	Answered
4: -1: Look back at the results you obtained - you don't need to rerun the stu...	4	Answered
5: 1: Consider the "SheetMetal1.sldprt" model in the attached zip file.	20	Answered
6: What is the modulus of elasticity?	3	Answered
7: Which of the following assumptions are true for static analysis in SolidWor...	3	Answered
8: To improve accuracy of the results, Adaptive methods can be used. The H-...	3	Answered
9: Is it possible to simulate the impact of a bullet made out of lead onto a st...	3	Answered
10: The statement that the displacement at a particular location in a specific di...	5	Answered
11: Where will the maximum stress be?	5	Answered
12: A 20 inches long beam of cross section C is fixed at one end. A tensile for...	5	Answered
13: After running a study, some material properties, loads and fixtures were ...	3	Answered
14: A Remote Load applied on a face with a Force component and no Mome...	3	Answered
15: Can a non uniform pressure or force be applied on a face in SolidWorks ...	3	Answered
16: Is cyclic symmetry another name for axisymmetry?	3	Answered
17: Is the part completely constrained by the fixture shown in the image?	3	Answered
18: In SolidWorks Simulation, what criteria are best suited to check the failure ...	3	Answered
19: To avoid any misinterpretation of the results, after performing a static ana...	3	Answered
20: In order to look at only portions of your model that lie between a range o...	3	Answered

← Previous Question	↻ Reset Question	7.4.5.436	ⓘ Show Summary	Next Question →

⊙ 20:16 ▮▮▮▮▮▮▮▮▮▮▮▮▮▮▮▮▮▮ -99:44

🔆 Download all needed model files from the SDC Publication website www.SDCpublications.com/downloads/978-1-63057-567-0.

🔆 Table 1 displays the User Interface updates for Contact terms between SOLIDWORKS 2020 and 2021 and newer versions. The CSWA-S exam uses the older contact terms. The older terms are used in this chapter.

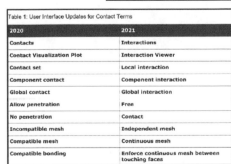

| ▮ SOLIDWORKS 2020 |
| ▮ SOLIDWORKS 2021 |
| ▮ SOLIDWORKS 2022 |
| ▮ SOLIDWORKS 2023 |

🔆 On the exam it will state, "Use default size, high quality elements". Set Jacobian points to 16. This could affect the results in the exam using an older SW version.

Table 1: User Interface Updates for Contact Terms

2020	2021
Contacts	Interactions
Contact Visualization Plot	Interaction Viewer
Contact set	Local interaction
Component contact	Component interaction
Global contact	Global interaction
Allow penetration	Free
No penetration	Contact
Incompatible mesh	Independent mesh
Compatible mesh	Continuous mesh
Compatible bonding	Enforce continuous mesh between touching faces

CSWA-S Exam Audience

The intended audience for the exam is anyone with a minimum of 3 - 6 months of SOLIDWORKS and SOLIDWORKS Simulation experience and knowledge in the following areas:

- Engineering Mechanics – Statics.

- Strength of Materials.

- Finite Element Method/Finite Element Analysis Theory.

- Define a Static Analysis Study.

- Knowledge of setting and modifying Unit systems.

- Knowledge of Simulation Options (System Options dialog box).

- Apply material to a part or assembly.

- Define a Beam, Solid or Shell element.

- Define Connector properties such as Contact Sets (Local Interaction), No Penetration (Contact) and Bonded.

- Apply and modify Connections, Contact Set (Local Interactions) and Contact Type (Interactions).

- Define Standard and Advanced Fixtures and External loads.

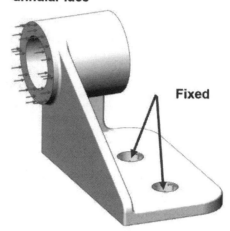

Normal Force applied on the annular face

Fixed

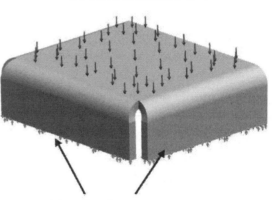

Force in direction -Y

4 bottom faces are immovable

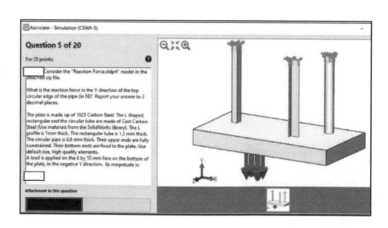

Sample CSWA-S model types

- Define Local and Global coordinate systems.

- Apply either the h-Adaptive or p-Adaptive method to the static study.

- Understand axial forces, sheer forces, bending moments and factor of safety.

- Set and modify plots to display in the correct results folders.

- Work with Multi-body parts as different solid bodies.

- Select different solvers as directed to optimize problems.

- Determine if the results are valid.

- Understand the types of problems SOLIDWORKS Simulation (Static Linear) can solve.

- Ability to use SOLIDWORKS Simulation Help.

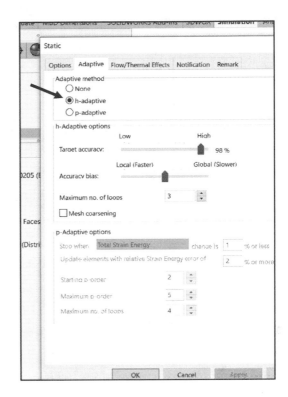

The CSWA-S **only** covers Static Linear analysis.

During the exam, you may be asked to apply a German material standard (DIN) to the model.

Note: Simulations using **SW 2023** are more accurate (than previous years) for curved surfaces that come into contact. Geometry corrections factors in the contact detection algorithms improve the representation of curved geometries such as cylindrical and spherical surfaces.

Illustrations will vary depending on your SOLIDWORKS version and operating system.

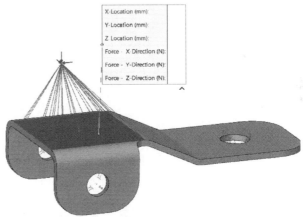

Sample CSWA-S model type

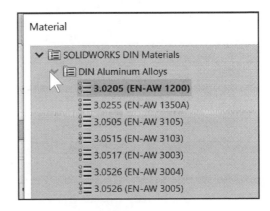

Basic FEA Concepts

SOLIDWORKS Simulation uses the Finite Element Method (FEM). FEM is a numerical technique for analyzing engineering designs. FEM is accepted as the standard analysis method due to its generality and suitability for computer implementation.

FEM divides the model into many small pieces of simple shapes called elements effectively replacing a complex problem by many simple problems that need to be solved simultaneously.

Elements share common points called nodes. The process of dividing the model into small pieces is called meshing.

The behavior of each element is well-known under all possible support and load scenarios. The finite element method uses elements with different shapes.

The response at any point in an element is interpolated from the response at the element nodes. Each node is fully described by a number of parameters depending on the analysis type and the element used.

For example, the temperature of a node fully describes its response in thermal analysis.

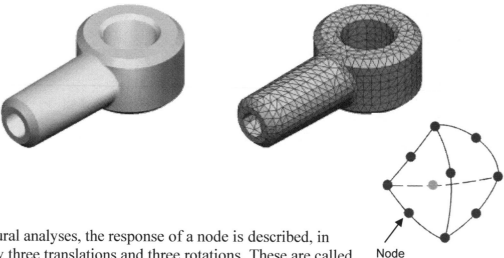

Node

Tetrahedral Element

For structural analyses, the response of a node is described, in general, by three translations and three rotations. These are called degrees of freedom (DOFs). Analysis using the FEM is called a Finite Element Analysis (FEA).

SOLIDWORKS Simulation formulates the equations governing the behavior of each element taking into consideration its connectivity to other elements. These equations relate the response to known material properties, restraints, and loads.

Next, SOLIDWORKS Simulation organizes the equations into a large set of simultaneous algebraic equations and solves for the unknowns.

In stress analysis, for example, the solver finds the displacements at each node and then the program calculates strains and finally stresses.

Static studies calculate displacements, reaction forces, strains, stresses, and factor of safety distribution. Material fails at locations where stresses exceed a certain level. Factor of safety calculations are based on one of the following failure criterion:

- **Maximum von Mises Stress**.

- **Maximum shear stress (Tresca)**.

- **Mohr-Coulomb stress**.

- **Maximum Normal stress**.

- **Automatic** (Automatically selects the most appropriate failure criterion across all element types).

Static studies can help avoid failure due to high stresses. A factor of safety less than unity indicates material failure. Large factors of safety in a contiguous region indicate low stresses and that you can probably remove some material from this region.

Simulation Advisor

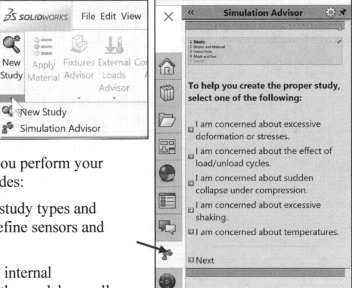

Simulation Advisor is a set of tools that guide you through the analysis process. By answering a series of questions, these tools collect the necessary data to help you perform your analysis. Simulation Advisor includes:

- **Study Advisor**. Recommends study types and outputs to expect. Helps you define sensors and creates studies automatically.

- **Interactions Advisor**. Defines internal interactions between bodies in the model as well as external interactions between the model and the environment. Interactions can include loads, fixtures, connectors, and contacts.

- **Mesh and Run Advisor**. Helps you specify the mesh and run the study.

- **Results Advisor**. Provides tips for interpreting and viewing the output of the simulation. Also, helps determine if frequency or buckling might be areas of concern.

Simulation Advisor works with the SOLIDWORKS Simulation interface by starting the appropriate PropertyManagers and linking to online help topics for additional information. Simulation Advisor leads you through the analysis workflow from determining the study type through analyzing the simulation output. By following the workflow, you use, depending on your requirements, each of the individual Advisors.

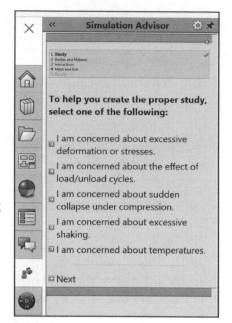

The purpose of this chapter is not to educate a new or intermediate user on SOLIDWORKS Simulation, but to cover and to inform on the types of questions, layout and what to expect when taking the CSWA-S exam.

Illustrations and values will vary slightly depending on your SOLIDWORKS release.

Simulation Help & Tutorials

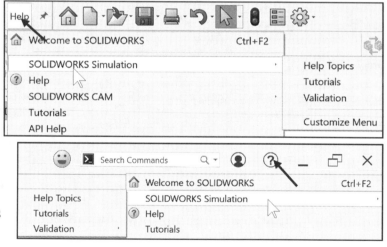

SOLIDWORKS Simulation is an Add-in. Use SOLIDWORKS Simulation during the CSWA-S exam to discover information during the exam. Utilize the Contents and Search tabs to locate subject matter.

Review the SOLIDWORKS Simulation Tutorials on Static parts and assemblies. Understand your options and setup parameters. Questions in these areas will be on the exam.

Review the SOLIDWORKS Simulation Validation, Verification Problems, Static section and the SOLIDWORKS Simulation, Verification, NAFEMS Benchmarks, Linear Static section.

Access SOLIDWORKS Simulation directly from the SOLIDWORKS Add-Ins tab in the CommandManager.

A model needs to be open to obtain access to SOLIDWORKS Simulation Help or SOLIDWORKS Simulation Tutorials.

Linear Static Analysis

This section provides the basic theoretical information required for a Static analysis using SOLIDWORKS Simulation. The CSWA-S only covers Static analysis. SOLIDWORKS Simulation and SOLIDWORKS Simulation Professional cover the following topics:

General Simulation:

- *Static Analysis (default)*
- *Use 2D Simplification*
- *Frequency Analysis*

Design Insight:

- *Topology Study*
- *Design Study*

Advanced Simulation:

- *Thermal*
- *Buckling*
- *Fatigue*
- *NonLinear*
- *Linear Dynamic*

Specialized Simulation:

- *Submodeling*
- *Drop Test*
- *Pressure Vessel Design*

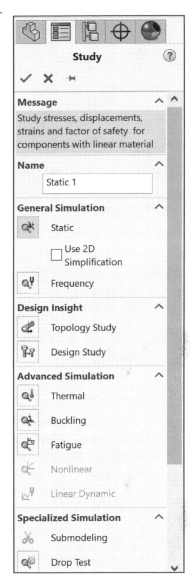

Linear Static Analysis

When loads are applied to a body, the body deforms and the effect of loads is transmitted throughout the body. The external loads induce internal forces and reactions to render the body into a state of equilibrium.

Linear Static analysis calculates displacements, strains, stresses, and reaction forces under the effect of applied loads.

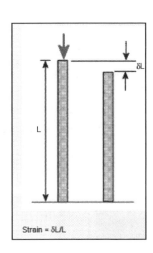

Strain ε is the ratio of change, δ L, to the original length, L. Strain, is a dimensionless quantity. Stress σ is defined in terms of Force per unit Area.

Linear Static analysis makes the following assumptions:

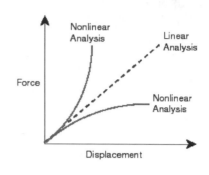

- **Static Assumption**. All loads are applied *slowly* and gradually until they reach their full magnitudes. After reaching their full magnitudes, loads *remain constant* (time-invariant). This assumption *neglects inertial and damping forces*.

  Time-variant loads that induce considerable inertial and/or damping forces and require dynamic analysis.

  Dynamic loads change with time and in many cases induce considerable inertial and damping forces that cannot be neglected.

- **Linearity Assumption**. The relationship between loads and induced responses is linear. For example, **if you double the loads, the response of the model (displacements, strains, and stresses) will also double**. Apply the linearity assumption if:

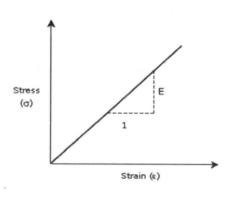

  - All materials in the model comply with Hooke's law; that is, stress is directly proportional to strain.
  - The induced displacements are small enough to ignore the change in stiffness caused by loading.
  - Boundary conditions do not vary during the application of loads. Loads must be constant in magnitude, direction, and distribution. They should not change while the model is deforming.

☀ In SOLIDWORKS Simulation for Static analysis, all displacements **are small relative** to the model geometry (unless the Large Displacements option is activated).

☀ FeatureManager and CommandManager tabs and tree folders will vary depending on system setup, year, and SOLIDWORKS Add-ins.

SOLIDWORKS Simulation assumes that the normals to contact areas do not change direction during loading. Hence, it applies the full load in one step. This approach may lead to inaccurate results or convergence difficulties in cases where these assumptions are not valid.

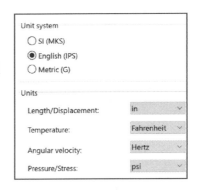

- Elastic Modulus, E, is the stress required to cause one unit of strain. The material behaves linearly at low stresses.

- Elastic Modulus (Young's Modulus) is the slope defined as stress divided by strain. *E = modulus of elasticity (Pa (N/m²), N/mm², psi).*

- Stress σ is proportional to strain in a Linear Elastic Material. Units: *(Pa (N/m²), N/mm², psi).*

You must be able to work in SI and English units for the CSWA-S exam within the same problem. For example, you apply a Force in Newtons and then you determine displacement in inches.

Different materials have different stress property levels. Mathematical equations derived from Elasticity theory and Strength of Materials are utilized to solve for displacement and stress. These analytical equations solve for displacement and stress for simple cross sections.

General Procedure to Perform a Linear Static Analysis

- Complete a Linear Static study by performing the following steps:

- Open the model (part or assembly).

- Add-in SOLIDWORKS Simulation.

- Select the Simulation tab.

- Create a new Static study.

- Apply material (part or assembly).

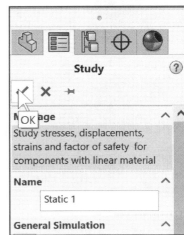

- Define and apply various Connections and Fixtures. Standard and Advanced options.

- Define and apply various External Loads. Standard and Advanced options.

- Knowledge in defining coordinates of the reference node location and remote forces and remote translations.

- Define and apply component contacts (Local interactions) and contact sets (interactions) for assemblies and multi-body parts.

- Mesh and Run the study. Select correct mesh type and parameters.

- Create Iso Clipping or Section Clipping plot.

- Define what criteria are best suited to check the failure of ductile or brittle material in a SOLIDWORKS Simulation.

- Assess failure based on a yield criterion (Define Factor of Safety Plot).

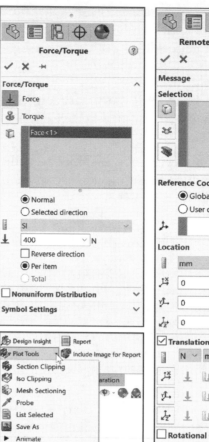

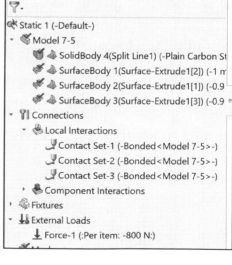

Sequence of Calculations in General

Given a meshed model with a set of displacement restraints and loads, the linear static analysis program proceeds as follows:

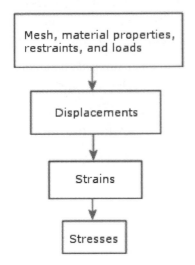

- The program constructs and solves a system of linear simultaneous finite element equilibrium equations to calculate displacement components at each node.

- The program then uses the displacement results to calculate the strain components.

- The program uses the strain results and the stress-strain relationships to calculate the stresses.

Stress Calculations in General

Stress results are first calculated at special points, called Gaussian points or Quadrature points, located inside each element.

💡 SOLIDWORKS Simulation utilizes a tetrahedral element (High quality mesh for a Solid). Each node contains a series of equations.

These points are selected to give optimal numerical results. The program calculates stresses at the nodes of each element by extrapolating the results available at the Gaussian points.

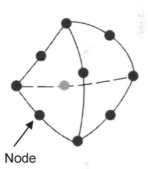

Node

Tetrahedral Element

After a successful run, nodal stress results at each node of every element are available in the database. Nodes common to two or more elements have multiple results. In general, these results are not identical because the finite element method is an approximate method. For example, if a node is common to three elements, there can be three slightly different values for every stress component at that node.

Overview of the Yield or Inflection Point in a Stress-Strain curve

When viewing stress results, you can ask for element stresses or nodal stresses. To calculate element stresses, the program averages the corresponding nodal stresses for each element.

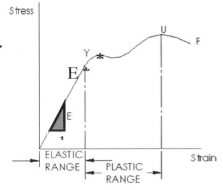

Stress vs. Strain Plot
Linearly Elastic Material

To calculate nodal stresses, the program averages the corresponding results from all elements sharing that node.

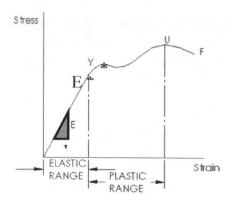

Stress vs. Strain Plot
Linearly Elastic Material

- The material remains in the Elastic Range until it reaches the elastic limit.

- The point E is the elastic limit. The material begins Plastic deformation.

- Yield Stress is the stress level at which the material ceases to behave elastically.

- The point Y is called the Yield Point. The material begins to deform at a faster rate. In the Plastic Range the material behaves non-linearly.

- The point U is called the ultimate tensile strength. Point U is the maximum value of the non-linear curve. Point U represents the maximum tensile stress a material can handle before a fracture or failure.

- Point F represents where the material will fracture.

- Designers utilize maximum and minimum stress calculations to determine if a part is safe. Simulation reports a recommended Factor of Safety during the analysis.

- The Simulation Factor of Safety is a ratio between the material strength and the calculated stress.

Brittle materials do not have a specific yield point and hence it is not recommended to use the yield strength to define the limit stress for the criterion.

Material Properties in General

Before running a study, you must define all material properties required by the associated analysis type and the specified material model. A material model describes the behavior of the material and determines the required material properties. Linear isotropic and orthotropic material models are available for all structural and thermal studies. Other material models are available for nonlinear stress studies. The von Mises plasticity model is available for drop test studies. Material properties can be specified as function of temperature.

- For solid assemblies, each component can have a different material.

- For shell models, each shell can have a different material and thickness.

- For beam models, each beam can have a different material.

- For mixed mesh models, you must define the required material properties for solid and shell separately.

Connections in General

A connection replaces a piece of hardware or fastener by simulating its effect on the rest of the model. Connections include Bolts, Springs, Flexible Support, Bearings, Bonding - Weld/Adhesives, Welds, etc.

The automatic detection tool in SOLIDWORKS Simulation defines Contact Sets **(SW 2021 and newer versions changed to Local interactions)**. Sometimes additional Contact Sets (Local Interactions) and types need to be defined. The SOLIDWORKS Simulation Study Advisor can help.

For example, the behavior of an adhesive depends on its strength and thickness. You can select the Type manually.

Fixtures: Adequate restraints to prevent the body from rigid body motion. If your model is not adequately constrained, check the Use soft springs to stabilize the model option in the Static dialog box.

When importing loads from SOLIDWORKS Motion, make sure that Use inertial relief option is checked. These options are available for the Direct Sparse solver and FFEPlus solver.

See SOLIDWORKS Simulation Help for additional information.

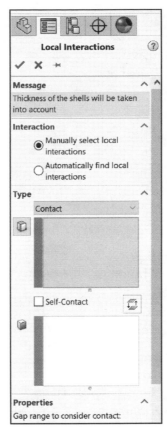

Restraint Types

The Fixture PropertyManager provides the ability to prescribe zero or non-zero displacements on vertices, edges, or faces for use with static, frequency, buckling, dynamic and nonlinear studies. This section will only address standard restraint types, namely Fixed Geometry and Immovable (No translation).

The Immovable (No translation) option is displayed for Sheet Metal parts.

Fixed: For solids, this restraint type sets all translational degrees of freedom to zero. For shells and beams, it sets the translational and the rotational degrees of freedom to zero. For truss joints, it sets the translational degrees of freedom to zero. When using this restraint type, no reference geometry is needed.

View the illustrated table for the attributes and input needed for this restraint.

Attribute	Value
DOFs restrained for solid meshes	3 translations
DOFs restrained for shells and beams	3 translations and 3 rotations
DOFs restrained for truss joints	3 translations
3D symbol (the arrows are for translations and the discs are for rotations)	
Selectable entities	Vertices, edges, faces and beam joints
Selectable reference entity	N/A
Translations	N/A
Rotations	N/A

Immovable (No translation): This restraint type sets all translational degrees of freedom to zero. It is the same for shells, beams and trusses. No reference geometry is used.

To access the immovable restraint, right-click on Fixtures in the Simulation study tree and select Fixed Geometry. Under Standard, select Immovable (No translation). View the illustrated table for the attributes and input needed for this restraint.

Attribute	Value
DOFs restrained for shell meshes	3 translations
DOFs restrained for beam and truss meshes	3 translations
3D symbol	
Selectable entities	Vertices, edges, faces and beam joints
Selectable reference entity	N/A
Translations	N/A
Rotations	N/A

There are differences for Shells and Beams between Immovable (No translation) and Fixed restraint types.

The Immovable option is not available for Solids.

The Fixture PropertyManager allows you to prescribe zero or non-zero displacements on vertices, edges or faces for use with static, frequency, buckling, dynamic and nonlinear studies.

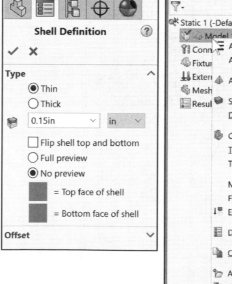

The Fixed Geometry Fixture allows for additional Advanced options: Symmetry, Circular Symmetry, User Reference Geometry, On Flat Faces, On Cylindrical Faces, On Spherical Faces.

Attributes of each option are available in SOLIDWORKS Simulation Help.

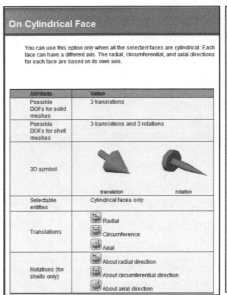

Loads and Restraints in General

Loads and restraints are necessary to define the service environment of the model. The results of analysis directly depend on the specified loads and restraints.

Loads and restraints are applied to geometric entities as features that are fully associative to geometry and automatically adjust to geometric changes.

For example, if you apply a pressure **P** to a face of area A_1, the equivalent force applied to the face is PA_1.

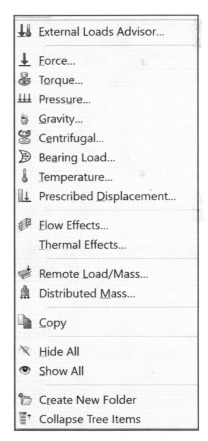

If you modify the geometry such that the area of the face changes to A_2, then the equivalent force automatically changes to PA_2. Re-meshing the model is required after any change in geometry to update loads and restraints.

The types of loads and restraints available depend on the type of the study. A load or restraint is applied by the corresponding Property

Manager accessible by right-clicking the Fixtures or External Loads folder in the Simulation study tree, or by clicking Simulation, Loads/Fixture.

Loads: At least one of the following types of loading is required:

- Concentrated force.

- Pressure.

- Prescribed nonzero displacements.

- Body forces (gravitational and/or centrifugal).

- Thermal (define temperatures or get the temperature profile from thermal analysis).

- Imported loads from SOLIDWORKS Motion.

- Imported temperature and pressure from Flow Simulation.

In a linear static thermal stress analysis for an assembly, it is possible to input different temperature boundary conditions for different parts.

Under the External Loads folder you can define Remote Load/Mass and Distributed Mass. In the Remote Load/Mass you define Load, Load/Mass or Displacement. Input values are required for the Remote Location and the Force.

By default, the Location is set to x=0, y=0, z=0. The Force is set to $F_x=0$, $F_y=0$, $F_z=0$. The Force requires you to first select the direction and then enter the value.

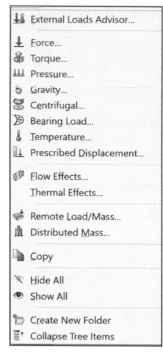

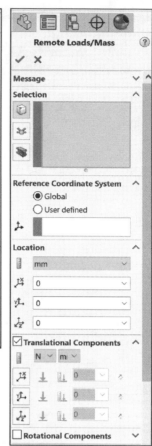

Meshing in General

Meshing splits continuous mathematical models into finite elements. The types of elements created by this process **depend on the type of geometry meshed**. SOLIDWORKS Simulation offers three types of elements:

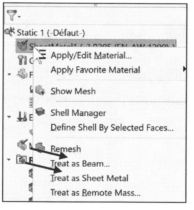

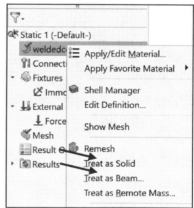

- **Solid elements** - solid geometry.

- **Shell elements** - surface geometry.

- **Beam elements** - wire frame geometry.

In CAD terminology, "Solid" denotes the type of geometry. In FEA terminology, "Solid" denotes the type of element used to mesh the solid CAD Geometry.

Meshing Types

Meshing splits continuous mathematical models into finite elements. Finite element analysis looks at the model as a network of interconnected elements.

Meshing is a crucial step in design analysis. SOLIDWORKS Simulation automatically creates a mixed mesh of:

- **Solid**: The Solid mesh is appropriate for bulky or complex 3D models. In meshing a part or an assembly with solid elements, Simulation generates one of the following types of elements based on the active mesh options for the study:

  - Draft quality mesh. The automatic mesher generates linear tetrahedral solid elements **(4 nodes)**.

  - High quality mesh. The automatic mesher generates parabolic tetrahedral solid elements **(10 nodes)**.

Linear elements are also called first-order, or lower-order elements. Parabolic elements are also called second-order, or higher-order elements.

A linear tetrahedral element is defined by four corner nodes connected by six straight edges.

A parabolic tetrahedral element assigns 10 nodes to each solid element: four corner nodes and one node at the middle of each edge (a total of six mid-side nodes).

Linear solid element

In general, for the same mesh density (number of elements), parabolic elements yield better results than linear elements because:

- They represent curved boundaries more accurately.

- They produce better mathematical approximations. However, parabolic elements require greater computational resources than linear elements.

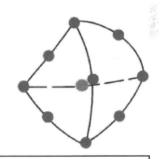

Parabolic solid element

For structural problems, each node in a solid element has three degrees of freedom that represent the translations in three orthogonal directions.

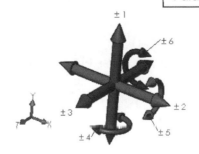

SOLIDWORKS Simulation uses the X, Y, and Z directions of the global Cartesian coordinate system in formulating the problem.

- **Shell**: Shell elements are suitable for thin parts (sheet metal models). Shell elements are 2D elements capable of resisting membrane and bending loads. When using shell elements, Simulation generates one of the following types of elements depending on the active meshing options for the study:

  Linear triangular element

  - **Draft quality mesh**. The Draft quality mesher generates linear triangular shell elements Specifies 4 corner nodes for each solid element and 3 corner nodes for each shell element.

  - **High quality mesh**. The High quality mesher generates parabolic triangular shell elements. Sets the number of integration points for the Jacobian ratio calculations. The Jacobian ratio calculations consider the Gaussian points located within each element. Select 4, 16, & 29 Jacobian points, or At Nodes. Default Jacobian points: 16.

  Parabolic triangular element

A linear triangular shell element is defined by three corner nodes connected by three straight edges.

A parabolic triangular element is defined by three corner nodes, three mid-side nodes, and three parabolic edges.

The Shell Definition PropertyManager is used to define the thickness of thin and thick shell elements. The program automatically extracts and assigns the thickness of the sheet metal to the shell. You cannot modify the thickness. You can select between the thin shell and thick shell formulations. You can also define a shell as a composite for static, frequency, and buckling studies. In general use thin shells when the thickness-to-span ratio is less than 0.05.

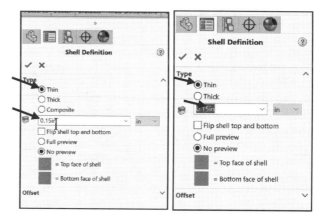

SW 2020 Version   SW 2021, 2022, 2023 version

 Surface models can only be meshed with shell elements.

- **Beam or Truss**: Beam or Truss elements are suitable for extruded or revolved objects and structural members with constant cross-sections. Beam elements can resist bending, shear, and torsional loads. The typical frame shown is modeled with beams elements to transfer the load to the supports. Modeling such frames with truss elements fails since there is no mechanism to transfer the applied horizontal load to the supports.

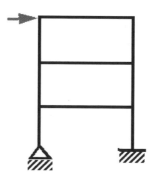

A truss is a special beam element that can resist axial deformation only.

Beam elements require defining the exact cross section so that the program can calculate the moments of inertia, neutral axes and the distances from the extreme fibers to the neutral axes. The stresses vary within the plane of the cross-section and along the beam. A beam element is a line element defined by two end points and a cross-section.

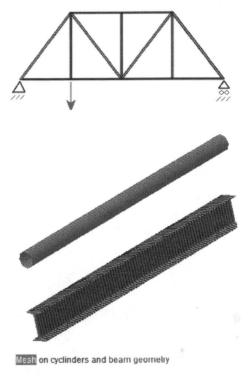

Consider a 3D beam with cross-sectional area (A) and the associated mesh. Beam elements can be displayed on actual beam geometry or as hollow cylinders regardless of their actual cross-section.

Beam elements are capable of resisting axial, bending, shear, and torsional loads. Trusses resist axial loads only. When used with weldments, the software defines cross-sectional properties and detects joints.

Mesh on cylinders and beam geometry

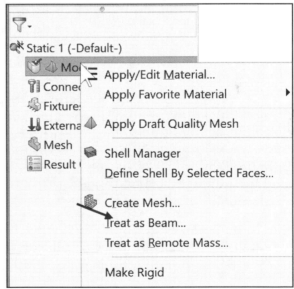

Beam and truss members can be displayed on actual beam geometry or as hollow cylinders regardless of their actual cross-sectional shape.

A Beam element has 3 nodes (one at each end) with 6 degrees of freedom (3 translational and 3 rotational) per node plus one node to define the orientation of the beam cross section.

A Truss element has 2 nodes with 3 translational degrees of freedom per node.

The accuracy of the solution depends on the quality of the mesh. In general, the finer the mesh the better the accuracy.

A compatible mesh is a mesh where elements on touching bodies have overlaying nodes.

The curvature-based mesher supports multi-threaded surface and volume meshing for assembly and multi-body part documents. The standard mesher supports only multi-threaded volume meshing.

It is possible to mesh a part or assembly with a combination of solids, shells and beam elements (mixed mesh) in SOLIDWORKS Simulation.

SOLIDWORKS Simulation Meshing Tips

SOLIDWORKS Simulation Help lists the following Meshing tips that you should know for the CSWA-S exam.

- When you mesh a study, the SOLIDWORKS Simulation meshes all unsuppressed solids, shells and beams:

- Use Solid mesh for bulky objects.

- Use Shell elements for thin objects like sheet metals.

- Use Beam or Truss elements for extruded or revolved objects with constant cross-sections.

- Simplify structural beams to optimize performance in Simulation to be modeled with beam elements. The size of the problem and the resources required are dramatically reduced in this case. For the beam formulation to produce acceptable results, the length of the beam should be 10 times larger than the largest dimension of its cross section.

- Compatible meshing (a mesh where elements on touching bodies have overlaying nodes) is more accurate than incompatible meshing in the interface region. Requesting compatible meshing can cause mesh failure in some cases. Requesting incompatible meshing can result in successful results. You can request compatible meshing and select Re-mesh failed parts with incompatible mesh so that the software uses incompatible meshing only for bodies that fail to mesh.

- Check for interferences between bodies when using a compatible mesh with the curvature-based mesher. If you specify a bonded contact condition between bodies, they should be touching. If interferences are detected, meshing stops, and you can access the Interference Detection PropertyManager to view the interfering parts. Make sure to resolve all interferences before you mesh again.

- If meshing fails, use the Failure Diagnostics tool to locate the cause of mesh failure. Try the proposed options to solve the problem. You can also try different element size, define mesh control, or activate Enable automatic looping for solids.

- The SOLIDWORKS Simplify utility lets you suppress features that meet a specified simplification factor. In the Simulation study tree, right-click Mesh and select Simplify Model for Meshing. This displays the Simplify utility.

- It is good practice to check mesh options before meshing. For example, the Automatic transition can result in generating an unnecessarily large number of elements for models with many small features. The high quality mesh is recommended for most cases. The Automatic looping can help solve meshing problems automatically, but you can adjust its settings for a particular model. The Curvature-based mesher automatically uses smaller element sizes in regions with high curvature.

- To improve results in important areas, use mesh control to set a smaller element size. When meshing an assembly with a wide range of component sizes, default meshing results in a relatively coarse mesh for small components. Component mesh control offers an easy way to give more importance to the selected small components. Use this option to identify important small components.

- For assemblies, check component interference. To detect interference in an assembly, click Tools, Interference Detection. Interference is allowed only when using **shrink fit**. The Treat coincidence as interference and Include multi-body part interferences options allow you to detect touching areas. These are the only areas affected by the global and component contact settings.

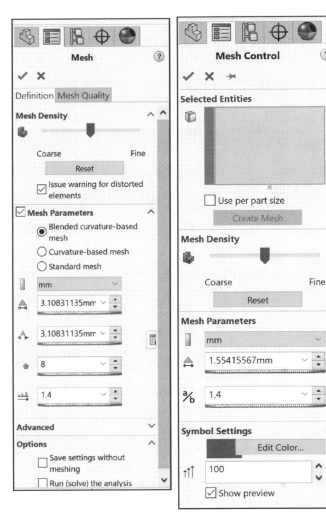

Use the mesh and displacement plots to calculate the distance between two nodes using SOLIDWORKS Simulation.

The Global element size parameter provides the ability to **set the global average element size**. SOLIDWORKS suggests a default value based on the model volume and surface area. This option is only available for a standard mesh.

The Ratio value in Mesh Control provides the geometric growth ratio from one layer of elements to the next. To access Mesh Control, right-click the Mesh folder in the Simulation study tree and click Apply Mesh Control.

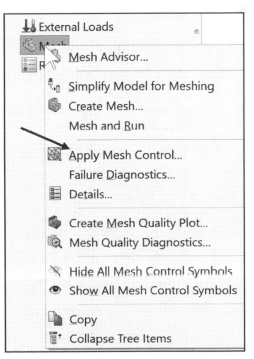

Running the Study

When you run a study, Simulation calculates the results based on the specified input for materials, restraints, loads and mesh.

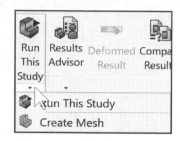

Set the default plots that you want to see in your Simulation Study tree under **Simulation**, **Options** from the Main menu.

When you run one or multiple studies, they run as background processes.

In viewing the results after running a study, you can generate plots, lists, graphs, and reports depending on the study and result types.

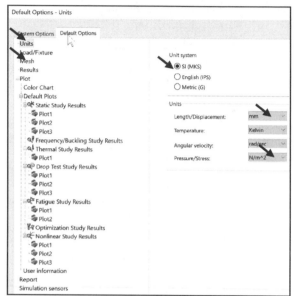

If you modify the study (force, material, etc.) you only need to **re-run the study** to update the results. You do not need to re-mesh unless you modified contact conditions.

Displacement Plot - Output of Linear Static Analysis

The Displacement Plot PropertyManager allows you to plot displacement and reaction force results for static, nonlinear, dynamic, drop test studies, or mode shapes for bucking and frequency studies. By default, directions X, Y and Z refer to the **global coordinate system**.

If you choose a reference geometry, these directions refer to the selected reference entity. Displacement components are:

UX = Displacement in the X-direction.

UY = Displacement in the Y-direction.

UZ = Displacement in the Z-direction.

URES = Resultant displacement.

RFX = Reaction force in the X-direction.

RFY = Reaction force in the Y-direction.

RFZ = Reaction force in the Z-axis.

RFRES = Resultant reaction force.

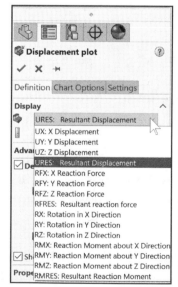

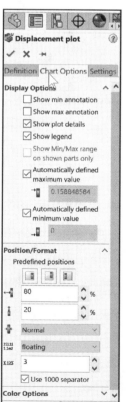

The **Probe** function allows you to query a plot and view the values of plotted quantities at **defined nodes or centers of elements**. When you probe a mesh plot, Simulation displays the node or element number and the global coordinates of the node. When you probe a result plot, SOLIDWORKS Simulation displays the node or element number, the value of the plotted result, and the global coordinates of the node or center of the element. For example, in a nodal stress plot, the node number, the stress value, and the global x, y, and z coordinates appear.

Adaptive Methods for Static Studies

Adaptive methods help you obtain an accurate solution for static studies. There are two types of adaptive methods: h-adaptive and p-adaptive method.

The concept of the h-method (available for solid part and assembly documents) is to use smaller elements (increase the number of elements) in regions with high relative errors to improve accuracy of the results. After running the study and estimating errors, the software automatically refines the mesh where needed to improve results.

The p-adaptive method (available for solid part and assembly documents) increases the polynomial order of elements with high relative errors. The p-method does not change the mesh. It changes the order of the polynomials used to approximate the displacement field using a unified polynomial order for all elements. See SOLIDWORKS Help for additional information.

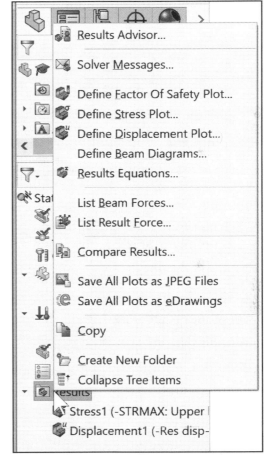

Understand how to select either p-adaptive or h-adaptive methods for the CSWA- S exam.

Table 1 displays the User Interface updates for Contact terms between SOLIDWORKS 2020 and 2021 and newer versions. The CSWA-S exam uses the older contact terms. The older terms are used in this chapter.

Table 1: User Interface Updates for Contact Terms	
2020	**2021**
Contacts	Interactions
Contact Visualization Plot	Interaction Viewer
Contact set	Local interaction
Component contact	Component interaction
Global contact	Global interaction
Allow penetration	Free
No penetration	Contact
Incompatible mesh	Independent mesh
Compatible mesh	Continuous mesh
Compatible bonding	Enforce continuous mesh between touching faces

Sample Exam Questions

These questions are examples of what to expect on the certification exam. The multiple-choice questions should serve as a check for your knowledge of the exam materials.

1. What is the Modulus of Elasticity?

- The slope of the Deflection-Stress curve.

- The slope of the Stress-Strain curve in its linear section.

- The slope of the Force-Deflection curve in its linear section.

- The first inflection point of a Strain curve.

2. What is Stress?

- A measure of power.

- A measure of strain.

- A measure of material strength.

- A measure of the average amount of force exerted per unit area.

3. Which of the following assumptions are true for a static analysis in SOLIDWORKS Simulation with small displacements?

- Inertia effects are negligible and loads are applied slowly.

- The model is not fully elastic. If loads are removed, the model will not return to its original position.

- Results are proportional to loads.

- All the displacements are small relative to the model geometry.

4. What is Yield Stress?

- The stress level beyond which the material becomes plastic.

- The stress level beyond which the material breaks.

- The strain level above the stress level which the material breaks.

- The stress level beyond the melting point of the material.

5. A high quality Shell element has _____ nodes.

- 4
- 5
- 6
- 8

6. Stress σ is proportional to _____ in a Linear Elastic Material.

- Strain.
- Stress.
- Force.
- Pressure.

7. The Elastic Modulus (Young's Modulus) is the slope defined as _____ divided by _____.

- Strain, Stress.
- Stress, Strain.
- Stress, Force.
- Force, Area.

8. Linear static analysis assumes that the relationship between loads and the induced response is _____.

- Flat.
- Linear.
- Doubles per area.
- Translational.

9. In SOLIDWORKS Simulation, the Factor of Safety (FOS) calculations are based on one of the following failure criteria.

- Maximum von Mises Stress.
- Maximum shear stress (Tresca).
- Mohr-Coulomb stress.
- Maximum Normal stress.

10. The Yield point is the point where the material begins to deform at a faster rate than at the elastic limit. The material behaves _____ in the Plastic Range.

- Flatly.

- Linearly.

- Non-Linearly.

- Like a liquid.

11. What are the Degrees of Freedom (DOFs) restrained for a Solid?

- None.

- 3 Translations.

- 3 Translations and 3 Rotations.

- 3 Rotations.

12. What are the Degrees of Freedom (DOFs) restrained for Truss joints?

- None.

- 3 Translations.

- 3 Translations and 3 Rotations.

- 3 Rotations.

13. What are the Degrees of Freedom (DOFs) restrained for Shells and Beams?

- None.

- 3 Translations.

- 3 Translations and 3 Rotations.

- 3 Rotations.

14. Which statements are true for Material Properties using SOLIDWORKS Simulation?

- For solid assemblies, each component can have a different material.

- For shell models, each shell cannot have a different material and thickness.

- For shell models, the material of the part is used for all shells.

- For beam models, each beam cannot have a different material.

15. A Beam element has _____nodes (one at each end) with _____degrees of freedom per node plus_____ node to define the orientation of the beam cross section.

- 6, 3, 1
- 3, 3, 1
- 3, 6, 1
- None of the above.

16. A Truss element has _____ nodes with _____ translational degrees of freedom per node.

- 2, 3
- 3, 3
- 6, 6
- 2, 2

17. In general, the finer the mesh the better the accuracy of the results.

- True.
- False.

18. How does SOLIDWORKS Simulation automatically treat a Sheet metal part with uniform thickness?

- Shell.
- Solid.
- Beam.
- Mixed Mesh.

19. Use the mesh and displacement plots to calculate the distance between two _____ using SOLIDWORKS Simulation.

- Nodes.
- Elements.
- Bodies.
- Surfaces.

20. Surface models can only be meshed with _____ elements.

- Shell.

- Beam.

- Mixed Mesh.

- Solid.

21. The shell mesh is generated on the surface (located at the mid-surface of the shell).

- True.

- False.

22. In general, use Thin shells when the thickness-to-span ratio is less than _____.

- 0.05

- .5

- 1

- 2

23. The model (a rectangular plate) has a length to thickness ratio of less than 5. You extracted its mid-surface to use it in SOLIDWORKS Simulation. You should use a _____.

- Thin Shell element formulation.

- Thick Shell element formulation.

- Thick or Thin Shell element formulation, it does not matter.

- Beam Shell element formulation.

24. The model, a rectangular sheet metal part, uses SOLIDWORKS Simulation. You should use a:

- Thin Shell element formulation.

- Thick Shell element formulation.

- Thick or Thin Shell element formulation, it does not matter.

- Beam Shell element formulation.

25. The Global element size parameter provides the ability to set the global average element size. SOLIDWORKS Simulation suggests a default value based on the model volume and _____ area. This option is only available for a standard mesh.

- Force.

- Pressure.

- Surface.

- None of the above.

26. A remote load applied on a face with a Force component and no Moment can result in: Note: Remember (DOFs restrain).

- A Force and Moment of the face.

- A Force on the face only.

- A Moment on the face only.

- A Pressure and Force on the face.

27. There are _____ DOFs restrained for a Solid element.

- 3

- 1

- 6

- None

28. There are _____ DOFs restrained for a Beam element.

- 3

- 1

- 6

- None

29. What best describes the difference(s) between a Fixed and Immovable (No translation) boundary condition in SOLIDWORKS Simulation?

- There are no differences.

- There are no difference(s) for Shells but it is different for Solids.

- There is no difference(s) for Solids but it is different for Shells and Beams.

- There are only differences(s) for a Static Study.

30. Can a non-uniform pressure or force be applied on a face using SOLIDWORKS Simulation?

- No.

- Yes, but the variation must be along a single direction only.

- Yes. The variation can be in two directions and is described by a binomial equation.

- Yes, but the variation must be linear.

31. You are performing an analysis on your model. You select five faces, 3 edges and 2 vertices and apply a force of 20lbf. What is the total force applied to the model using SOLIDWORKS Simulation?

- 100lbf

- 1600lbf

- 180lbf

- 200lbf

32. Yield strength is typically determined at _____ strain.
- 0.1%

- 0.2%

- 0.02%

- 0.002%

33. There are four key assumptions made in Linear Static Analysis: 1. Effects of inertia and damping are neglected, 2. The response of the system is directly proportional to the applied loads, 3. Loads are applied slowly and gradually, and_____ .

- Displacements are very small. The highest stress is in the linear range of the stress-strain curve.

- There are no loads.

- Material is not elastic.

- Loads are applied quickly.

34. How many degrees of freedom does a physical structure have?

- Zero.
- Three - Rotations only.
- Three - Translations only.
- Six - Three translations and three rotational.

35. What criteria are best suited to check the failure of brittle material in SOLIDWORKS Simulation?

- Maximum von Mises Stress and Maximum Shear Stress criteria.
- Maximum von Misses Stress and Maximum Shear Strain criteria.
- Maximum von Misses Stress and Maximum Normal Stress criteria.
- Mohr-Coulomb Stress and Maximum Normal Stress criteria.

36. You are performing an analysis on your model. You select three faces and apply a force of 40lb. What is the total force applied to the model using SOLIDWORKS Simulation?

- 40lb
- 20lb
- 120lb
- Additional information is required.

37. A material is orthotropic if its mechanical or thermal properties are not unique and independent in three mutually perpendicular directions.

- True
- False

38. An increase in the number of elements in a mesh for a part will:

- Decrease calculation accuracy and time.
- Increase calculation accuracy and time.
- Have no effect on the calculation.
- Change the FOS below 1.

39. SOLIDWORKS Simulation uses the von Mises Yield criteria to calculate the Factor of Safety of many ductile materials. According to the criteria:

- Material yields when the von Mises stress in the model equals the yield strength of the material.

- Material yields when the von Mises stress in the model is 5 times greater than the minimum tensile strength of the material.

- Material yields when the von Mises stress in the model is 3 times greater than the FOS of the material.

- None of the above.

40. SOLIDWORKS Simulation calculates structural failure on:

- Buckling.

- Fatigue.

- Creep.

- Material yield.

41. Apply a uniform total force of 200lb on two faces of a model. The two faces have different areas. How do you apply the load using SOLIDWORKS Simulation for a Linear Static Study?

- Select the two faces and input a normal to direction force of 200lb on each face .

- Select the two faces and a reference plane. Apply 100lb on each face.

- Apply equal force to the two faces. The force on each face is the total force divided by the total area of the two faces.

- None of the above.

42. Maximum and Minimum value indicators are displayed on Stress and Displacement plots in SOLIDWORKS Simulation for a Linear Static Study.

- True.

- False.

43. What SOLIDWORKS Simulation tool should you use to determine the result values at specific locations (nodes) in a model using SOLIDWORKS Simulation?

- Section tool.

- Probe tool.

- Clipping tool.

- Surface tool.

44. What criteria are best suited to check the failure of ductile materials in SOLIDWORKS Simulation?

- Maximum von Mises Strain and Maximum Shear Strain criteria.

- Maximum von Misses Stress and Maximum Shear Stress criteria.

- Maximum Mohr-Coulomb Stress and Maximum Mohr-Coulomb Shear Strain criteria.

- Mohr-Coulomb Stress and Maximum Normal Stress criteria.

45. Set the scale factor for plots_____ to avoid any misinterpretation of the results, after performing a Static analysis with gap/contact elements.

- Equal to 0.

- Equal to 1.

- Less than 1.

- To the Maximum displacement value for the model.

46. It is possible to mesh _____ with a combination of Solids, Shells and Beam elements in SOLIDWORKS Simulation.

- Parts and Assemblies.

- Only Parts.

- Only Assemblies.

- None of the above.

47. SOLIDWORKS Simulation supports multi-body parts. Which of the following is a true statement?

- You can employ different mesh controls to each Solid body.
- You can classify Contact conditions between multiple Solid bodies.
- You can classify a different material for each Solid body.
- All of the above are correct.

48. Which statement best describes a Compatible mesh?

- A mesh where only one type of element is used.
- A mesh where elements on touching bodies have overlaying nodes.
- A mesh where only a Shell or Solid element is used.
- A mesh where only a single Solid element is used.

49. The Ratio value in Mesh Control provides the geometric growth ratio from one layer of elements to the next.

- True.
- False.

50. The structures displayed in the following illustration are best analyzed using:

- Shell elements.
- Solid elements.
- Beam elements.
- A mixture of Beam and Shell elements.

51. The structure displayed in the following illustration is best analyzed using:

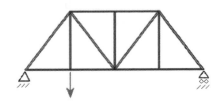

- Shell elements.

- Solid elements.

- Beam elements.

- A mixture of Beam and Shell elements.

52. The structure displayed in the following illustration is best analyzed using:

- Shell elements.

- Solid elements.

- Beam elements.

- A mixture of Beam and Shell elements.

Sheet metal model

53. The structure displayed in the following illustration is best analyzed using:

- Shell elements.

- Solid elements.

- Beam elements.

- A mixture of Beam and Shell elements.

54. Surface models can only be meshed with _____ elements.

- Shell elements.

- Solid elements.

- Beam elements.

- A mixture of Beam and Shell elements.

55. Use the _____ and _____ plots to calculate the distance between two nodes using SOLIDWORKS Simulation.

- Mesh and Displacement.

- Displacement and FOS.

- Resultant Displacement and FOS.

- None of the above.

56. You can simplify a large assembly in a Static Study by using the _____ or _____ options in your study.

- Make Rigid, Fix.

- Shell element, Solid element.

- Shell element, Compound element.

- Make Rigid, Load element.

57. A force "F" applied in a static analysis produces a resultant displacement URES. If the force is now 2X and the mesh is not changed, then the resultant displacement URES will:

- Double if there are no contacts specified and there are large displacements in the structure.

- Be divided by 2 if contacts are specified.

- The analysis must be run again to find out.

- Double if there is no source of nonlinearity in the study (like contacts or large displacement options).

58. To compute thermal stresses on a model with a uniform temperature distribution, what type/types of study/studies are required?

- Static only.

- Thermal only.

- Both Static and Thermal.

- None of these answers is correct.

59. In an h-adaptive method, use smaller elements in mesh regions with high errors to improve the accuracy of results.

- True.

- False.

60. In a p-adaptive method, use elements with a higher order polynomial in mesh regions with high errors to improve the accuracy of results.

- True.

- False.

61. Where will the maximum stress be in the illustration?

- A

- B

- C

- D

62. Is axisymmetric the same as cyclic symmetry?

- Yes.

- No.

63. Can you perform a dynamic simulation using a static SOLIDWORKS Simulation study?

- Yes.

- No.

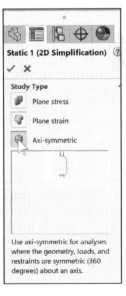

64. The concept of the h-method (available for solid part and assembly documents) is to use smaller elements (increase the number of elements) in regions with high relative errors to improve accuracy of the results.

- True.

- False.

65. Various adaptive methods can be applied to improve accuracy of the results. The p-method increases the _____ with each iteration.

- Number of elements.

- Meshing speed.

- Polynomial order.

- None of the above.

```
Static

Options   Adaptive   Flow/Thermal Effects

Adaptive method
  ● None
  ○ h-adaptive
  ○ p-adaptive

h-Adaptive options
                Low
```

66. Various adaptive methods can be applied to improve accuracy of the results. The h-adaptive method increases the _____ with each iteration.

- Number of elements.

- Meshing speed.

- Polynomial order.

- None of the above.

67. In SOLIDWORKS Simulation, how is a sheet metal part treated in a static study with uniform thickness?

- Solids.

- Beams.

- Shells.

- Mixed mesh.

68. Global size in the Mesh PropertyManager refers to the _____.

- Smallest element size of the mesh.

- Largest element size of the mesh.

- Average element size of the mesh.

- None of the above.

69. Global size in the Mesh PropertyManager is only available for a Standard Mesh.

- True.

- False.

70. Can you use a Static study in SOLIDWORKS Simulation to simulate the deflection of a bridge with a uniform force?

- Yes.

- No.

71. Can you calculate the distance between two nodes in a SOLIDWORKS Simulation results plot using the Probe tool?

- Yes, for all parts.

- No.

- Yes, only strain plots.

- Yes, only for mesh and displacement plots.

72. True or False: Use the probe tool to display the numerical value of the plotted field at the closest node or element's center to the selected model location. You can graph the results or save them to a file.

A. True.

B. False.

73. Which of the following assumptions are correct for a SOLIDWORKS static Simulation analysis? There can be more than a single answer.

A. All displacements are small relative to the model geometry; unless Large Displacements are activated.

B. Inertia effects are negligible and loads are applied slowly.

C. Results are proportional to loads. With the exception of when the following conditions are present: No penetration, Large Displacements, Shrink fit or Virtual wall contacts.

D. None of the above.

74. Is this model fixed at the two holes?

A. Yes.

B. No.

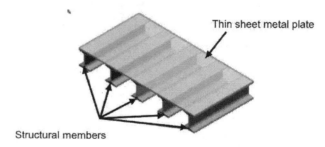

75. True or False: Check for interferences between bodies when using a compatible mesh with the curvature-based mesher. If you specify a bonded contact condition between bodies, they should be touching.

A. True.

B. False.

76. True or False: If interferences are detected, meshing stops, and you can access the Interference Detection PropertyManager to view the interfering parts. Make sure to resolve all interferences before you mesh again.

A. True.

B. False.

77. True or False: For static studies, you should define either **gravity** or **centrifugal loads** in order for the distributed mass to be effective. Simulation calculates the static load based on the acceleration due to gravity or the angular velocity and angular acceleration parameters defined in the Centrifugal PropertyManager

A. True.

B. False.

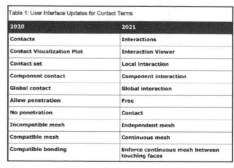

Thin sheet metal plate

Structural members

78. A thin sheet metal plate (Shell) is stiffened by structural members (H-beams). What SOLIDWORKS Simulation contact type or connector would you apply between the shell surface and each structural member?

A. Bonded Contact (Interactions).

B. Rigid Connector.

C. Thermal Resistance.

D. None of the above.

Table 1: User Interface Updates for Contact Terms

2020	2021
Contacts	Interactions
Contact Visualization Plot	Interaction Viewer
Contact set	Local Interaction
Component contact	Component interaction
Global contact	Global interaction
Allow penetration	Free
No penetration	Contact
Incompatible mesh	Independent mesh
Compatible mesh	Continuous mesh
Compatible bonding	Enforce continuous mesh between touching faces

79. In a SOLIDWORKS linear static study; which is true about a "Distributed Mass" load? You can have more than a single answer.

A. Moment effects due to mass are considered. Thus, the X, Y, and Z location of the mass needs to be specified.

B. A "Distributed Mass" increases the stiffness of the face to which it is applied.

C. It has to be used in conjunction with gravity or centrifugal load.

D. The mass is assumed to be uniformly distributed on selected faces based on their areas.

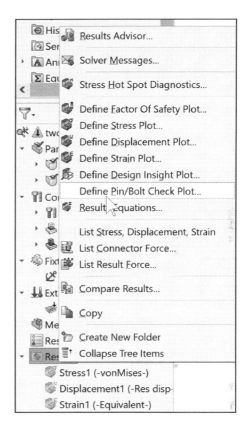

80. Which results are available for a Bolt connector in a SOLIDWORKS linear static study?

A. Stresses.

B. Torque.

C. Axial forces.

D. Shear forces.

E. Bending Moments.

F. Calculated Factor of Safety.

81. If you modify the study (force, material, etc.) you only need to re-run to update the results. You do not need to re-mesh unless you modified contact conditions.

A. True.

B. False.

82. What criteria are best suited to check the failure of ductile material in SOLIDWORKS Simulation?

- Maximum von Mises Stress and Maximum Shear Stress criteria.

- Maximum von Misses Stress and Maximum Shear Strain criteria.

- Maximum von Misses Stress and Maximum Normal Stress criteria.

- Mohr Coulomb Stress and Maximum Normal Stress criteria.

83. You have a 10-inch-long beam of cross section A. The beam is fixed at one end. A tensile force of 200 lbf is applied to the other end. The beam material is Alloy Steel. The tensile stress if found to be approximately 1000 psi. You modify the beam material to titanimum Ti-SAI-2.55n. Everything else in the study remains the same. What would the tensile force be?

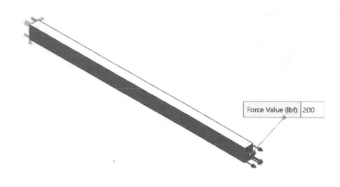

A. Approximately the same original stress.

B. Approximately half of the original stress.

C. Approximately 2X of the original stress.

D. Approximately 3X of the original stress.

84. A force "F" applied in a static linear analysis produces a URES: Resultant Displacement. If you double the force and the mesh is not changed, then the URES: Resultant Displacement will:

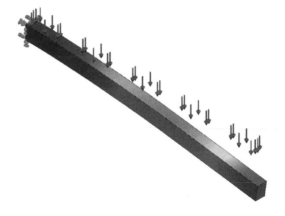

A. Be divided by 2, if contacts (Interactions) are specified.

B. Double if there is no source of nonlinearity in the study—like contacts (Interactions) or large deployment options.

C. Double if there are no contacts specified and there are large displacements in the structure.

D. The analysis must be run again to find out.

Table 1: User Interface Updates for Contact Terms

2020	2021
Contacts	Interactions
Contact Visualization Plot	Interaction Viewer
Contact set	Local interaction
Component contact	Component interaction
Global contact	Global interaction
Allow penetration	Free
No penetration	Contact
Incompatible mesh	Independent mesh
Compatible mesh	Continuous mesh
Compatible bonding	Enforce continuous mesh between touching faces

85. If you want to view a portion of stress values (range) on your model, which of the following would you do?

A. Change the model and re-run the study.

B. Create a section plot with two section clipping planes. The first plane could be the upper limit range. The second plane would be the lower limit range.

C. Create an iso plot with two iso values. The first value could be the upper limit. The second value would be the lower limit.

D. None of the above.

86. Which of the following, best describes the Ratio value in the Mesh Control PropertyManager when using Mesh control? Use Simulation help if needed.

A. The maximum depth to which the mesh control affects the mesh size, expressed a fraction of the overall model size.

B. The geometric growth ratio from one layer of elements to the next.

C. The maximum ratio between the largest element and the smallest element in the entire model.

D. The maximum ratio of the longest edge to the shortest edge of any element (i.e. maximum aspect ratio).

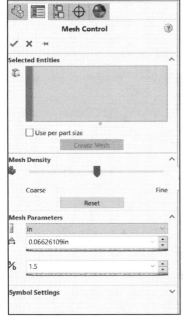

87. Which of the following geometrics can be meshed using beam elements? There can be more than one answer. Note: Beam or Truss elements are suitable for extruded or revolved objects and structural members with constant cross-sections.

A. A

B. B

C. C

D. None of them.

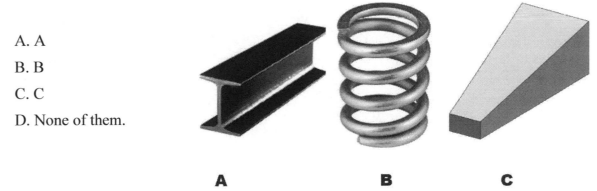

A   B   C

88. The cylindrical faces A are only allowed to rotate about their axis. What Fixture type would you apply to the cylindrical faces?

A. Fixed.

B. Roller/Slider.

C. Fixed Hinge.

D. On Spherical Face.

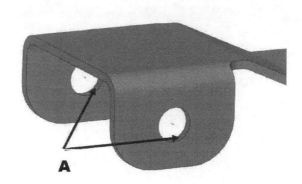

A

Table 1 displays the User Interface updates for Contact terms between SOLIDWORKS 2020 and 2021 and newer versions. The CSWA-S exam uses the older contact terms.

Note: Simulations using **SW 2023** are more accurate (than previous years) for curved surfaces that come into contact. Geometry corrections factors in the contact detection algorithms improve the representation of curved geometries such as cylindrical and spherical surfaces.

Table 1: User Interface Updates for Contact Terms	
2020	**2021**
Contacts	Interactions
Contact Visualization Plot	Interaction Viewer
Contact set	Local interaction
Component contact	Component interaction
Global contact	Global interaction
Allow penetration	Free
No penetration	Contact
Incompatible mesh	Independent mesh
Compatible mesh	Continuous mesh
Compatible bonding	Enforce continuous mesh between touching faces

Download all needed model files from the SDC Publication website www.SDCpublications.com/downloads/978-1-63057-567-0.

	SOLIDWORKS 2020
	SOLIDWORKS 2021
	SOLIDWORKS 2022
	SOLIDWORKS 2023

FEA Modeling Section

Tutorial FEA Model 7-1

An exam question in this category could read:

Calculate the vertical displacement in the Global Y direction at the location of the red dot? Decimal place: 3. Units: inch. Use default size, high quality elements.

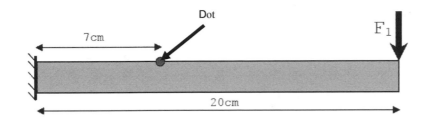
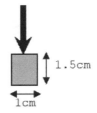

Given Information:

Material: Alloy Steel

Elastic modulus = $2.1e11 \, N/m^2$

Poisson's ratio = 0.28

$F_1 = 200lbf$

Use the default high quality element size to mesh.

Use the models from the CSWA-S model folder for this section.

Let's start.

1. **Open** Model 7-1 from the CSWA-S model folder.

The bar was created on the Front Plane.

The upper left corner of the rectangle is located at the origin. This simplifies the actual deformation of the part. The height dimension references zero in the Global Y direction.

Split Lines were created to provide the ability to locate the needed Joints in this problem.

To add the force at the center of the beam, a split line in the shape of a small circle on the top face of the right end of the beam is used.

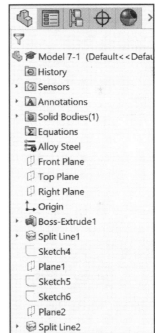

💡 Download all needed model files from the SDC Publication website www.SDCpublications.com/downloads/978-1-63057-567-0.

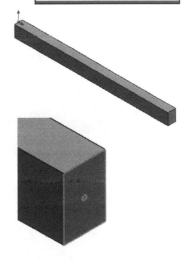

Proper model setup is very important to obtain the correct mesh and final results.

Create a Static SOLIDWORKS Simulation Study. Set Simulation Options.

2. **Add-In** SOLIDWORKS Simulation.

3. Click the **Simulation** tab in the CommandManager.

4. Click **New Study** from the drop-down menu.

5. Accept the default name (Static #). Click **OK** ✓ from the Study PropertyManager.

Set unit system, units, and mesh type.

6. Click **Simulation**, **Options** from the Main menu. The System Options - General dialog box is displayed.

7. Click the **Default Options** tab.

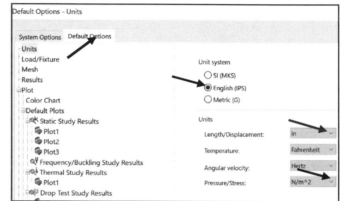

Set unit system.

8. Select **English (IPS)**.

Set units.

9. Select **in** for Length/Displacement.

10. Select **(N/m²)** for Pressure/Stress.

Select mesh quality.

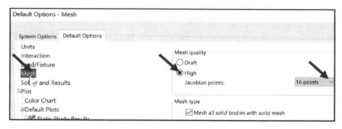

11. Click the **Mesh folder**.

12. Select **High** for Mesh quality.

13. Set **Jacobian** points to 16.

14. Click **OK** from the Default Options - Mesh dialog box.

Treat the model as a Beam.

15. Right-click **Model 7-1** in the Study.

16. Click **Treat as Beam**. A Joint group folder is created.

17. Right-click the **Joint group** folder. A joint is identified at the free end of a structural member or at the intersection of two or more structural members.

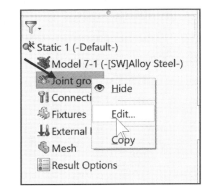

18. Click **Edit**. The Edit Joints PropertyManager is displayed. All is selected by default. The software considers all structural members when calculating joints. Joint 1 and Joint 2 are displayed in the Results box.

19. Click **OK** ✓ from the Edit Joints PropertyManager.

The Results box displays the joints that the software finds in the model. Select a joint from the list to highlight it in the graphics area. Pink joints are highlighted with orange, and light green joints with a bright green color.

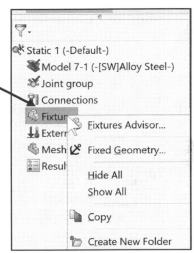

Set Fixed Geometry as Fixture type.

20. Right-click the **Fixtures** folder.

21. Click **Fixed Geometry**. The Fixture PropertyManager is displayed. View your options. Fixed Geometry is selected by default.

22. Click the **joint on the left side** of the beam. Joint<1,1> is displayed in the Joints box.

23. Click **OK** ✓ from the Fixture PropertyManager. Fixed-1 is created in the Static Study.

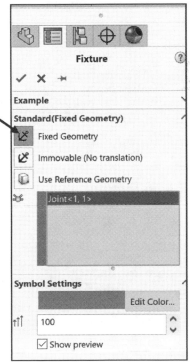

Set the Force at the end of the beam.

24. Right-click the **External Loads** folder.

💡 Illustrations will vary depending on your SOLIDWORKS version and operating system.

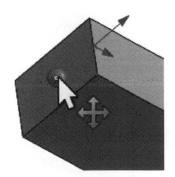

25. Click **Force**. The Force/Torque PropertyManager is displayed.

26. Click the **Joints** option.

27. Click the **joint on the right side** of the beam. Joint<2,1> is displayed in the Joints box.

28. Click **inside** the Face, Edge, Plane for Direction 1 box.

29. Click the **end face** of the beam as the plane for direction as illustrated.

30. Select **English (IPS)** for units.

31. Click the **Along Plane Dir 1 Force** box.

32. Enter **200**lbf Along Plane Direction1. The arrow is displayed downwards; reverse direction if needed.

33. Click **OK** ✓ from the Force/Torque PropertyManager. Force-1 is created in the Static Study.

Meshing generates 3D tetrahedral solid elements, 2D triangular shell elements, and 1D beam elements. A mesh consists of one type of elements unless the mixed mesh type is specified. Solid elements are naturally suitable for bulky models. Shell elements are naturally suitable for modeling thin parts (sheet metals), and beams and trusses are suitable for modeling structural members.

Apply the standard default setting for the mesh. Mesh and run the study.

34. Right-click the **Mesh** folder.

35. Click **Mesh and Run**.

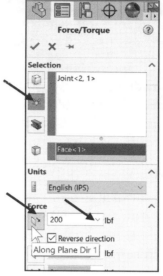

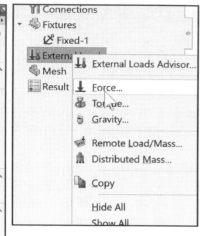

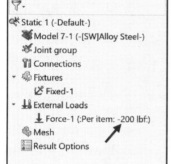

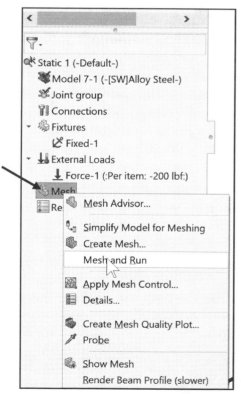

36. **View** the results.

37. Double-click the **Displacement1 (-Res disp-)** folder.

38. **View** the results.

Set plot results. Set number format to floating. Set number of decimals to 3. Display URES: Resultant Displacement.

39. Right-click the **Displacement1** folder.

40. Click **Edit Definition**. The Displacement plot FeatureManager is displayed.

41. Click the **Definition tab**.

42. Click the **Display drop-down** arrow.

43. Select **URES: Resultant Displacement**.

44. Select **in**.

45. Click the **Chart Options** tab.

46. Click the **show max annotation** box.

47. Select **floating** from the number format box.

48. Select **3** for number of decimal places.

49. Click **OK** from the Displacement plot FeatureManager.

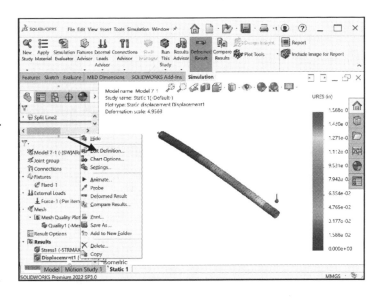

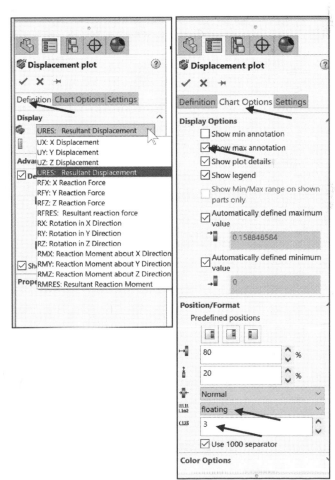

50. **View** the results in the graphic window.

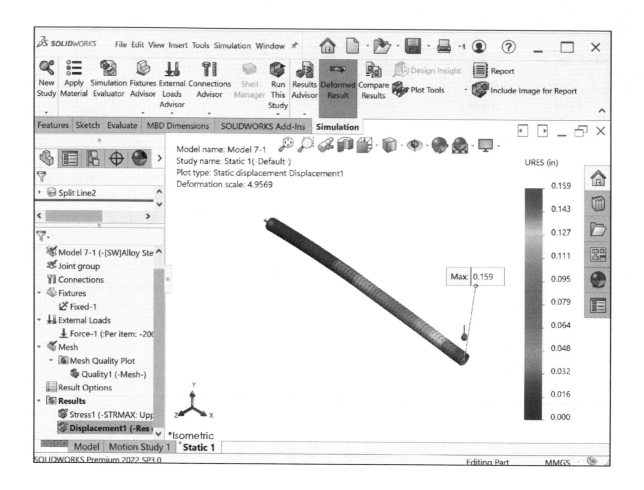

Calculate the displacement at 7cm.

51. Click **List Stress,
    Displacement and Strain** from
    the Results Advisor drop-down
    menu. The List Results
    PropertyManager is displayed.

52. Select **Displacement** for
    Quantity.

53. Select **UY: Y Displacement**
    for Component.

54. Select **in** for Units.

55. Click **OK** ✓ from the List
    Results PropertyManager. The
    List Results dialog box is
    displayed.

56. **View** the results.

57. **Scroll down** until you see values
    around 70mm for the distance along the
    X direction. See the value of
    displacement (UY) (in).

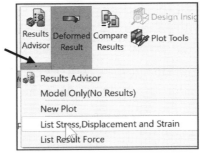

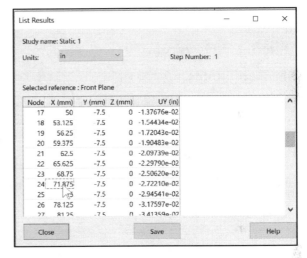

To find the exact value at 70mm, use linear
interpolation.

This method is shown below and uses the
values greater and less than the optimal one
to find the actual displacement.

$$\frac{X_U - X_L}{UY_U - UY_L} = \frac{X_U - X_O}{UY_U - UY_O}$$

$$\frac{71.875 - 68.75}{-0.0272206 - -0.0250617} = \frac{71.875 - 70}{-0.02718 - UY_O}$$

$$UY_o = 0.0258$$

At the distance of 7cm (70mm) the displacement is found to be approximately 0.026in.

The correct answer is **B**.

A = 0.034in

B = **0.026**in

C – 0.043in

D = 0.051in

Tutorial FEA Model 7-2

Below is a second way to address the first problem (Tutorial FEA Model 7-1) with a different model, using the **Study Advisor** and the **Probe** tool.

Calculate the vertical displacement in the Global Y direction in (inches) at the location of the red dot. Decimal place: 3. Use default size, high quality elements.

Given Information:

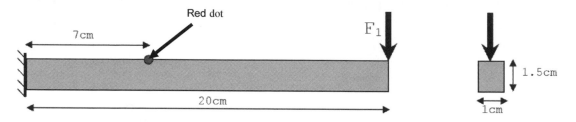

Material: Alloy Steel

Elastic modulus = 2.1e11 N/m^2

Poisson's ratio = 0.28

F₁ = 200lbf

Use the default high quality element size to mesh.

Use the models from the CSWA-S model folder for this section.

Let's start.

1. **Open** Model 7-2 from the CSWA-S model folder.

2. **Show** Sketch 2.

The bar was created on the Front Plane. The bar was created so that the origin is the point at which the force is applied. A construction line is created across the part at a distance of 7cm from the end that is to be fixed.

Illustrations may vary depending on your SOLIDWORKS version and system setup.

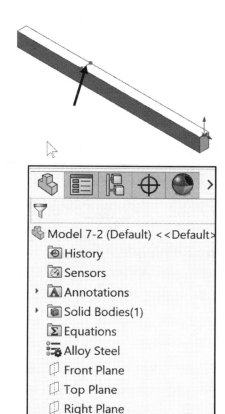

Create a Static SOLIDWORKS Simulation Study. Set Simulation Options.

3. **Add-In** SOLIDWORKS Simulation.

4. Click the **Simulation** tab in the CommandManager.

5. Click **New Study** from the drop-down menu.

6. Accept the default name (Static #). Click **OK** ✓ from the Study PropertyManager.

7. Click **Simulation**, **Options** from the Main menu. The System Options - General dialog box is displayed.

8. Click the **Default Options** tab.

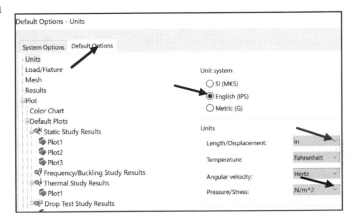

Set unit system, units and mesh quality.

9. Select **English (IPS)** for unit system.

10. Select **in** for Length/Displacement for units.

11. Select **N/m²** for units.

Set mesh quality.

12. Click the **Mesh folder**.

13. Select **High** for Mesh quality. Set **Jacobian points** to 16.

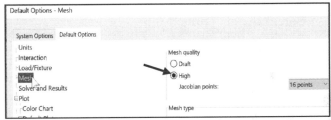

14. Click **OK** from the Default Options - Mesh dialog box.

Set Fixture type. Use the Study Advisor.

15. Right-click the **Fixtures** folder from the Simulation study tree.

16. Click **Fixtures Advisor**. The Simulation Advisor is displayed to the right of the graphics area.

17. Click **Add a fixture**. The Fixture PropertyManager is displayed. SOLIDWORKS 2021 and newer versions replace the word contact with interaction.

18. Select the **left face** as illustrated.

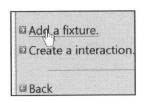   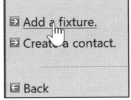

2023, 2022, 2021 version   2020 version

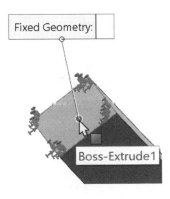

19. Set as **Fixed Geometry**.

20. Click **OK** ✓ from the Fixture PropertyManager.

Set the Force.

21. Right-click the **External Loads** folder from the Simulation study tree.

22. Select **Remote Load/Mass**. The Remote Loads/Mass PropertyManager is displayed.

23. Select the **top face of the part** - for Select faces.

24. Set the Reference Coordinate system to **Global**.

25. Leave all of the location boxes at **zero** value.

26. Select the **Translational Components** box.

27. Select **lbf** for Force unit.

28. Select **in** for Translational unit.

29. Click the **Y-Direction** button.

30. Enter **200 in Y-Direction**. The force points downward. Reverse direction if necessary.

31. Click **OK** ✓ from the Remote Loads/Mass PropertyManager. -200 lbf is displayed in the Remote load folder.

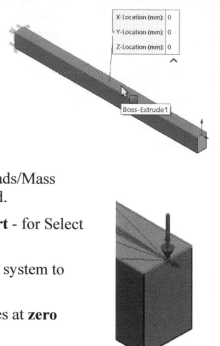

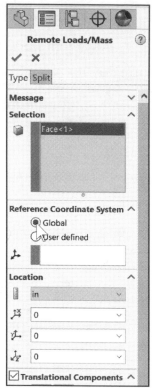

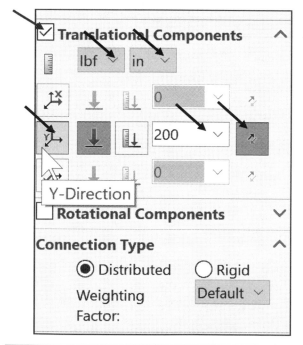

Meshing generates 3D tetrahedral solid elements, 2D triangular shell elements, and 1D beam elements. A mesh consists of one type of elements unless the mixed mesh type is specified. Solid elements are naturally suitable for bulky models. Shell elements are naturally suitable for modeling thin parts (sheet metals), and beams and trusses are suitable for modeling structural members.

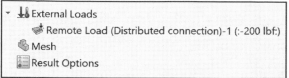

Use the standard default setting for the mesh. Mesh and run the study.

32. Right-click the **Mesh** folder.

33. Click **Mesh and Run**. View the results. Results may vary depending on SOLIDWORKS version (less than 1%).

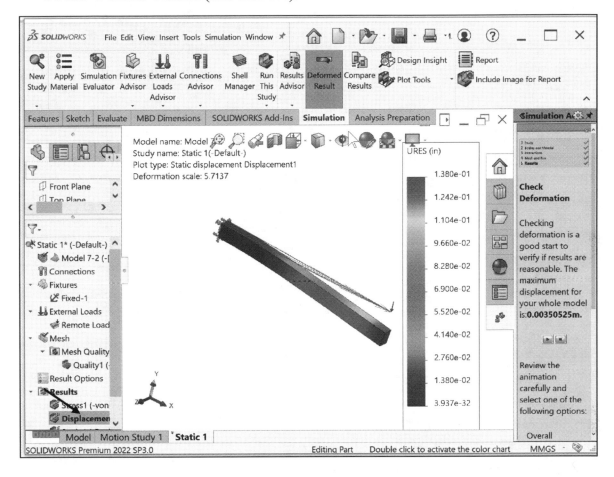

34. Click the **Play Animation** button. View the model. Explore the below options in the Simulation Advisor dialog box.

35. Click **Stop**.

36. Double-click the **Displacement1 (-Res-disp1)** folder.

37. **View** the results.

Set plot results. Set number format to floating. Decimal place: 3. Display URES: Resultant Displacement.

38. Right-click the **Displacement1** folder.

39. Click **Edit Definition**. The Displacement plot FeatureManager is displayed.

40. Click the **Definition tab**.

41. **View** your options.

42. Click the **Display drop-down** arrow.

43. Select **URES: Resultant Displacement**.

44. Select **in**.

45. Click the **Chart Options** tab. View your options.

46. Click the **show max annotation** box.

47. Select **floating** from the Number format box.

48. Select **3** for number of decimal places.

49. Click **OK** from the Displacement plot FeatureManager.

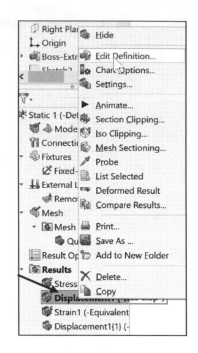

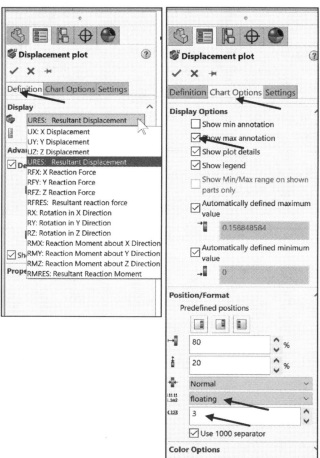

50. **View** the results in the graphic window.

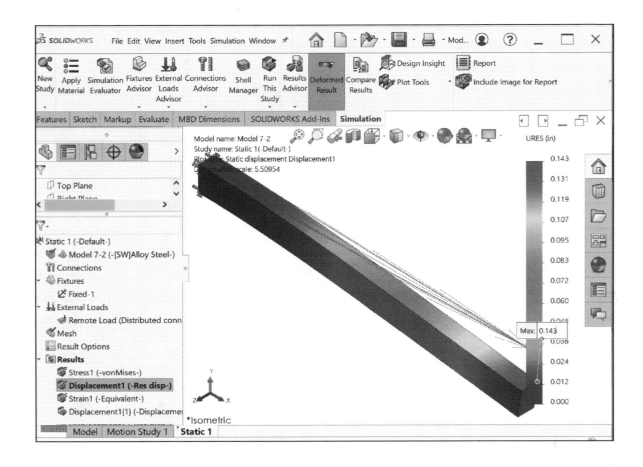

Locate the displacement at 7cm using the Probe tool.

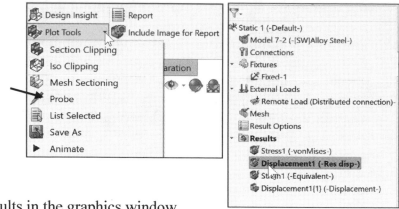

51. Click the **Displacement1 (-Res disp-)** Results folder.

52. Click **Probe** from the Plot Tools drop-down menu.

53. Click the **Split line** at 7cm on the model as illustrated. View the results in the graphics window.

54. Click **OK** from the Probe Results PropertyManager.

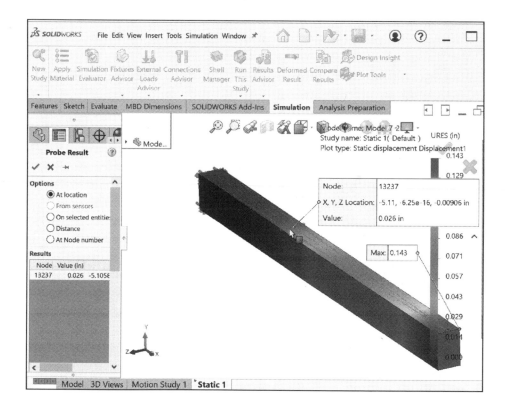

At the distance of 7cm (70mm) the displacement is found to be approximately 0.026in.

The correct answer is **B**.

A = 0.034in

B = 0.026in

C = 0.043in

D = 0.051in

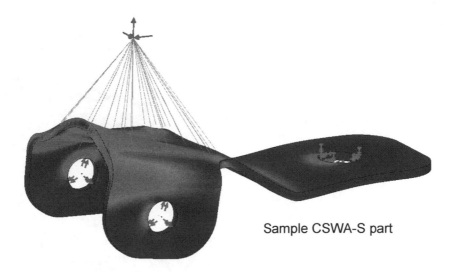

Sample CSWA-S part

In the CSWA-S exam:

- Comprehend Remote loads/mass to apply an external load.

- Knowledge in defining coordinates of the reference, node location, remote forces, and remote translations.

- Awareness of how to apply Fixed Geometry, Fixed Hinge, and Advanced Fixtures.

- Familiarity with how to apply h-adaptive or p-adaptive method during a mesh.

- Ability to apply the German material standard (DIN).

- Ability to apply different materials to verious components in an assembly.

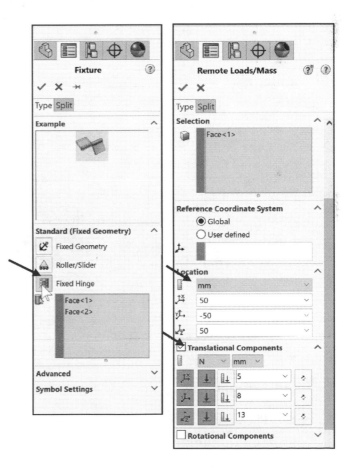

Tutorial FEA Model 7-3

An exam question in the Solid category could read:

Calculate the maximum URES: Resultant Displacement in mm on the annular face of the model. A normal force is applied on the top annular face as shown. The three cylindrical faces are fixed. Use default size, high quality elements. Decimal place: 3.

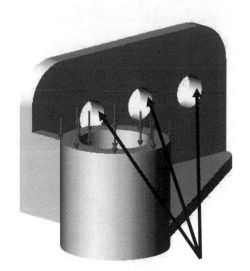

Given Information:

Material: Alloy Steel

A normal force **F₁** is applied to the annular face.
F_1 = 3000lbf.

The three cylindrical faces are **fixed**.

Use the default high quality element size to mesh.

Use the models from the CSWA-S model folder for this section.

Let's start.

1. **Open** Model 7-3 from the CSWA-S model folder.

Create a Static SOLIDWORKS Simulation Study. Set Simulation Options.

2. **Add-In** SOLIDWORKS Simulation.

3. **Create** a new Static Study. Accept the default name (Static #).

4. Click **Simulation**, **Options** from the Main menu.

5. Click the **Default Options** tab.

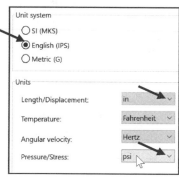

Set unit system, units, and mesh quality.

6. Select **English (IPS)** for unit system.

7. Select **in** for units.

8. Select **psi** for pressure/stress units.

9. Click the **Mesh folder**.

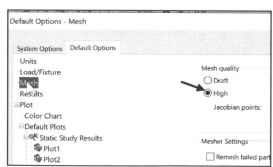

10. Select **High** for mesh quality.

11. Set **Jacobian points** to 16.

12. Click **OK** from the Default Options - Mesh dialog box.

Apply Fixed Geometry to the three cylindrical faces.

13. Right-click the **Fixtures** folder.

14. Click **Fixed Geometry**. The Fixture PropertyManager is displayed.

15. Select the **three cylindrical faces** of the hole pattern as illustrated.

16. Click **OK** ✓ from the Fixtures PropertyManager. Fixed-1 is created.

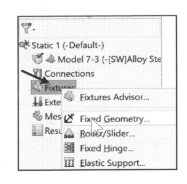

Apply an External Load.

17. Right-click the **External Loads** folder.

18. Click **Force**. The Force/Torque PropertyManager is displayed.

19. Click the **face** of the model as illustrated. Face<1> is displayed in the Faces and Shell edges for Normal Force box.

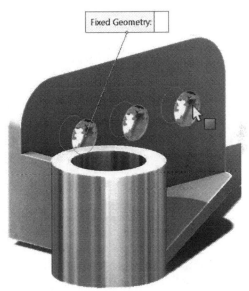

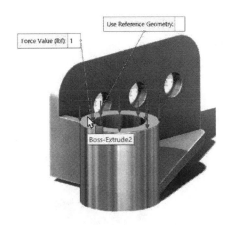

20. Select **English (IPS)** for units.

21. Enter **3000**lbf normal to the annular face.

22. Click **OK** ✓ from the Force/Torque PropertyManager. Force-1 is created in the Study tree.

23. **Mesh and Run** the study.

Set plot results. Set number format to floating. Set number of decimals to 3. Display URES: Resultant Displacement in mm.

24. Right-click the **Displacement1 (-Res disp-)** folder.

25. Click **Edit Definition**. The Displacement plot FeatureManager is displayed.

26. Click the **Definition tab**.

27. Select **mm** for units.

28. Click the **Chart Options** tab.

29. Click the **show max annotation** box.

30. Select **floating** from the Number format box.

31. Select **3** for number of decimal places.

32. Click **OK** from the Displacement plot FeatureManager.

33. **View** the results in the graphic window.

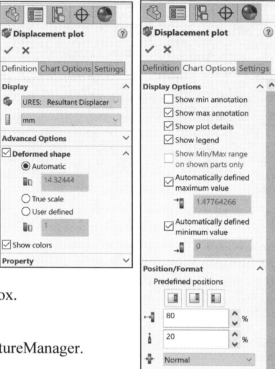

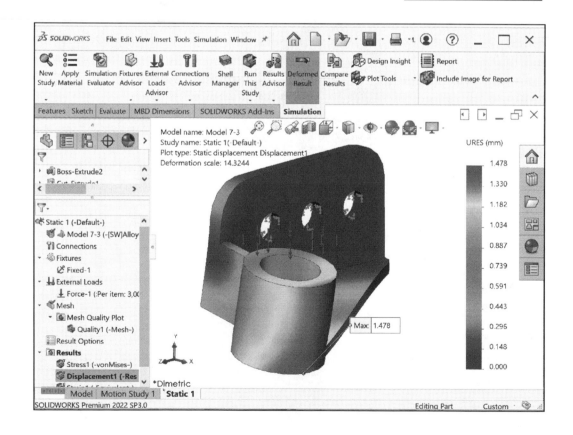

The correct answer is **±1%** of this value.

The correct answer is **C**.

A = 1.112mm

B = 1.014mm

C = 1.478mm

D = 1.734mm

Displacement components are: UX = Displacement in the X-direction, UY = Displacement in the Y-direction, UZ = Displacement in the Z-direction, URES = Resultant displacement.

After you calculate displacement or other parameters in a Simulation Study, the CSWA-S exam will also deliver a series of successive questions to test the understanding of the results.

In the first question you calculate Displacement; in the second question you calculate Resultant Force. Do not re-mesh and re-run. Create the required parameter in the Results folder.

In the third question, you are asked to determine if the results are valid or invalid. If the materials yield strength was passed, then the results are invalid.

Use the Define Factor of Safety Plot to determine if your results are valid or invalid. The CSWA-S exam requires you to apply Finite Element Method theory and review displacement values, factory of safety, mesh refinement, and material properties such as yield strength.

The following are some statements you will encounter in the CSWA-S exam:

- The results are invalid because the material's yield strength was passed.

- The results area is invalid because the displacement was more than ½ the plate's thickness.

- The results are valid as they are, even if mesh refinement was better.

- The results are invalid because a dynamic study is required.

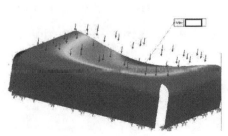

In general use Thin Shells when thickness to span ratio < 0.05.

Tutorial FEA Model 7-4

An exam question in the Sheet Metal category could read:

Calculate the **maximum UX: X displacement** in mm. Use default size, high quality elements. Use the default study properties. There is a normal force on face B. It's value is 400N. Decimal place: 4.

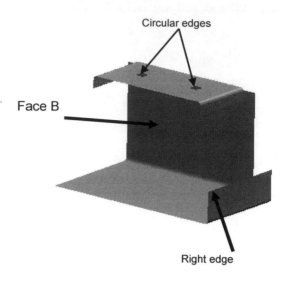

Circular edges

Face B

Given Information:

Material: Alloy Steel.

Normal force of **400N** to face B.

The thickness of the materials is **0.15in**.

The right edge as illustrated and two circular edges are **immovable**.

Right edge

Use the default high quality element size to mesh.

Let's start.

You need to define the thickness. The Fixture option Immovable is added for shell models. Thin models created with no Sheet Metal feature require you to define the use of **Shell elements** in SOLIDWORKS Simulation.

Use the models from the CSWA-S folder for this section. Models created with the Sheet Metal feature automatically create Shell elements in SOLIDWORKS Simulation.

1. **Open** Model 7-4 from the CSWA-S model folder.

Create a Static SOLIDWORKS Simulation Study. Set Simulation Options.

2. **Add-In** SOLIDWORKS Simulation.

3. Click the **Simulation** tab in the CommandManager.

4. Click **New Study** from the drop-down menu.

5. Accept the default name (Static #).
   Click **OK** ✓ from the Study PropertyManager.

6. Click **Simulation**, **Options** from the Main menu. The System Options - General dialog box is displayed.

7. Click the **Default Options** tab.

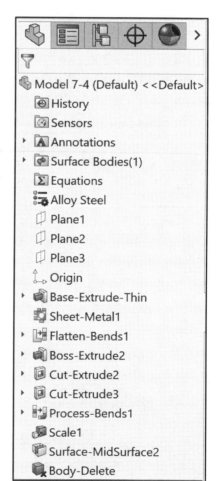

Set unit system, units, and mesh quality.

8. Select **SI (MKS)** for unit system.

9. Select **mm** for length/Displacement units.

10. Select **N/m²** for Pressure/Stress units.

11. Click the **Mesh** folder.

12. Click **High** for Mesh quality.

13. Set **Jacobian** points to 16.

14. Click **OK** from the Default Options – Mesh dialog box. A check mark is displayed next to the model name in the Study Simulation tree. Note the comment.

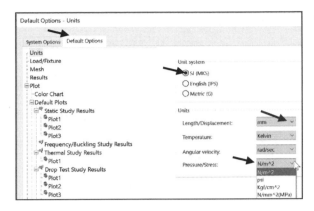

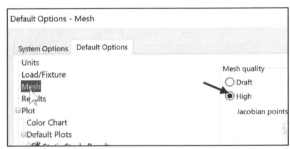

The Shell Definition PropertyManager is used to define the thickness of thin and thick shell elements. You can also define a shell as a composite for static, frequency, and buckling studies.

There are three shell types in a linear static study:

- Thin: In general use thin shells when the thickness-to-span ratio is less than 0.05.

- Thick: Use the thick shells when the thickness-to-span ratio is greater than 0.05.

- Composite: Available only for surface geometries in static, frequency and buckling studies. Defines the shell as a composite laminate.

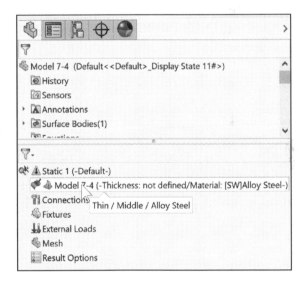

Define Shell thickness.

15. Right-click **Model 7-4** in the Study tree.

16. Click **Edit Definition** from the drop-down menu. The Shell Definition PropertyManager is displayed.

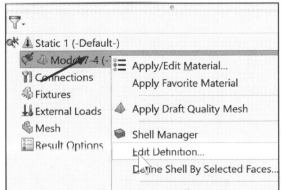

17. Click **Thin** for Type.

18. Enter **0.15in** for shell thickness.

19. Click **OK** ✓ from the Shell Definition plot PropertyManager. Note: Composite is available only for surface geometries in static, frequency and buckling studies.

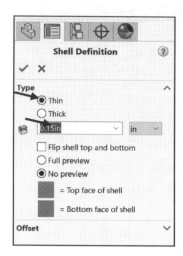

Apply Fix Geometry Fixtures. The right edge as illustrated and two circular edges are immovable. This restraint type sets all translational degrees of freedom to zero. It is the same for shells, beams and trusses. No reference geometry is used.

20. Right-click **Fixtures** folder.

21. Click **Fixed Geometry**. The Fixture PropertyManager is displayed.

22. Click the **Immovable (No translation)** option.

23. Select the **edge** and the **two circular edges** of the model as illustrated.

24. Click **OK** ✓ from the Fixture PropertyManager. Immovable-1 is created.

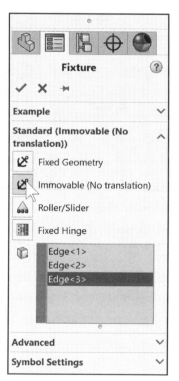

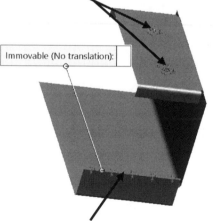

Immovable (No translation):

Apply a normal load (Force) of 400N to face A.

25. Right-click the **External Loads** folder from the study.

26. Click **Force**. The Force/Torque PropertyManager is displayed.

27. Click **face A** of the model as illustrated.

28. Select **Normal** for Force component.

29. Select **SI** for units.

30. Enter **400N** for force value.

31. Click **OK** ✓ from the Force/Torque PropertyManager.

Mesh and Run the model. Use the standard default setting for the mesh.

32. Right-click the **Mesh** folder.

33. Click **Mesh and Run**.

34. **View** the results.

Note: Using SW 2021 and newer versions, the answer will be different, but they will be within 1% for the exam.

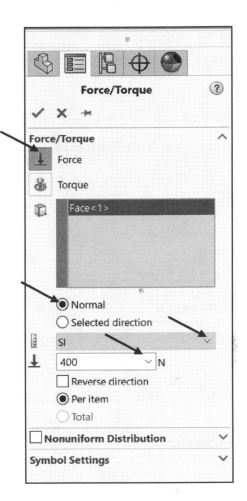

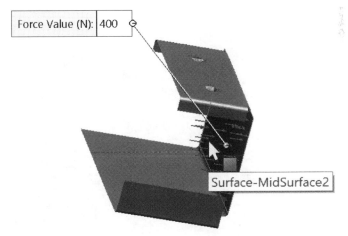

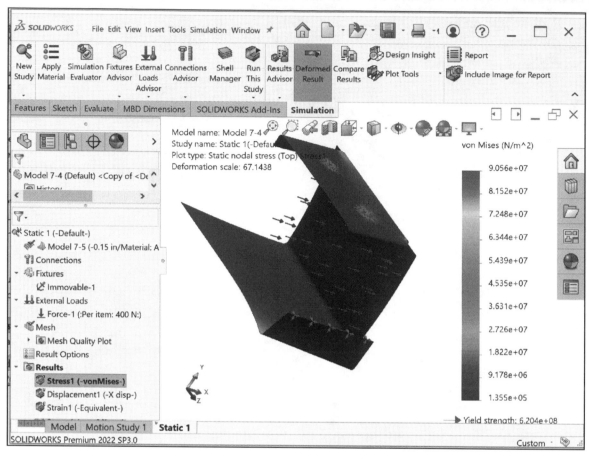

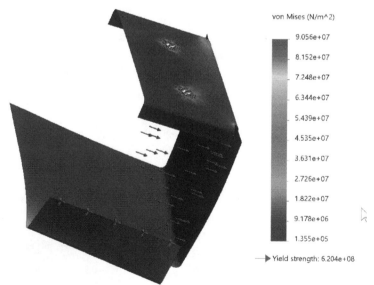

Set plot results. Set number format to floating. Set number of decimals to 4. Calculate maximum global displacement in X. Change units to mm.

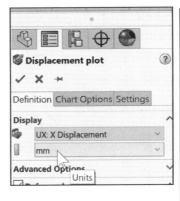

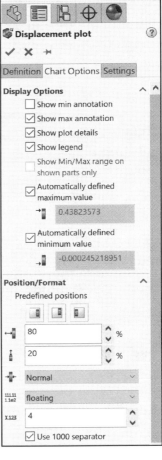

35. Right-click the **Displacement1 (-Res disp-)** folder.

36. Click **Edit Definition**. The Displacement plot FeatureManager is displayed.

37. Click the **Definition tab**.

38. Select **UX: X Displacement** for display.

39. Select **mm** for units.

40. Click the **Chart Options** tab.

41. Click the **show max annotation** box.

42. Select **floating** from the Number format box.

43. Select **4** for number of decimal places.

44. Click **OK** from the Displacement plot FeatureManager. View the results in the graphic window. Note: Using SW 2021 and newer versions, the answer will vary, but they should be within 1%.

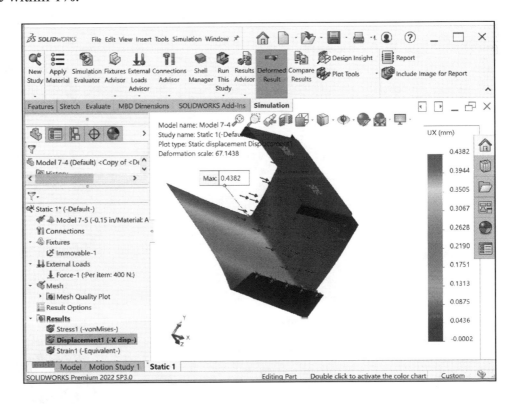

Tutorial FEA Model 7-4 Part 2

An exam question in the Understanding your knowledge category could read:

Given Information:

From the study you just completed, can you modify the applied force from **400N** to **4000N** without creating a new mesh and still calculate the correct maximum von Mises stress?

Think about the problem. Think about the model.

Answer: In SOLIDWORKS Simulation you need to only select **Run**.

In a Yes/No question, the answer would be: **Yes**.

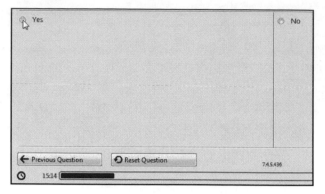

Screen shot from the exam

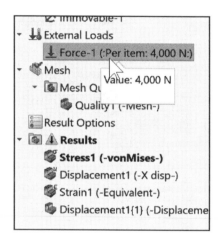

Tutorial FEA Model 7-4 Part 3

An exam question in the understanding your knowledge category could read:

Given Information:

From the study you just completed with a force of 4000N applied, are the results for the maximum von Mises stress valid?

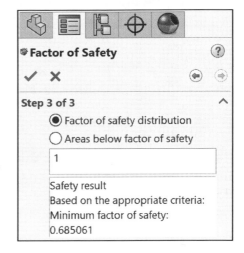

Answer: In SOLIDWORKS 2020 and older versions, the result is not valid because the maximum von Mises stress (960 MPa) is greater than the Yield strength (620 MPa) of the material. In a Yes/No question, the answer would be: **No**.

Answer: In SOLIDWORKS 2021 and newer versions, the result is not valid because the maximum von Mises stress (905 MPa) is greater than the Yield strength (620 MPa) of the material. In a Yes/No question, the answer would be: **No**.

Note: Simulations using **SW 2023** are more accurate (than previous years) for curved surfaces that come into contact. Geometry corrections factors in the contact detection algorithms improve the representation of curved geometries such as cylindrical and spherical surfaces.

Tutorial FEA Model 7-5

An exam question in the Connections category could read:

Calculate the maximum URES: Resultant Displacement in mm Decimal place: 4. Use default size, high quality elements.

Given Information:

Two cylindrical tubes, with a **thickness of 0.9mm** and one rectangular tube with a **thickness of 1.0mm** are **fixed** to the rectangular plate.

The 3 tubes are manufactured from **Alloy Steel**.

Their upper ends are **fully** constrained.

The bottom plate is manufactured from **Plain Carbon Steel**.

A load is applied to the circular face at the bottom of the rectangular plate facing downward (**negative direction**).

The load magnitude is **800N**.

Use the models from the CSWA-S model folder for this section. Models created with the Sheet Metal feature automatically create Shell elements in SOLIDWORKS Simulation.

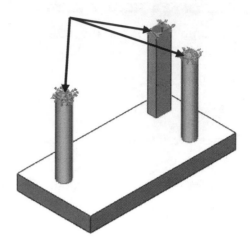

In general use Thin Shells when thickness to span ratio < 0.05.

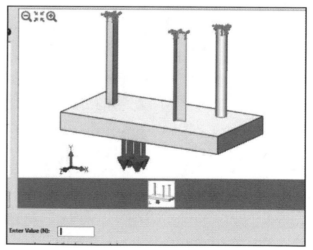

Screen shot from a CSWA-S exam

Let's start.

1. **Open** Model 7-5 from the CSWA-S model folder.

Create a Static SOLIDWORKS Simulation Study. Set Simulation Options.

2. **Add-In** SOLIDWORKS Simulation.

3. **Create** a Static Study. Accept the default name (Static #).

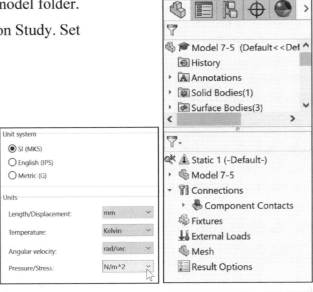

Set unit system, units, and mesh quality.

4. Click **Simulation**, **Options** from the Main menu.

5. Click **SI (MKS)** for unit system.

6. Click **mm** for Length/Displacement units.

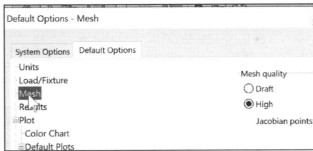

7. Click **N/m²** for Pressure stress units.

8. Click the **Mesh folder**.

9. Select **High** for Mesh quality.

10. Set **Jacobian** points to 16.

11. Click **OK** from the Default Options - Mesh dialog box.

12. Apply **Material**. The tubes are manufactured from **Alloy Steel**. The bottom plate is manufactured from **Plain Carbon Steel**. A check mark is displayed next to the SurfaceBody.

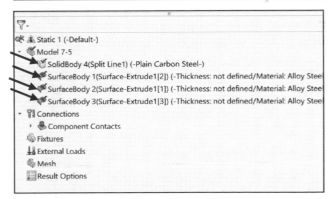

Define the Shell thickness for each SurfaceBody.

13. Select **Edit Definition** for each SurfaceBody.

14. Type: **Thin**.

15. Enter **0.9mm** for the two cylinders thickness.

16. Enter **1.0mm** for the rectangle thickness.

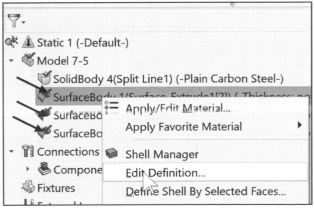

Create three Bonded Type Connections to
the top face of the plate.

17. Right-click the **Connections** folder
    from the study tree.

18. Click **Contact Set (Local
    Interaction)**. Note: The word **Contact**
    was replaced by **Interaction** in SW
    2021.

19. Create a **Bonded Content Set
    (Interaction)** between the **bottom 4
    edges** of the rectangle and the **top face**
    of the plate.

20. **Repeat** for the two bottom circular
    edges.

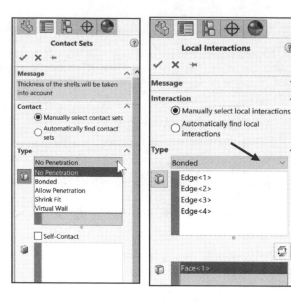

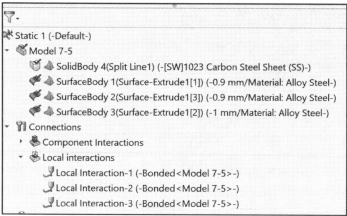

SOLIDWORKS 2021, 2022, 2023 version

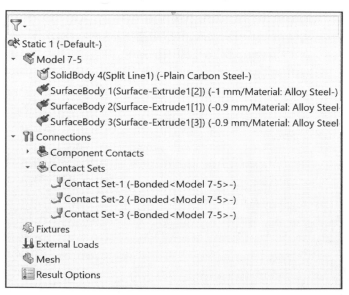

SOLIDWORKS 2020 version

Apply Immovable fixed geometry.

21. Right-click the **Fixtures** folder from the study.

22. Click **Fixed Geometry**.

23. Select the **Immovable (No translation)** option.

24. Select the **top rectangular edges** and the **two top circular edges**. The tops of each tube are fixed.

Apply a load to the circular face at the bottom of the rectangular plate facing downward (**negative direction**).

25. Apply an External Load of **800N** normal to the bottom circular face of the model. The direction points downward.

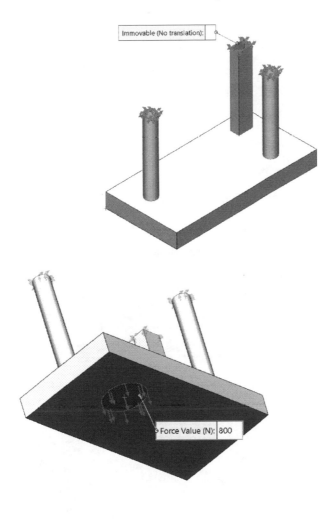

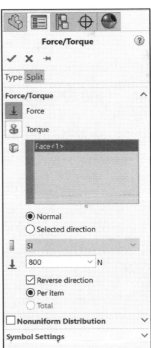

Mesh and Run the model. Use the standard default setting for the mesh.

26. Right-click the **Results** folder.

27. Click **Mesh and Run**. View the results.

Set plot results. Set number format to floating. Set number of decimals to 4. Calculate Maximum URES: Resultant Displacement in mm.

28. Right-click the **Displacement1 (-Res disp-)** folder in the study.

29. Click **Edit Definition**. The Displacement plot FeatureManager is displayed.

30. Click the **Definition tab**.

31. Select **URES: Resultant Displacement** for display.

32. Select **mm** for units.

33. Click the **Chart Options** tab.

34. Click the **show max annotation** box.

35. Select **Floating** from the Number format box.

36. Select **4** for number of decimal places.

37. Click **OK** from the Displacement plot FeatureManager.

38. **View** the results in the Graphic window. Note your answer should be within 1%.

Answer: 0.0233mm.

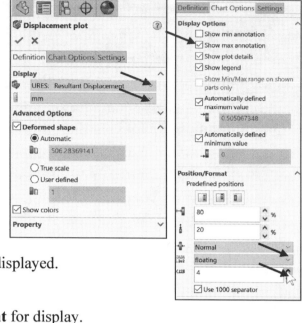

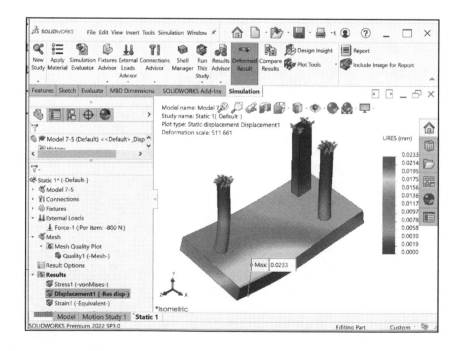

Note you need to input trailing zeroes for four decimal places in the exam.

Tutorial FEA Model 7-5 Part 2

An exam question in the understanding your knowledge category could read:

Given Information:

Without rerunning the study, what is the Resultant Force of the **top circular edge** in the illustration?

Think about the problem. Think about the model.

1. Right-click the **Results folder** in the study.

2. Click **List Result Force**. The Result Force PropertyManager is displayed.

The Result Force PropertyManager provides the ability to list reaction forces on selected entities for an active static, nonlinear, drop test and dynamic study.

The software lists the X-, Y-, Z-components of the reaction force, and the resultant reaction force on the selected entities and on the entire model as well.

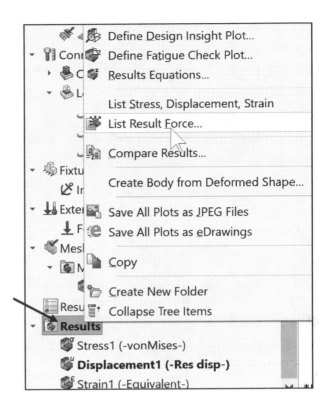

3. Click **Reaction force** under Options.

4. Set units to **SI**.

5. Click the **top edge of the cylinder** as illustrated.

6. Click **Update**.

7. **View** the results in the graphics area.

Note: In SW 2021 and newer versions, the answer is **572.37**. In SW 2020 and older versions, the answer is **572.33N**. Both are within 1% for the exam.

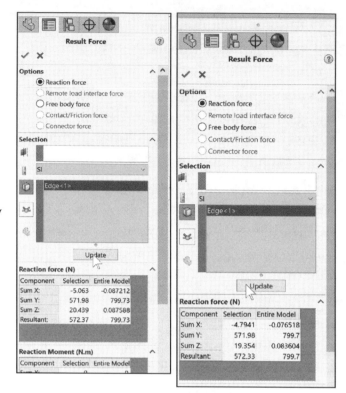

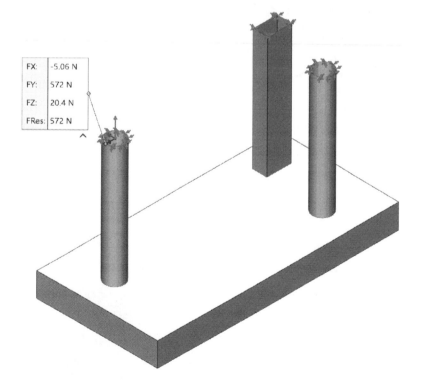

Tutorial FEA Model 7-5 Part 3

An exam question in the understanding your knowledge category could read:

Given Information:

What force component, Fx, Fy, or Fz contributes the lease to the overall Resultant Force?

Answer: **FX**.

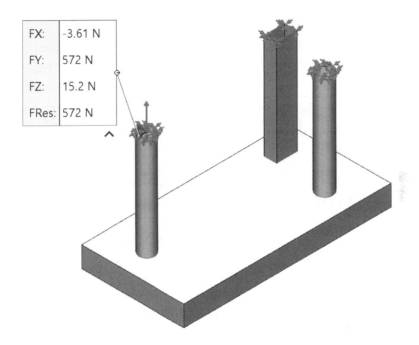

FX:	-3.61 N
FY:	572 N
FZ:	15.2 N
FRes:	572 N

Tutorial FEA Model 7-6

Part of an exam question could read:

Use a solid mesh. Start with a default mesh created with the Standard mesher and then use the H-Adaptive method to the model.

Note: There is no part for this section. Just the procedure.

1. Right-click the **Static 1 (-Default-)** folder in the Study tree.

2. Click **Properties** from the drop-down menu. The Static dialog box is displayed.

3. Click the **Adaptive** tab. The settings take place when you run the study. The options are ignored if you run design scenarios. The p-method does not refine the mesh but uses progressively higher element order to improve results. The h-method **refines the mesh** and does not change the element order. In addition to having better accuracy by using more elements in critical regions, the refined mesh represents the geometry in critical regions more closely in each loop.

4. Click **h-adaptive** as illustrated.

5. Click **OK** from the Static dialog box.

6. Right-click **Mesh**.

7. Click **Mesh and Run**.

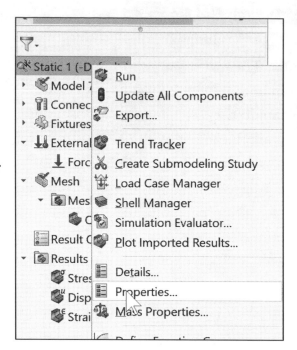

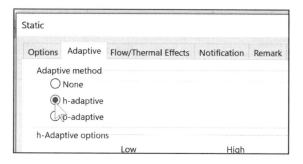

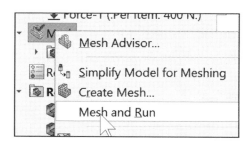

Tutorial FEA Model 7-7

Part of an exam question could read:

A remote load with direct transfer is applied to the illustrated face. The location in the global coordinate system is (50mm, -50mm, 50mm). Its magnitude is Fx = 2N, Fy = -17N, and Fz = 4N.

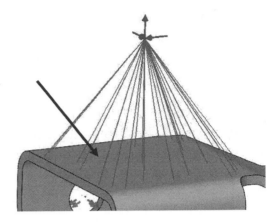

Create and apply the external remote load with direct transfer. **Note:** There is no part for this section. Just the procedure.

1. Right-click the **External Load** folder from the Static study tree.

2. Click **Remote Load/Mass**. The Remote Load/Mass PropertyManager is displayed. The Remote Loads/Mass PropertyManager allows you to apply remote loads, remote masses, and remote displacements for static, nonlinear static, and topology studies.

3. Click the selected **face**.

4. Click **Global** for reference coordinate system.

5. Enter the **X (50mm)**, **Y (-50mm)**, **Z (50mm)** location. The coordinates of the reference node location. As an example: X-coordinate of the point of application of the remote load, mass, or translation with reference to the selected coordinate system (or the global coordinate system).

6. Click the **Translational Components** box.

7. Select **Newtons**.

8. Enter the **illustrated** information. Note the direction of the triad.

9. Click **OK** ✓ from the Remote Loads/Mass PropertyManager.

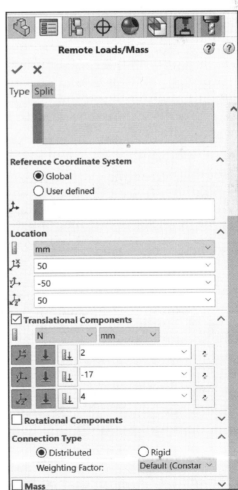

Tutorial FEA Model 7-8

Part of an exam question could read:

The cylindrical face noted A is fixed.

The cylindrical faces noted B are only allowed to rotate about their axis.

Create and apply the two Fixtures for the model.

Address cylindrical face A. Create fixed geometry. **Note:** There is no part for this section. Just the procedure.

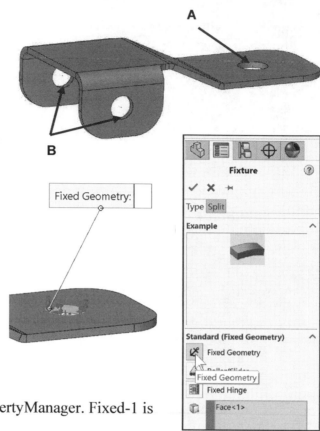

1. Right-click the **Fixtures** folder.

2. Click **Fixed Geometry**. The Fixture PropertyManager is displayed.

3. Click **Fixed Geometry**.

4. Click **cylindrical face A** in the graphics area.

5. Click **OK** ✓ from the Fixture PropertyManager. Fixed-1 is created.

Address cylindrical faces B. The two cylindrical faces are only allowed to rotate around their axis.

6. Right-click the **Fixtures** folder.

7. Click **Fixed Geometry**. The Fixture PropertyManager is displayed.

8. Click **Fixed Hinge**. The Hinge restraint specifies that a cylindrical face can only rotate about its own axis. The radius and the length of the cylindrical face remain constant under loading. This condition is similar to selecting the On cylindrical face restraint type and setting the radial and axial components to zero.

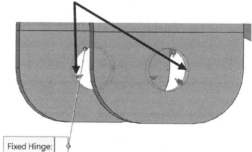

9. Click the **two cylindrical faces** in the graphics area.

10. Click **OK** ✓ from the Fixture PropertyManager. Fixed-Hinge-1 is created.

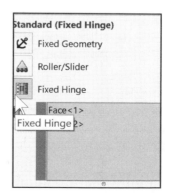

Definitions:

The following are a few key definitions for the exam:

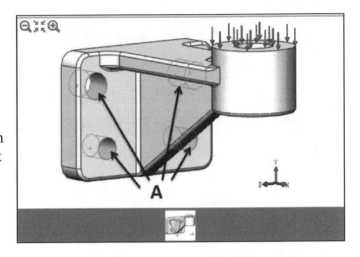

Axisymmetry: Having symmetry around an axis.

Brittle: A material is brittle if, when subjected to stress, it breaks without significant deformation (strain). Brittle materials, such as concrete and carbon fiber, are characterized by failure at small strains. They often fail while still behaving in a linear elastic manner, and thus do not have a defined yield point. Because strains are low, there is negligible difference between the engineering stress and the true stress. Testing of several identical specimens will result in different failure stresses; this is due to the Weibull modulus of the brittle material.

Compatible meshing: A mesh where elements on touching bodies have overlaying nodes.

Cyclic Symmetry: To define the number of sectors and the axis of symmetry in a cyclic symmetric structure for use in a cyclic symmetry calculation.

Deflection: is a term to describe the magnitude to which a structural element bends under a load.

Deformation: is the change in geometry created when stress is applied (in the form of force loading, gravitational field, acceleration, thermal expansion, etc.). Deformation is expressed by the displacement field of the material.

Distributed Mass Load: Distributes a specified mass value on the selected faces for use with static, frequency, buckling, and linear dynamic studies. Use this functionality to simulate the effect of components that are suppressed or not included in the modeling when their mass can be assumed to be uniformly distributed on the specified faces. The distributed mass is assumed to lie directly on the selected faces, so rotational effects are not considered.

Ductile Material: In materials science, ductility is a solid material's ability to deform under tensile stress; this is often characterized by the material's ability to be stretched into a wire. Stress vs. Strain curve typical of aluminum.

Maximum Normal Stress criterion: The maximum normal stress criterion also known as Coulomb's criterion is based on the Maximum normal stress theory. According to this theory failure occurs when the maximum principal stress reaches the ultimate strength of the material for simple tension.

This criterion is used for brittle materials. It assumes that the ultimate strength of the material in tension and compression is the same. This assumption is not valid in all cases. For example, cracks decrease the strength of the material in tension considerably while their effect is far smaller in compression because the cracks tend to close.

Brittle materials do not have a specific yield point and hence it is not recommended to use the yield strength to define the limit stress for this criterion.

This theory predicts failure to occur when:

$$\sigma_1 \geq \sigma_{limit}$$

where σ_1 is the maximum principal stress.

Hence:

Factor of safety = $\sigma_{limit} / \sigma_1$

Maximum Shear Stress criterion: The maximum shear stress criterion, also known as Tresca yield criterion, is based on the Maximum Shear stress theory.

This theory predicts failure of a material to occur when the absolute maximum shear stress (τ_{max}) reaches the stress that causes the material to yield in a simple tension test. The Maximum shear stress criterion is used for ductile materials.

$$\tau_{max} \geq \sigma_{limit} / 2$$

τ_{max} is the greatest of τ_{12}, τ_{23} and τ_{13}

Where:

$$\tau_{12} = (\sigma_1 - \sigma_2)/2; \ \tau_{23} = (\sigma_2 - \sigma_3)/2; \ \tau_{13} = (\sigma_1 - \sigma_3)/2$$

Hence:

Factor of safety (FOS) = $\sigma_{limit} / (2 * \tau_{max})$

Maximum von Mises Stress criterion: The maximum von Mises stress criterion is based on the von Mises-Hencky theory, also known as the Shear-energy theory or the Maximum distortion energy theory.

In terms of the principal stresses s1, s2, and s3, the von Mises stress is expressed as:

$$\sigma_{vonMises} = \{[(s1 - s2)2 + (s2 - s3)2 + (s1 - s3)2]/2\}(1/2)$$

The theory states that a ductile material starts to yield at a location when the von Mises stress becomes equal to the stress limit. In most cases, the yield strength is used as the stress limit. However, the software allows you to use the ultimate tensile or set your own stress limit.

$$\sigma_{vonMises} \geq \sigma_{limit}$$

Yield strength is a temperature-dependent property. This specified value of the yield strength should consider the temperature of the component. The factor of safety at a location is calculated from:

Factor of Safety (FOS) = $\sigma_{limit} / \sigma_{vonMises}$

Modulus of Elasticity or Young's Modulus: The Elastic Modulus (Young's Modulus) is the slope defined as stress divided by strain. E = modulus of elasticity (Pa (N/m^2), N/mm^2, psi). The Modulus of Elasticity can be used to determine the stress-strain relationship in the linear-elastic portion of the stress-strain curve. The linear-elastic region is either below the yield point, or if a yield point is not easily identified on the stress-strain plot it is defined to be between 0 and 0.2% strain, and is defined as the region of strain in which no yielding (permanent deformation) occurs.

Force is the action of one body on another. A force tends to move a body in the direction of its action.

Mohr-Coulomb: The Mohr-Coulomb stress criterion is based on the Mohr-Coulomb theory, also known as the Internal Friction theory. This criterion is used for brittle materials with different tensile and compressive properties. Brittle materials do not have a specific yield point and hence it is not recommended to use the yield strength to define the limit stress for this criterion.

Mohr-Coulomb Stress criterion: The Mohr-Coulomb stress criterion is based on the Mohr-Coulomb theory, also known as the Internal Friction theory. This criterion is used for brittle materials with different tensile and compressive properties. Brittle materials do not have a specific yield point and hence it is not recommended to use the yield strength to define the limit stress for this criterion.

This theory predicts failure to occur when:

$$\sigma_1 \geq \sigma_{TensileLimit} \quad \text{if } \sigma_1 > 0 \text{ and } \sigma_3 > 0$$

$$\sigma_3 \geq -\sigma_{CompressiveLimit} \quad \text{if } \sigma_1 < 0 \text{ and } \sigma_3 < 0$$

$$\sigma_1 / \sigma_{TensileLimit} + \sigma_3 / -\sigma_{CompressiveLimit} \geq 1 \quad \text{if } \sigma_1 \geq 0 \text{ and } \sigma_3 \leq 0$$

The factor of safety is given by:

Factor of Safety (FOS) = $\{\sigma_1 / \sigma_{TensileLimit} + \sigma_3 /- \sigma_{CompressiveLimit}\}^{(-1)}$

Stress: Stress is defined in terms of Force per unit Area: $Stress = \dfrac{f}{A}$.

Stress vs. Strain diagram: Many materials display linear elastic behavior, defined by a linear stress-strain relationship, as shown in the figure up to point 2, in which deformations are completely recoverable upon removal of the load; that is, a specimen loaded elastically in tension will elongate, but will return to its original shape and size when unloaded. Beyond this linear region, for ductile materials such as steel, deformations are plastic. A plastically deformed specimen will not return to its original size and shape when unloaded. Note that there will be elastic recovery of a portion of the deformation. For many applications, plastic deformation is unacceptable and is used as the design limitation.

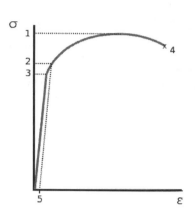

Stress vs. Strain curve typical of aluminum

1 - Ultimate Strength

2 - Yield Strength

3 - Proportional Limit Stress

4 - Rupture

5 - Offset Strain (usually 0.2%)

Tensile strength: Ultimate tensile strength (UTS), often shortened to tensile strength (TS) or ultimate strength, is the maximum stress that a material can withstand while being stretched or pulled before necking, which is when the specimen's cross-section starts to significantly contract. Tensile strength is the opposite of compressive strength and the values can be quite different.

Yield Stress: The stress level beyond which the material becomes plastic.

Yield Strength: The lowest stress that produces a permanent deformation in a material. In some materials, like aluminum alloys, the point of yielding is difficult to identify, thus it is usually defined as the stress required to cause 0.2% plastic strain. This is called a 0.2% proof stress.

Young's Modulus, or the "Modulus of Elasticity": The Elastic Modulus (Young's Modulus) is the slope defined as stress divided by strain. E = modulus of elasticity (Pa (N/m2), N/mm2, psi). The Modulus of Elasticity can be used to determine the stress-strain relationship in the linear-elastic portion of the stress-strain curve. The linear-elastic region is either below the yield point, or if a yield point is not easily identified on the stress-strain plot it is defined to be between 0 and 0.2% strain, and is defined as the region of strain in which no yielding (permanent deformation) occurs.

CHAPTER 8 - CERTIFIED SOLIDWORKS ASSOCIATE ADDITIVE MANUFACTURING (CSWA-AM)

The Certified SOLIDWORKS Associate Additive Manufacturing (CSWA-AM) exam indicates a foundation in and apprentice knowledge of today's 3D printing technology and market.

Go to the 3DEXPERIENCE® Certification Center at https://3dexperience.virtualtester.com/#home.

The 3DEXPERIENCE® Certification Center is where you are able to log in to manage your certificates, take certifications, and make changes to your account settings.

If your school is an academic certification provider, your instructor can allocate free exam credits for the CSWA (Segment 1 & 2), CSWA-SD, CSWA-S, and CSWA-AM certifications. The instructor will require your .edu email address.

The CSWA-AM exam is meant to be taken after the completion of the 10-part learning path located on MySOLIDWORKS.com.

Note: The book is meant to provide all of the needed information from the online MySOLIDWORKS lessons to pass the CSWA-AM exam.

The learning content is free, but the creation of a free MySOLIDWORKS account is needed in order to access the content.

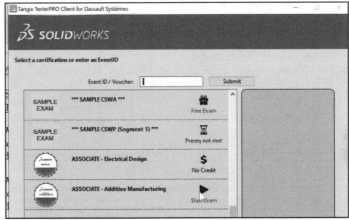

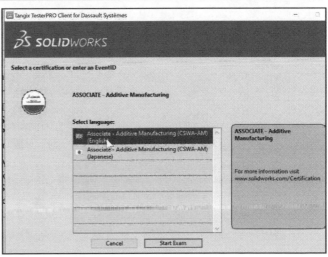

The lessons cover: **Introduction to Additive Manufacturing** (7 minutes), **Machine Types** (8 minutes), **Materials** (7 minutes), **Model Preparation** (9 minutes), **File Export Settings** (8 minutes), **Machine Preparation** (7 minutes), **Printing the Part** (6 minutes), **Post Printing** (7 minutes), **Part Finishing** (9 minutes) and **Software Options** (9 minutes).

The lessons are focused on two types of 3D printer technology: Fused Filament Fabrication (FFF) and STereoLithography (SLA). There are a few questions on Selective Laser Sintering (SLS) technology and available software-based printing aids.

The CSWA-AM (Additive Manufacturing) exam covers numerous areas: material types, printing technologies, machine types and processes, part design and orientation for 3D printing, printer preparation, post printing finishing STereoLithography (SLA), Slicer software features and functionality and available software-based printing aids.

There are 50 questions in the exam. Each question is worth 2 points. All questions are in a multiple-choice format.

You are allowed to answer the questions in any order. Total exam time is 60 minutes. You need a minimum passing score of 80 or higher. The exam is out of 100 points.

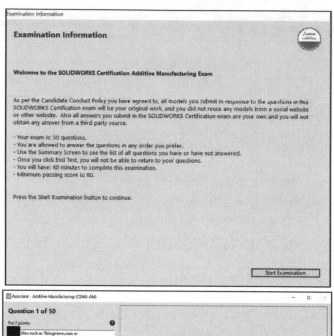

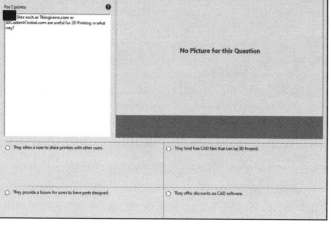

The exam covers the Ultimaker 3 (FFF) and the FormLabs Form 2 (SLA) machine as examples.

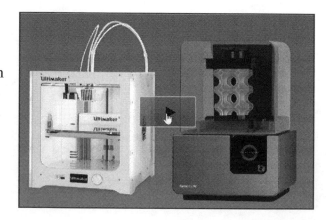

If you fail the exam, you will need to wait 14-days before retaking it again.

💡 Fused Filament Fabrication (FFF) and Fused Deposition Modeling (FDM) are used interchangeably in the certification exam.

Additive vs. Subtractive Manufacturing

In April 2012, *The Economist* published an article on 3D printing. In the article they stated that this was the "beginning of a third industrial revolution, offering the potential to revolutionize how the world makes just about everything."

Avi Reichental, who was President and CEO, 3D Systems stated, "With 3D printing, complexity is free. The printer doesn't care if it makes the most rudimentary shape or the most complex shape, and that is completely turning design and manufacturing on its head as we know it."

Over the past five years, companies are now using 3D printing to evaluate more concepts in less time to improve decisions early in product development. As the design process moves forward, technical decisions are iteratively tested at every step to guide decisions big and small, to achieve improved performance, lower manufacturing costs, delivering higher quality and more successful product introductions. In pre-production, 3D printing is enabling faster first article production to support marketing and sales functions, and early adopter customers. And in final production processes, 3D printing is enabling higher productivity, increased flexibility, reduced warehouse and other logistics costs, economical customization, improved quality, reduced product weight, and greater efficiency in a growing number of industries.

Technology for 3D printing continues to advance in three key areas: **printers** and **printing methods**, **design software**, and **materials** used in printing.

Already, 3D printing is being used in the medical industry to help save lives and in some space exploration efforts. But how will 3D printing affect the average, middle-class person in the future? Low-cost 3D printers are addressing this consumer market.

Additive manufacturing is the process of joining materials to create an object from a 3D model, usually adding layer over layer.

Subtractive manufacturing relies upon the removal of material to create something. The blacksmith hammered away at heated metal to create a product. Today, a Computer Numerical Control CNC machine cuts and drills and otherwise removes material from a larger initial block to create a product.

💡 Additive manufacturing, sometimes known as ***rapid prototyping***, can be slower than Subtractive manufacturing. Both take skill in creating the Geometric Code "G-code" and understanding the machine limitations.

3D Printer Technology

Stages of 3D printing

There are three basic stages to preparing files for 3D printing.

3D modeling

The process of developing a mathematical representation of any surface of an object (either inanimate or living) in three dimensions via software such as SOLIDWORKS. There are many other software applications for 3D modeling. These applications all have their own file format.

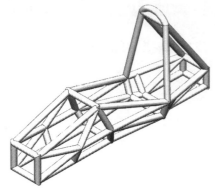

3D file export

Once the model is created, it needs to be saved and exported as a STereoLithography (*.stl), 3D Manufacturing Format (*.3mf), or an Additive Manufacturing File (*.amf). These file formats are recognized by many 3D printers (Slicers).

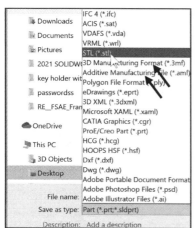

Slicing the file

The (*.3mf), (*amf), or (*.stl) file is imported into the 3D printer slicer. The slicer takes the file and translates the model into individual layers. It then generates the machine Geometric Code (G-code) that the 3D printer requires.

Slicer software allows the user to calibrate 3D printer settings: filament type, part orientation, extruder speed, extruder temperature, bed temperature, cooling fan rate, raft type, support type, percent infill, infill pattern type, etc.

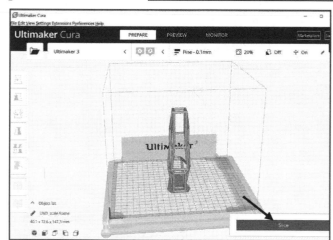

Ultimaker Cura slicer - Version: 4.4.1

Note: Unlike its predecessor STL (*.stl) format, 3D Manufacturing Format (*.3mf) and Additive Manufacturing File (*.amf) has native support for color, materials, lattices, and constellation (groups). Therefore, it requires less post-processing to define data such as the position of your model relative to the selected 3D printer, orientation, color, materials, etc.

An OBJ (*.obj) file is a standard 3D image format that can be exported and opened by various 3D image editing programs.

Fused Filament Fabrication (FFF)

Fused filament fabrication (FFF) is a relatively new method of Additive manufacturing (also known as FDM) technology used for building three-dimensional prototypes or models layer by layer with a range of thermoplastics.

FFF technology is the most widely used form of 3D desktop printing at the consumer level, fueled by students, hobbyists, and office professionals.

The technology uses a continuous filament of a thermoplastic material. The thermoplastic filament is provided in a spool (open filament area) or in a refillable auto loading cartridge. Thermoplastic filament comes in two standard diameters: 1.75mm and 3mm (true size of 2.85mm).

🔆 Ultimaker 3 and newer machines uses Near Field Communication technology with their 3mm filament. **NFC technology informs the printer of the filament type and color**. NFC spools tend to be more expensive than generic filament and you need to manually feed the filament correctly through the Bowden-tubes into the extruder (hot end). The spools are in the back of the printer. This increases the footprint of the printer.

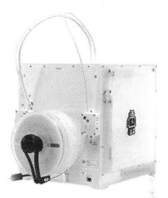

Ultimaker 3 3D printer

🔆 Sindoh 3DWOX printers uses a replacement filament spool (1.75mm) inside their cartridge with a smart chip. The benefit is the automatic loading and unloading feature along with informing the printer of the filament type and properties. Replacement spools are available. The Sindoh 3DWOX 1 printer provides the ability to either use their filament or open source filament.

Sindoh 3DWOX replacement spool and cartridge

The thermoplastic material is heated in the nozzle (hot end) (160C - 250ºC) to form liquid, which solidifies immediately when it's deposited onto the build plate or platform. The nozzle travels at a controlled rate and moves in the X and Y direction. The build plate moves in the Z direction. This creates mechanical adhesion (not chemical).

Fused Filament Fabrication (FFF) technology can be described as the inverse process of a computer numerical cutting (CNC) machine. This technology only uses the amount of material required for the part, as opposed to a CNC machine which requires significant amounts of scrap material.

Sindoh 3DWOX DP200 printer

Fused Filament Fabrication printers are available with multiple-head extruders. The most common usage for multiple-head extruders is to print in different colors or a different material for support or increase bed adhesion (raft).

A few advantages of Fused Filament Fabrication (FFF) Additive manufacturing:

Sindoh 3DWOX 2X

- Lower cost (different entry levels) into the manufacturing environment.

- Lowers the barriers (space, power, safety, and training) to traditional subtractive manufacturing.

- Reduce part count in an assembly from traditional subtractive manufacturing (complex parts vs. assemblies).

- Numerous thermoplastic filament types and colors (open and closed source) are available and affordable.

- Remote monitoring ability with a camera and LED lighting.

- Print more than one thermoplastic filament type and color during the print cycle (multiple-head extruders).

- Change or load new filament material during a print cycle.

- Customize percent infill and infill pattern type in the print.

- Only uses the amount of material required for the part. One exception, when using a raft and or support during the print.

Ultimaker S3

- No post-curing process required.

- Reduce prototyping time.

- Faster development cycle.

- Quicker customer feedback.

- Quicker product customization and configuration.

- Parallel verticals: develop and prototype at the same time.

- Most work on either the Window or Mac OSX platform.

- Open source slicing engines. To name a few: Slic3r, Skeinforge, Netfabb, KISSkice, and Cura.

- Open source filament.

Prusa i3 MK3

A slicer takes a 3D model, most often in STereoLithography (*.stl) file format and translates the model into individual layers. It then generates the machine code that the 3D printer uses. You can also use Additive Manufacturing file (*.amf) or a 3D Manufacturing Format (*.3mf) file if supported.

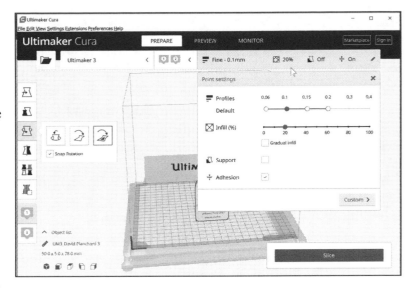

Ultimaker Cura slicer - Version: 4.4.1

Slicer software allows the user to calibrate 3D printer settings: filament type, part orientation, extruder speed, extruder temperature, bed temperature, cooling fan rate, raft type, support type, percent infill, infill pattern type, etc.

A few disadvantages of Fused Filament Fabrication (FFF) Additive manufacturing:

- Slow build rates. Many printers lay down material at a speed of one to five cubic inches per hour. Depending on the part needed, other manufacturing processes may be significantly faster.

- May require post-processing. The surface finish and dimensional accuracy may be lower quality than other manufacturing methods.

- Poor mechanical properties. Layering and multiple interfaces can cause defects in the product.

- Frequent calibration is required. Without frequent calibration, prints may not be the correct dimensions, they may not stick to the build plate, and a variety of other not-so-wanted effects can occur.

- Limited by the accuracy of the stepper motors, extruder nozzle diameter, user calibration as well as print speed.

- Print time increases linearly as part tolerances become tighter. In general, FFF print tolerances range from 0.05mm to 0.5mm.

- To print something, you require a CAD model. You either need to know how to design using CAD software (SOLIDWORKS) or download a CAD model (native format, or an (*.stl), (*.amf), or (*.3mf) file from a website (GrabCAD, Thingiverse, etc.).

StereoLithography (SLA)

Stereolithography (SLA) was the world's first 3D printing technology. It was introduced in 1988 by 3D Systems, Inc., based on work by inventor Charles Hull.

SLA is one of the most popular resin-based 3D printing technologies for professionals today.

SLA technology is a form of 3D printing in which a computer-controlled low-power, highly focused ultraviolet (UV) laser is used to create layers (layer by layer) from a liquid polymer (photopolymerization) that hardens on contact.

Markforged X7
SLS printer

When a layer is completed, a leveling blade is moved across the surface to smooth it before depositing the next layer. The platform moves by a distance equal to the layer thickness, and a subsequent layer is formed on top of the previously completed layers.

The process of tracing and smoothing is repeated until the part is complete. Once complete, the part is removed from the resin tank and drained of any excess polymer.

The part is in a "green" state. This green state differs from the completely cured state in one very important way: there are still polymerizable groups on the surface that subsequent layers can covalently bond to.

Through the application of heat and light, the strength and stability of printed parts improves beyond their original "green" state. However, each resin behaves slightly differently when post-cured, and requires different amounts of time and temperature to arrive at the material's optimum properties.

After the post-cure, you need to address the post-process. Post-processing is the removal of the supports and any needed polishing or sanding.

SLA prints are watertight and fully dense.

Resolution of SLA technology varies from 0.05mm to 0.15mm with the industry average tolerance around 0.1mm. On average, this is significantly more precise than FFF technology and is the preferred rapid prototyping solution when extremely tight tolerances are required.

As the ultraviolet (UV) laser traces the layer, the liquid polymer solidifies, and the excess areas are left as liquid in the tank.

(1micron = 1μm = 0.001mm).

One example of a popular SLA 3D desktop printer is the Formlabs Form 2. Formlabs was founded in September 2011 by three MIT Media Lab students.

The build area is 45mm × 145mm × 175mm with a layer thickness (Axis Resolution) of 25, 50, and 100 microns.

💡 Desktop area increases significantly if you include their Form Wash and Form Cure products.

The resin (liquid polymer) is provided in a cartridge. There are different cartridges for various colors and materials. Not all resin cartridges support the 25, 50, and 100 micron resolutions.

Formlabs Form 2
SLA 3D printer

Once the print is finished, remove it from the resin tank. Resin tanks are consumables and require replacement. Expect to replace a standard resin tank after 1,000-3,000 layers of printing (1-1.5 liters) of resin.

Wash the print with isopropyl alcohol (IPA). You should wait approximately 30 minutes for the IPA to fully evaporate after washing.

To ensure proper washing, Formlabs recommends their Form Wash. Form Wash automatically cleans uncured liquid resin printed part surfaces.

Formlabs Form 2
resin cartridges

IPA dissolves uncured resin. The part is covered in uncured resin when it's removed from the resin tank. Use IPA in a well-ventilated area. Always wear protective gloves and eyewear. Find a safe way to dispose of the used IPA.

After the IPA wash and dry, perform a post-cure. At a basic level, exposure to sunlight triggers the formation of additional chemical bonds within a printed part, making the material stronger and stiffer.

Formlabs
Form Wash

To ensure proper post-curing, Formlabs recommends their post-curing unit, Form Cure. Form Cure precisely combines temperature and 405 nm light to post-cure parts for peak material performance.

💡 You cannot use a camera or webcam inside the build area. Formlabs Dashboard feature provides the ability to keep track of what layer the printer is on, resin consumptions, and various other stats.

Formlabs
Form Cure

Formlabs uses their PreForm software. PreForm provides the ability to select the One-Click Print option, to automatically orient, support and layout the part.

Formlabs Form 2 printer works with the Window or Mac OSX platform.

💡 Clean the resin tank after a print fails. Remove any cured resin on the elastic layer, discard print failures, and filter out debris. Clean any contamination on the clear acrylic tank window.

A few advantages of Stereolithography (SLA) Additive manufacturing:

- Final parts are stronger than using FFF technology.

- Parts are watertight and fully dense. No infill is required.

- Higher resolution than FFF technology.

A few disadvantages of Stereolithography (SLA) Additive manufacturing:

- Higher cost (different entry levels) into the manufacturing environment than FFF.

- Material costs are significantly higher than FFF due to the proprietary nature and limited availability of the photopolymers.

- Significantly slower fabrication speed than FFF.

- Suitable for low volume production runs of small, precise parts.

- Print only a single material type (color) at a time.

- Requires an isopropyl alcohol (IPA) wash.

- Ability to safely dispose of the used isopropyl alcohol (IPA).

- Requires a post-cure.

- Parts are sensitive to long exposure to UV light.

- May need drain hole in the part to remove excess liquid polymer.

Selective Laser Sintering (SLS)

Selective Laser Sintering (SLS) technology is the most common additive manufacturing technology for industrial powder base applications.

SLS fuses particles together layer by layer through a high energy pulse laser. Similar to SLA, this process starts with a tank full of bulk material but is in a powder form vs. liquid. As the print continues, the bed lowers itself for each new layer as done in the SLS process.

3D Systems
ProX DMP 300

Both plastics and metals can be fused in this manner, creating much stronger and more durable prototypes.

Although the quality of the powders is dependent on the supplier's proprietary processes, the base materials used are typically more abundant than photopolymers, and therefore cheaper. However, there are additional costs in energy used for fabricating with this method which may reverse any savings realized in the material cost.

💡 Speed and resolution of SLS printers typically match that of SLA, with industry averages at around 0.1mm tolerances. Only suitable for low volume production runs of small, precise parts.

💡 SLS is the preferred rapid-prototyping method of metals and exotic materials.

Select the Correct Filament Material for FFF

There are many materials that are being explored for 3D printing; however, the two most dominant plastics for FFF technology are PLA (Polylactic acid) and ABS (Acrylonitrile-Butadiene-Styrene).

Both PLA and ABS are known as thermoplastics; that is, they become soft and moldable when heated and return to a solid when cooled.

There are three key printing stages for both thermoplastics.

1. Cold to warm: The thermoplastic starts in a hard state. It stays this way until heated to its glass transition temperature.

2. Warm to hot: The thermoplastic is now in a viscous state. It stays this way until heated to its melting temperature.

3. Hot to melting: The thermoplastic is now in a liquid state.

For PLA, the glass transition temperature is approximately 60°C. The melting temperature is approximately 155°C with a printing temperature range of 190°C - 220°C.

For ABS, the glass transition temperature is approximately 100°C with a printing temperature range between 210C - 250°C. Therefore, a heated print bed is needed for ABS and is optional for PLA. The print bed should be kept near the glass transition temperature and well below the melting temperature. It also helps to keep the part from cooling too fast and warping.

PLA and ABS are hygroscopic, meaning it attracts and absorbs moisture from the air. The effects of attracting water may result in one or more of the following problems: increased brittleness, diameter augmentation (potential problems with Bowden-tube printers), or filament bubbling once reaching the extruder (hot-end).

It is recommended that you store used filament in an airtight plastic bag (container) with a few silica gel packs. Place the bag in a dry, dark, controlled environment.

Silica gel pack

PLA

PLA (Polylactic Acid) is a biodegradable thermoplastic, made from renewable resources like corn starch or sugarcane. Relevant information:

- **Strength**: High | Flexibility: Low | Durability: Medium
- **Difficulty to use**: Low
- **Print temperature**: 190°C - 220°C
- **Print bed temperature**: 20°C - 40°C (not required)
- **Shrinkage/warping**: Minimal
- **Soluble**: No
- **Hot Head**: Standard Polytetrafluoroethylene, PTFE (Teflon)
- **Filament size**: 1.75mm and 2.85mm
- **Food safety**: Refer to manufacturer guidelines

PLA has a lower printing temperature (20°C - 30°C) than ABS, and it doesn't warp as easily. PLA does not require a heated bed. Most non-heated beds are made from glass or metal.

 A removable flexible bed makes it ideal to retrieve printed parts.

PLA is normally used for its nice finish, easy and fast printing characteristics and for the large amounts of colors and varieties available.

Avoid using PLA if the print is exposed to temperatures of 60°C or higher or might be bent, twisted or dropped repeatedly.

Outside of 3D printing, PLA is typically used in medical implants, food packaging, and disposable tableware.

Darker colors and glow in the dark materials often require higher extruder temperatures (5°C - 10°C).

PLA in general is more forgiving to temperature fluctuations and moisture than ABS and Nylon during a build cycle.

 Flex/Soft PLA - Common flexible filaments are polyester-based (non-toxic). Recommended print temperature range is 220°C - 250°C. It is highly recommended to drastically lower your printing speed to around 10-20mm/s. To take advantage of the filament's properties, print it with 10% infill or less. Most flexible filament adheres well to a heated bed.

PLA - Storage

PLA is mildly hygroscopic, meaning it attracts and absorbs moisture from the air.

PLA responds somewhat differently to moisture than ABS. Over time, it can become very brittle. In addition to bubbles or spurting at the nozzle (hot end), you may see discoloration and a reduction in 3D printed part properties.

Store the filament in an airtight plastic bag (container) with a few silica gel packs. Place the bag (container) in a dry, dark, temperature controlled environment. As an extra precaution, filament manufacturers often recommend using up rolls as soon as possible.

PLA can be dried using something as simple as a food dehydrator. It is important to note that this can alter the crystallinity ratio in the PLA and will lead to changes in extrusion temperature and other extrusion characteristics.

PLA - Part Accuracy

Compared to ABS, PLA demonstrates much less part warping. PLA is less sensitive to changes in temperature than ABS.

PLA undergoes more of a phase-change when heated and becomes much more liquid. If actively cooled, sharper details can be seen on printed corners without the risk of cracking or warping. The increased flow can also lead to stronger binding between layers, improving the strength of the printed part.

In a small enclosed space, it is recommended to have your printer enclosed with a HEPA filtration filter.

ABS

ABS (Acrylonitrile-Butadiene-Styrene) is an oil-based thermoplastic, commonly found in (DWV) pipe systems, automotive trim, bike helmets, and toys (LEGO). Relevant information:

- **Strength**: High | Flexibility: Medium | Durability: High
- **Difficulty to use**: Medium
- **Print temperature**: 210°C - 250°C
- **Print bed temperature**: 80°C - 110°C (required)
- **Shrinkage/warping**: Medium
- **Soluble**: In esters, ketones, and acetone
- **Hot Head**: Standard Polytetrafluoroethylene, PTFE (Teflon)
- **Filament size**: 1.75mm and 2.85mm
- **Food safety**: Not food safe

ABS boast slightly higher strength, flexibility, and durability. ABS is more sensitive to changes in temperature than PLA, which can result in cracking and warping if the print cools too quickly.

A heated bed plate is required for ABS. Most heated beds are made from glass or metal.

It is recommended to have a ventilated printing area when using ABS material (oil-based thermoplastic). It is also recommended to have the printer enclosed with a HEPA filtration filter.

ABS is better suited for items that are frequently handled, dropped, or heated. It can be used for mechanical parts, especially if they are subjected to stress or must interlock with other parts.

Sindoh 3DWOX 1, 2X HEPA filter is built into the printer.

🔅 For high temperature applications, ABS (glass transition temperature of 105°C) is more suitable than PLA (glass transition temperature of 60°C). PLA can rapidly lose its structural integrity as it approaches 60°C.

ABS - Storage

ABS is mildly hygroscopic. Diameter augmentation (potential problems with Bowden-tube printers) can be an issue.

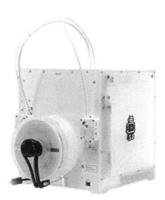

Store the filament in an airtight plastic bag (container) with a few silica gel packs. Place the bag in a dry, dark, temperature controlled environment. As an extra precaution, filament manufacturers often recommend using up rolls as soon as possible. ABS can be easily dried using a source of hot (preferably dry) air such as a food dehydrator.

Ultimaker 3 using Bowden-tubes

ABS - Part Accuracy

For most, the single greatest hurdle for accurate parts is good bed (platform) adhesion. Start with a clean, level, heated bed. Check the bed and extruder (hot end) are set to the correct temperature.

Eliminate all build area drafts (open windows, air conditioning vents, etc.). Use dry filament. Wet filament during printing prevents good layer adhesion and greatly weakens the part.

When printing on a glass plate, a (Polyvinyl Acetate) PVA based glue stick applied to the bed helps with bed adhesion. Elmer's or Scotch permanent glue sticks are inexpensive and easily found. Remember, less is more when applying the glue stick to the build plate.

You may need to add a raft (a horizontal latticework of filament located underneath the part) to the build. Over time, a heated metal bed can warp. Check for flatness.

🔅 ABS provides a more matte appearance than PLA, but it can become very shiny after acetone vapor smoothing.

Nylon

Nylon (618, 645) is a popular family of synthetic polymers used in many industrial applications. Compared to most other filaments, it ranks as the number one contender when together considering strength, flexibility, and durability.

Relevant information for Nylon 618:

- **Strength**: High | Flexibility: High | Durability: High
- **Difficulty to use**: Medium
- **Print temperature**: 240°C - 255°C
- **Print bed temperature**: 50°C - 60°C (required)
- **Shrinkage/warping**: Medium
- **Soluble**: No
- **Hot Head**: Metal
- **Filament size**: 1.75mm and 2.85mm
- **Food safety**: Refer to manufacturer guidelines

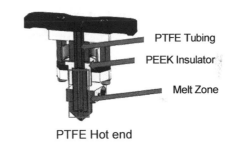

PTFE Hot end

Relevant information for Nylon 645:

- **Strength**: High | Flexibility: High | Durability: High
- **Difficulty to use**: Medium
- **Print temperature**: 255°C - 265°C
- **Print bed temperature**: 85°C - 95°C (required)
- **Shrinkage/warping**: Considerable
- **Soluble**: No
- **Hot Head**: Metal
- **Filament size**: 1.75mm and 2.85mm
- **Food safety**: Refer to manufacturer guidelines

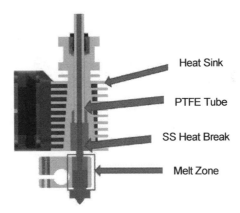

All metal Hot end

Nylon filament requires temperatures above 240°C to extrude. Most low-end 3D printers come standard with hot ends that use Polyether ether Ketone (PEEK) and Polytetrafluoroethylene (Teflon). Both PEEK and PTFE begin to breakdown above 240°C and will burn and emit noxious fumes. You should only use an all metal hot end.

☀ PLA and ABS are less likely to get stuck in the inner wall PTFE (Teflon) than an all metal hot end.

When should I use Nylon? Taking advantage of nylon's strength, flexibility, and durability use this 3D printer filament to create tools, functional prototypes, or mechanical parts (like hinges, buckles, or gears). Dry nylon prints buttery smooth and has a glossy finish.

Nylon - Storage

Nylon is very hygroscopic, more so than PLA or ABS. Nylon can absorb more than 10% of its weight in water in less than 24 hours. Successful 3D printing with nylon requires dry filament. When you print with nylon that isn't dry, the water in the filament explodes causing air bubbles during printing that prevents good layer adhesion and greatly weakens the part. It also ruins the surface finish. To dry nylon, place it in an oven at 50C - 60°C for 6-8 hours. After drying, store in an airtight container (Vacuum bag), preferably with dry silica gel packets. Place the container in a dry, dark, temperature controlled environment.

Nylon - Part Accuracy

Compared to ABS and PLA, Nylon and ABS warp approximately the same. PLA demonstrates much less part warping. A heated plate (50°C - 90°C) is required for Nylon. When printing on a glass plate, a (Polyvinyl Acetate) PVA based glue stick applied to the bed is the best method of bed adhesion. Elmer's or Scotch permanent glue sticks are inexpensive and easily found. Remember, less is more when applying the glue stick to the plate. You will also have to clean it up after your build.

PVA (Polyvinyl Alcohol)

PVA (Polyvinyl Alcohol) filament is a water-soluble synthetic polymer. PVA filament dissolves in water. Many multi-head extruder users find PVA to be a useful support material because of its dissolvable properties. In general, PVA filament is used in conjunction with PLA not ABS. PVA adheres well to PLA and not ABS. Moreover, the extrusion temperature difference between PVA and ABS can be problematic.

PVA extrusion temperatures range between 160°C - 190°C. A heated bed is recommended. Bed temperatures range between 40°C - 50°C.

PVA is highly hygroscopic and is costly. PVA should be stored in an airtight box or container and may need to be dried before use.

Submerge the finished part in a bath of cold circulating water, until the PVA support structure is completely dissolved. This can be time consuming and messy.

Do not expose PVA filament to temperatures higher than 200°C for an extended period of time. An irreversible degradation of the material will occur, known as pyrolysis. It will jam the extruder nozzle (hot end). Unlike PLA and ABS, you cannot remove the jam by increasing the temperature. Clearing the jam in the nozzle will often require it to be re-drilled or replaced altogether.

A few filament companies provide breakaway material to replace the high cost of PVA. Breakaway is a support material used with multi head 3D printers. It is quick to remove and does not need further post-processing.

STereoLithography (*.stl) file

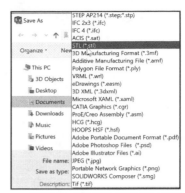

STereoLithography (*.stl) is a file format native to the Stereolithography CAD software created by 3D Systems. STL has several after-the-fact backronyms such as "Standard Triangle Language" and "Standard Tessellation Language".

An STL file describes only the surface geometry of a three-dimensional object without any representation of color, texture, or other common CAD model attributes. The STL format specifies both ASCII and Binary representations.

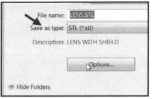

Binary files are more common, since they are more compact. An STL file describes a raw unstructured triangulated (point cloud) surface by the unit normal and vertices (ordered by the right-hand rule) of the triangles using a three-dimensional Cartesian coordinate system.

STL Save options allow you to control the number and size of the triangles by setting the various parameters in the CAD software.

Save an STL (*.stl) file in SOLIDWORKS

To save a SOLIDWORKS model as a STL file, click **File**, **Save As** from the Main menu or **Save A**s from the Main menu toolbar. The Save As dialog is displayed. Select **STL(*.stl)** as the Save as type.

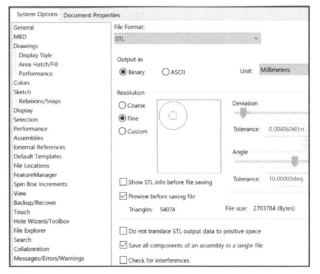

Click the **Options** button. The dialog box is displayed. View your options and the three types of Resolution: **Coarse**, **Fine** and **Custom**. The resolution options provide the ability to control the number and size of the triangles by setting various parameters in the CAD software. The Custom setting provides the ability to control the deviation and angle for the triangles. Click **OK**. View the generate point cloud of the part. Click **Yes**. The STL file is now ready to be imported into your 3D printer software.

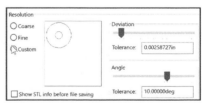

In SOLIDWORKS for a smoother STL file, change the Resolution to Custom. Change the deviation to 0.0005in (0.01mm). Change the angle to 5. Smaller deviations and angles produce a smoother file but increase the file size and print time.

Additive Manufacturing (*.amf) file

Additive Manufacturing (*.amf) is a file format that includes the materials that have been applied to the parts or bodies in the 3D model.

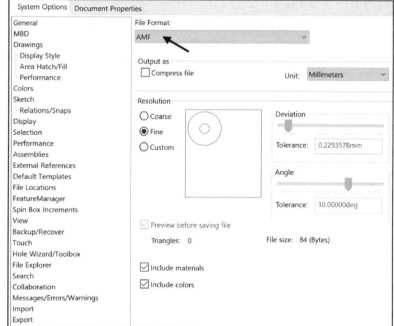

Additive Manufacturing (*.amf) is an open standard for describing objects for additive manufacturing processes such as 3D printing. The official ISO/ASTM 52915:2013 standard is an XML-based format designed to allow any computer-aided design software to describe the shape and composition of any 3D object to be fabricated on any 3D printer. Unlike its predecessor STL format, AMF has native support for color, materials, lattices, and constellation (groups). Therefore, it requires less post-processing to define data such as the position of your model relative to the selected 3D printer, orientation, color, materials, etc.

Save an Additive Manufacturing (*.amf) file in SOLIDWORKS

To save a SOLIDWORKS model as an Additive Manufacturing (*.amf) file, Click **File**, **Save As** from the Main menu or **Save As** from the Main menu toolbar. The Save As dialog is displayed. Select **Additive Manufacturing (*.amf)** as the Save as type.

Click the **Options** button. The dialog box is displayed. View your options. For most parts, utilize the default setting.

Close the dialog box.

Click **OK**. View the generate point cloud of the part.

Click **Yes**. The file is now ready to be imported into your 3D printer software or print directly from SOLIDWORKS.

3D Manufacturing Format (*.3mf) file

3D Manufacturing Format (*.3mf) is a file format developed and published by the 3MF Consortium. This format became natively supported in all Windows operating systems since Windows 8.1. 3MF has since garnered considerable support from large companies such as HP, 3D Systems, Stratasys, GE, Siemens, Autodesk and Dassault Systems although it is unknown how many actively use this file format. 3D Manufacturing Format file has similar native support for color, materials, lattices, and constellation (groups).

VRML (*.wrl)
STL (*.stl)
3D Manufacturing Format (*.3mf)
Additive Manufacturing File (*.amf)
eDrawings (*.eprt)
3D XML (*.3dxml)
Microsoft XAML (*.xaml)
CATIA Graphics (*.cgr)

Save a 3D Manufacturing Format (*.3mf) file in SOLIDWORKS

To save a SOLIDWORKS model as a 3D Manufacturing Format (*.3mf) file, Click **File**, **Save As** from the Main menu or **Save As** from the Main menu toolbar. The Save As dialog is displayed. Select **3D Manufacturing Format (*.3mf)** as the Save as type.

Click the **Options** button. The dialog box is displayed. View your options. For most parts, utilize the default setting.

Close the dialog box. Click **OK**. View the generate point cloud of the part. Click **Yes**. The file is now ready to be imported into your 3D printer software or print directly from SOLIDWORKS.

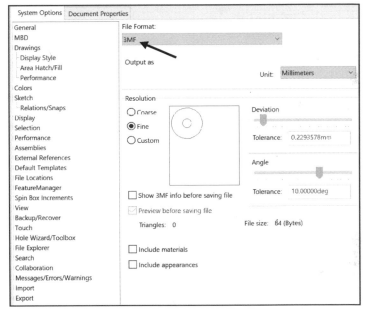

The include materials and include appearance option is not selected by default.

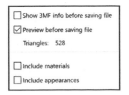

In SOLIDWORKS for a smoother STL file, change the Resolution to Custom. Change the deviation to 0.0005in (0.01mm). Change the angle to 5. Smaller deviations and angles produce a smoother file but increase the file size and print time.

What is a Slicer? How does a Slicer Work?

The (*.3mf), (*amf), or (*.stl) file is imported into the printer slicer. The slicer takes the file and translates the model into individual layers. It then generates the machine code (G-code) that the printer requires.

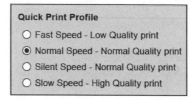

Slicing is the process of turning the 3D model into a toolpath for the printer. Most people call it slicing because the first thing the slicing engine does is cut the 3D model into thin horizontal layers.

The G-code file contains the instructions based on settings you choose, and calculates how much material the printer will need and how long it will take to print.

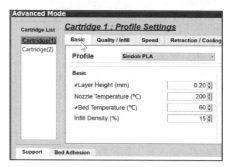

Slicer Parameters

Proper slicer parameters can mean the difference between a successful print, and a failed print. That's why it's important to know how slicers work and how various settings will affect the final print. There are open and close source slicers.

Sindoh 3DWOX slicer - Version: 1.4.2102.0

Layer Height

Layer height controls the resolution of the print. The setting specifies the height of each filament layer.

The higher values produce faster prints in lower resolution. Lower values produce slower prints in higher resolution.

The default value for (PLA) using the Sindoh 3DWOX 2X - Version: 1.4.2102.0 is .20mm. There are four default settings: Fast, Normal, Silent, and Slow. Settings range from .05mm - .4mm. Sindoh uses a .4mm nozzle with 1.75mm filament.

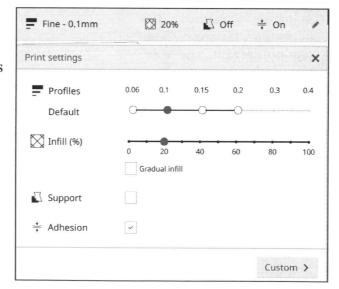

The default value for (PLA) using in the Ultimaker 3 Cura slicer - Version: 4.4.1 is .1mm. There are four default

Ultimaker Cura slicer - Version: 4.4.1

settings: Fast, Normal, Fine, and Extra Fine. Settings range from .06mm - .2mm.

Shell (Wall) Thickness

Wall thickness refers to the distance between one surface of the model and the opposite sheer surface. Set the wall thickness of the outside shell in the horizontal direction. Use in combination with the nozzle size to define the number of perimeter lines and the thickness of these perimeter lines.

The default value for (PLA) using the Sindoh 3DWOX 2X - Version: 1.4.2102.0 is .80mm. The minimum is .40mm.

The default value for (PLA) using the Ultimaker 3 Cura - Version: 3.4.1 is 1mm.

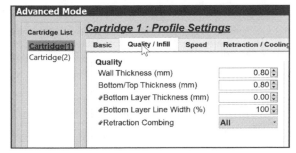

Sindoh 3DWOX slicer - Version: 1.4.2102.0

Shell		
Wall Thickness	1	mm
Wall Line Count	3	
Top/Bottom Thickness	1	mm
Top Thickness	1	mm
Top Layers	10	
Bottom Thickness	1	mm
Bottom Layers	10	

Ultimaker Cura slicer - Version: 3.4.1

Infill Density/Overlap

Infill is the internal structure of your object, which can be as sparse or as substantial as you would like it to be. A higher percentage will result in a more solid object, so 100% (not recommended) infill will make your object completely solid, while 0% infill will give you something completely hollow.

The higher the infill percentage, the more material and longer the print time. It will also increase weight and material cost.

When using any infill percentage, a pattern is used to create a strong and durable structure inside the print. A few standard patterns are: Rectilinear, Honeycomb, Circular, Tri-hexagon, Cubic, Octet, and Triangular. In general, use a 10% - 15% infill with a maximum infill of 60%. 100% infill is not recommended. Part warping can be a concern.

Shells are the outer layers of a print which make the walls of an object, prior to the various infill levels being printed within. The number of shells affect stability and translucency of the model.

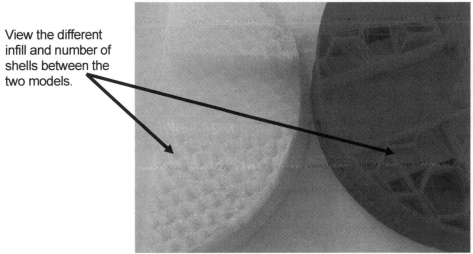

View the different infill and number of shells between the two models.

The default value for (PLA) using the Sindoh 3DWOX 2X - Version: 1.4.2102.0 is 15%.

The default value for (PLA) using the Ultimaker 3 Cura - Version: 3.4.1 is 20%.

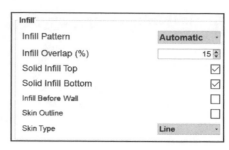

Strength corresponds to the maximum stress the print can take before breaking.

Sindoh 3DWOX slicer - Version: 1.4.2102.0

Print Speed

Print speed refers to the speed at which the extruder travels while it lays down filament.

Print speed affects the following areas: Infill speed, Wall speed, Top/Bottom speed, Initial layer travel speed, Raft print speed, and Maximum travel resolution.

The default value for (PLA) using the Sindoh 3DWOX 2X - Version: 1.4.2102.0 is 40mm/s.

The default value for (PLA) using the Ultimaker 3 Cura - Version: 3.4.1 is 70mm/s.

Supports

Generate structures to support parts (features) of the model which have overhangs to the build plate.

Without these structures, such parts would collapse during the print process.

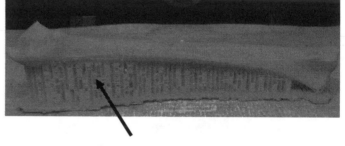

Support material

In order to create an overhang at any angle less than vertical, your printer offsets each successive layer. The lower the angle gets to horizontal, or 90°, the more each successive layer is offset.

General 45-degree rule. If your model has overhangs greater than 45 degrees, you need support material. If the part has numerous holes, sharp edges, long run (bridge), or thin bodies, support material may also be required.

Support Types

Touching Buildplate

The Touching Buildplate option provides supports only where the part touches the build plate. This reduces build time, clean up and support material.

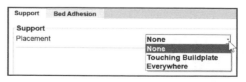

Sindoh 3DWOX slicer - Version: 1.4.2102.0

Use the touching Buildplate option when you have overhangs and tricky angles toward the bottom of a design, but do not wish to plug up holes, hollow spaces, or arches in the rest of the design.

Everywhere

The Everywhere option provides support material everywhere on the part, not only where the part touches the build plate.

Bed (Platform) Adhesion

Bed adhesion is one of the most important elements for getting a good 3D print. Set various options to ensure good bed adhesion and to prevent part warping. If needed, use blue painter's tape, hair spray or a glue stick.

Bed Adhesion Type- Raft

A Raft is a horizontal latticework of filament located underneath the part. A Raft is used to help the part stick to the build plate (heated or non-heated).

Rafts are also used to help stabilize thin tall parts with small build plate footprints.

When the print is complete, remove the part from the build plate. Peel the raft away from the part. If needed, use a scraper or spatula.

Bed Adhesion Type- Skirt

A Skirt is a layer of filament that surrounds the part with a 3mm - 4mm offset. The layer does not connect the part directly to the build plate. The Skirt primes the extruder and establishes a smooth flow of filament. In some slicers, the skirt is added automatically when you select the None option, for bed adhesion type.

Bed Adhesion Type- Brim

A Brim is basically like a Skirt for the part. A Brim has a zero offset from the part. It is a layer of filament laid down around the base of the part to increase its surface area. A Brim, however, does not extend underneath the part, which is the key difference between a brim and a raft.

Touching Build plate

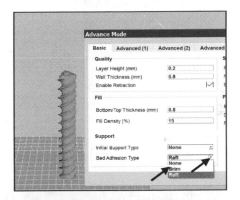

Part with a Raft on the Build plate

Part with a Skirt on the Build plate

Part Orientation

Insert the file into your printer's slicer. The model is displayed in the build plate area.

Depending on your printer's slicer software, you may or may not receive a message indicating the object is too large for the current build plate.

If the object is too large, you will need to scale it down or redesign it into separate parts.

Use caution when scaling critical features if you require fasteners or a minimum wall thickness.

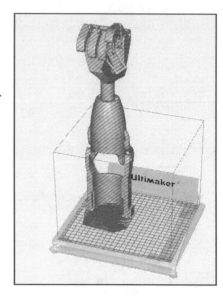

You should always center the part and have it lay flat on the build plate. Bed adhesion is one of the most important elements for getting a good 3D print.

If you are printing more than one part, space them evenly on the plate or position them for a single build.

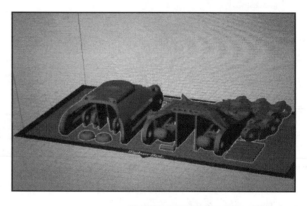

In SOLIDWORKS, lay the parts out in an assembly. Save the assembly as a part. Save the part file as an STL (*.stl), or Additive Manufacturing (*.amf), or 3D Manufacturing format (*.3mf).

Consideration should be used when printing an assembly. If the print takes 20 hours, and a failure happens after 19 hours, you just wasted a lot of time versus printing each part individually.

STL (*.stl)
3D Manufacturing Format (*.3mf)
Additive Manufacturing File (*.amf)
eDrawings (*.eprt)
3D XML (*.3dxml)
Microsoft XAML (*.xaml)

Example 1: Part Orientation

Part orientation is very important on build strength and the amount of raft and support material required for the build. Incorrect part orientations can lead to warping, curling, and delamination.

If maximizing strength is an issue, select the part orientation on the build plate so that the "grain" of the print is oriented to maximize the strength of the part.

Example 1: First Orientation - Vertical

In the first orientation (vertical), due to the number of holes and slots, additional support material is required (with minimum raft material) to print the model.

Removing the material in these geometrics can be very time consuming.

First orientation - vertical

Example 1: Second Orientation - Horizontal

In the second orientation (horizontal), additional raft material is used, and the support material is reduced.

The raft material can be easily removed with a pair of needle-nose pliers and no support material clean-up is required for the holes and slots. Note: In some cases, raft material is not needed.

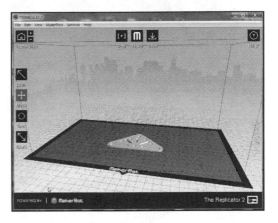

Example 2: Part Orientation

The lens part is orientated in a vertical position with the large face flat on the build plate. This reduces the required support material and ensures proper contact (maximum surface area) with the build plate.

Second orientation - horizontal

Optimize Print Direction

Some slicers (Sindoh 3DWOX desktop) provide the ability to run an optimization print direction analysis of the part under their Advanced Mode. The areas in evaluations are: **Thin Region**, **Area of Overhang Surface** and the **Amount of needed support** material. This can be very useful when you are unsure of the part orientation.

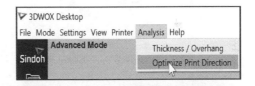

🔅 Suppress mates in an assmbly to have the model lay flat on the build plate. If the model does not lay flat, you will require a raft and additional supports. This will increase build time and material cost.

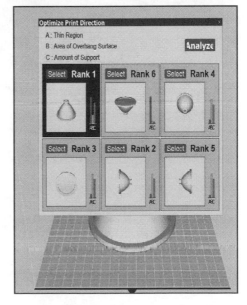

Sindoh 3DWOX slicer - Version: 1.4.2102.0

The needed support material is created mainly internal to the part to print the CBORE feature. Note: There is some outside support material on the top section of the part.

Raft material   Internal support material for the CBORE

💡 Proper part orientation for thin parts will make the removal of the raft easier.

Remove the Model from the Build Plate

Non-heated Build Plate

Most non-heated build plates are made from glass or metal. If needed, use blue painter's tape, glue stick or hair spray to assure good model adhesion. After the build, remove the plate from the printer.

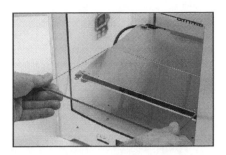

Ultimaker 3 glass build plate

Utilize a flat edge tool (thin steel spatula). Gently work under the part, and lift the part directly away from yourself. Clean the build plate. Return the plate. Re-level the plate after every build.

The Sindoh 3DWOX DP201 printer (non-heated) has a flexible magnet removable plate that does not require blue painter's tape, glue stick or hair spray to assure good model adhesion. This eliminates the need for scrapers or any sharp tools to remove the print. After the build, remove the plate and bend.

Sindoh 3DWOX DP201
flexible magnet build plate

 Bed adhesion is one of the most important elements for getting a good 3D print.

Heated Build Plate

Most heated build plates are made from glass or metal. If needed, use a glue stick or hair spray to assure good model adhesion. If you have a heated build plate your temperatures can range between 30°C - 90°C depending on the material, so be careful. After the build, remove the plate from the printer. Utilize a flat edge tool (thin steel spatula). Gently work under the part, and lift the part directly away from yourself. Clean the build plate. Return the plate. Re-level the plate after every build.

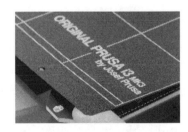

Prusa i3 MK3 flexible metal
magnet build plate

A few manufacturers (Sindoh 3DWOX 1 and 2X and Prusa i3 MK3) provide a flexible heated magnet metal build plate.

This eliminates the need for scrapers or any sharp tools most of the time. After the build, remove the flexible magnet plate and bend.

 Most heated metal bed plates are coated. Over time, the coating wears and the plate needs to be replaced.

Sindoh 3DWOX 1 and 2X
flexible metal magnet build plate

Know your Printer's Limitations

Overall part size can be an issue. Most affordable 3D printers typically are small enough to fit on your desktop. Typical build volumes range between 200 x 200 x 185 mm and 228 x 200 x 300 mm. The Sindoh 3DWOX 2X has one of the largest build volumes for an affordable desktop FFF 3D printer.

Sindoh 3DWOX 2X

There are features that are too small to be printed on a desktop 3D printer. An important, but often overlooked variable in what the printer can achieve is thread width. Thread width is determined by the diameter of the extruder nozzle. Most printers have a 0.4mm nozzle. A circle created by the printer is approximately two thread widths deep: 0.8mm thick with a 0.4mm nozzle to 1mm thick for a 0.5mm nozzle. A good rule of thumb is "The smallest feature you can create is double the thread width".

Tolerance for Interlocking Parts

For objects with multiple interlocking parts, design for a tolerance fit. Getting tolerances correct can be difficult using FFF technology.

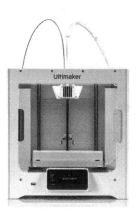

Ultimaker S3

In general, use the below suggested guidelines.

- Use ±0.1mm (±.004 in.) tolerance for a tight fit (press fit parts, connectors).

- Use ±0.2mm (±.008 in.) tolerance for a print in place (hinge).

- Use ±0.3mm (±.012 in.) tolerance for loose fit (pin in hole).

Test the fit yourself with the particular model to determine the right tolerance for the items you are creating and material you are using.

Tolerance may vary depending on filament type, manufacturer, color, humidity, build plate flatness, bed temperature, extruder temperature, etc.

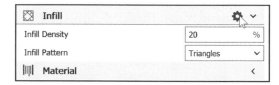

Ultimaker Cura slicer - Version: 3.4.1

General Printing Tips

Reduce Infill (Density/Overlap)

Infill is a settable variable in most Slicers. The amount of infill can affect the top layers, bottom layers, infill line distance and infill overlap.

Reduced infill can have a negative effect on part strength. There are always trade-offs.

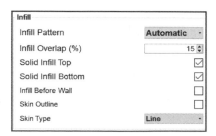

Sindoh 3DWOX slicer - Version: 1.4.2102.0

The material inside the part (infill) exerts a force on the entire printed part as it cools. More material increases cost of the part and build time.

Parts with a lower percentage of infill should have a lower internal force between layers and can reduce the chance of curling, cracking, and layer delamination along with a low build cost and time.

Sindoh 3DWOX 1 - Standard enclosed build area

Control Build Area Temperature

For a consistent quality build, control the build area and environment temperature. Eliminate all drafts and control air flow that may cause a temperature gradient within the build area.

Changes in temperature during a build cycle can cause curling, cracking and layer delamination, especially on long thin parts. From 1000s of hours in 3D printing experience, we have found that having a top cover and sides along with a consistent room temperature provides the best and repeatable builds.

FlashForge 3D Printer Creator Pro

Some printer companies like Ultimaker and FlashForge, sell additional enclosure kits (doors, sides, and tops). This helps regulate the temperature and reduce drafts for improved print quality.

The kits also lower the sound of the printer and provide a more secure print area.

When troubleshooting issues with your printer, it is always best to know if the nozzle and heated bed are achieving the desired temperatures. A thermocouple and a thermometer come in handy. I prefer a Type-K Thermocouple connected to a multi-meter. A non-contact IR or Laser-based thermometer also works well.

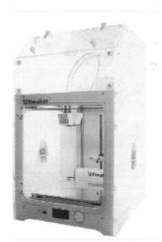

One important thing to remember is that IR and Laser units are not 100% accurate when it comes to shiny reflective surfaces. A Type-K thermocouple can be taped to the nozzle or heated bed using Kapton tape, for a very accurate temperature measurement.

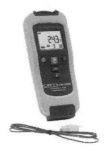

Cover/Door kit for Ultimaker 3

Add Pads

Sometimes, when you are printing a large flat object, you may view warping at the corners or extremities. One way to address this is to create small pads to your part during the modeling process. Create the model for the print. Think before you print and know your printer limitations. The pads can be any size and shape, but generally, diameter 10mm cylinders that are 1-2 layers thick work well. After the part is printed, remove them.

Makerbot Image

Unique Shape or a Large Part

If you need to make parts larger than your build area or create parts that have intricate projections, here are a few suggestions:

- Fuse smaller sections together using acetone (if using ABS). Glue if using PLA.

- Design smaller parts to be attached together (without hardware).

- Design smaller parts to be screwed together (with hardware).

Safe Zone Rule

Parts may have a safe zone. The safe zone is called "self-supporting" and no support material is required to build the part.

The safe zone can range between 30° to 150°. If the part's features are below 30° or greater than 150°, it should have support material during the build cycle. This is only a rule. Are there other factors to consider? Yes. They are layer thickness, extrusion speed, material type, length of the overhang along with the general model design of features.

🔅 Design your part for your printer. Use various modeling techniques (ribs, fillets, pads, etc.) during the design process to eliminate or to minimize the need for supports and clean up.

First Layer Not Sticking

One of the toughest aspects of 3D printing is to get your prints to stick to the build surface or bed platform. Investigate the following:

- Clean the bed. Remove any residue of tape, glue, hair spray, etc.

- Apply new blue painter's tape (non-heated), hair spray or glue to the build plate if needed.

- Level the bed (build platform). Perform an automatic (Assisted Bed Leveling) or manual leveling.

- Check extruder (hot end) temperature. Different filaments require different hot end temperatures.

- Check heated build plate temperature. Different filaments require different bed temperatures.

- Control the build area temperature around the printer. Eliminate all drafts.

- Layer height. Min layer height = 1/4 nozzle diameter. Max layer height = 1/2 nozzle diameter. Layer height too low might cause the filament to be pushed back into the nozzle (plugging). If layer height is too high, the layers won't stick to the build plate.

Level Build Platform

An unleveled build platform will cause many headaches during a print. You can quickly check the platform by performing the business card test: use a single business card to judge the height of the extruder nozzle over the build platform. Achieve a consistent slight resistance when you position the business card between the tip of the extruder and the bed platform for all leveling positions.

Most 3D printers have an automatic Assisted Bed Leveling feature as illustrated.

Sindoh 3DWOX Printers

Minimize Internal Support

Design the part for the printer. Use various modeling techniques in SOLIDWORKS (ribs, fillets, pads, etc.) during the design process to eliminate or to minimize the need for support and final part clean up.

Design a Water Tight Mesh

A water-tight mesh is achieved by having closed edges creating a solid volume. If you were to fill your geometry with water, would you see a leak? You may have to clean up any internal geometry that could have been left behind accidentally from Booleans.

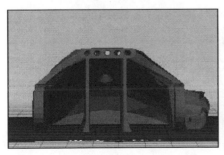

Clearance

If you are creating separate or interlocking parts, make sure there is a large enough distance between tight areas. 3D printing production makes moving parts without assembly a possibility. Take advantage of this strength by creating enough clearance that the model's pieces do not fuse together or trap support material inside.

In General (FFF Printers)

- Keep your software and firmware up to date.

- Think before you print. Design the model for your printer.

- Understand the printer's limitations. Adjust one thing at a time between prints and keep notes about the settings and effect on the print. Label test prints and take photographs.

- Control the build area and environment temperature. Eliminate all drafts and control air flow that may cause a temperature gradient within the build area.

- Level and clean the build plate before a build.

- Select the correct filament (material) for the application. Materials are still an area of active exploration.

- Set the correct extruder (hot end) and bed plate temperature.

- Most low-end 3D printers use an extruder (hot end) with Polyether ether Ketone (PEEK) or Polytetrafluoroethylene (Teflon). Both PEEK and PTFE begin to break down above 240°C and will burn and emit noxious fumes. Use an all metal hot end above 240°C.

- Select the correct part orientation.

- Suppress mates in an assembly to have the model lay flat on the build plate.

- Most parts can be printed successfully with 15 - 20% infill.

- Select the correct settings for your Slicer. If in doubt, use the factory default settings.

- Control the filament storage environment (temperature, humidity, etc.).

- General 45 degree rule. If your model doesn't have any overhangs greater than 45 degrees, you should not need support material. If the part has numerous holes, sharp edges, long run (bridge), or thin bodies, support material may be needed.

- If in doubt, create your first build with a raft and support.

- If needed, orientate the printed part on the bed plate for maximum strength (lines perpendicular to the force being applied).

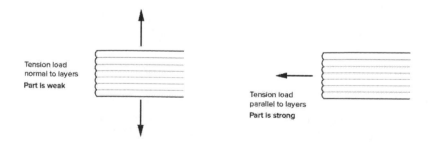

Print Directly from SOLIDWORKS

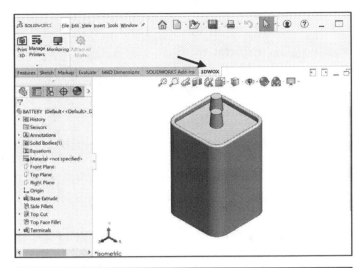

Download and install the printer drivers.

Download and install the SOLIDWORKS slicer Add-in. In this case, the SOLIDWORKS 3DWOX.

💡 Screen shots and procedure will vary depending on the 3D printer manufacture's slicer Add-in.

When the 3DWOX Add-in installation is complete, a 3DWOX tab is displayed in the CommandManager.

Open a SOLIDWORKS model.

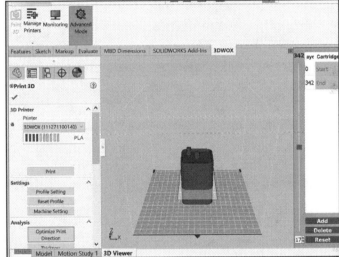

Click the 3DWOX tab. View your options:

- **Print 3D**

- **My Printers**

- **Monitoring**

- **Advanced Mode**

The **Print 3D** button provides the ability to access the slicer within SOLIDWORKS. The SOLIDWORKS part model is automatically converted into an STL file and appears on the selected (Settings button) printer bed. The Sindoh 3DWOX slicer is displayed.

The **My Printers** button provides the ability to manage your printers (printer name, IP and availability) over a network.

The **Monitoring** button provides the ability to connect to the network and view your printing real time with an internal camera.

The **Advanced Mode** button provides the ability to access the default settings, or the Advanced settings to customize the print.

Taking the Exam

Go to the 3DEXPERIENCE® Certification Center at https://3dexperience.virtualtester.com/#home.

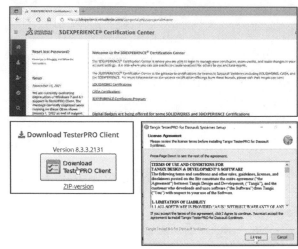

The 3DEXPERIENCE® Certification Center is where you are able to log in to manage your certificates, take certifications, and make changes to your account settings.

If your school is an academic certification provider, your instructor can allocate a free exam credit.

The instructor will require your .edu email address.

Download the TesterPRO Client.

Un-zip the Tester PRO Client.

Read the License Agreement. Agree to the Candidate Conduct Policy. Click I Agree.

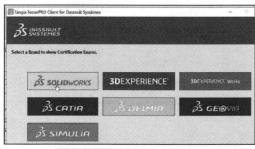

Install the Tester PRO Client. Click install. Click Finish.

Select Test Language. Click Continue.

Create an account. If you already have an account, select "I already have a VirtualTester UserID and password".

Login. Click Continue.

Click SOLIDWORKS.

Select ASSOCIATE - Additive Manufacturing.

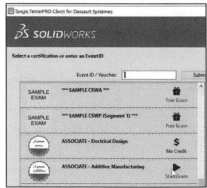

This will depend on the exam credit that your instructor set up. Click Start Exam. Select language. Click Start Examination.

Read the instructions. Agree to the Confidentiality Agreement and Candidate Conduct Policy. Click Start Examination.

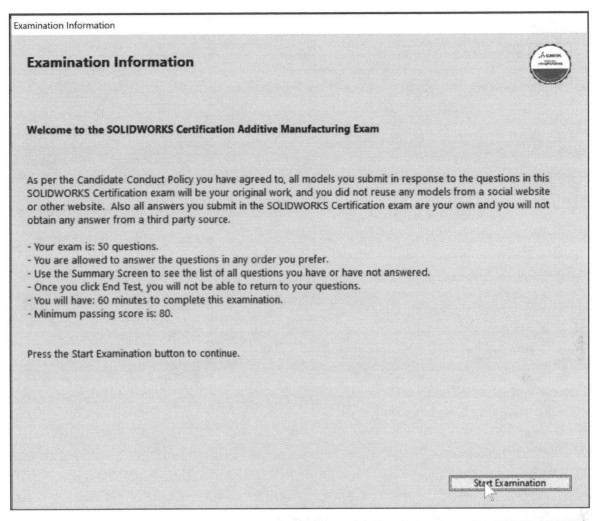

Examination Information

Examination Information

Welcome to the SOLIDWORKS Certification Additive Manufacturing Exam

As per the Candidate Conduct Policy you have agreed to, all models you submit in response to the questions in this SOLIDWORKS Certification exam will be your original work, and you did not reuse any models from a social website or other website. Also all answers you submit in the SOLIDWORKS Certification exam are your own and you will not obtain any answer from a third party source.

- Your exam is: 50 questions.
- You are allowed to answer the questions in any order you prefer.
- Use the Summary Screen to see the list of all questions you have or have not answered.
- Once you click End Test, you will not be able to return to your questions.
- You will have: 60 minutes to complete this examination.
- Minimum passing score is: 80.

Press the Start Examination button to continue.

Start Examination

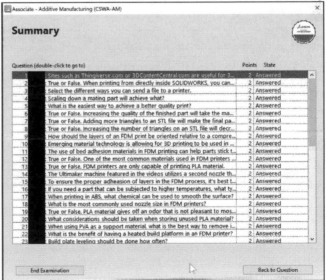

Summary

Stereolithography (SLA) was the world's first 3D printing technology. It was introduced in 1988 by 3D Systems, Inc., based on work by inventor Charles Hull.

SLA is one of the most popular resin-based 3D printing technologies for professionals today.

Selective Laser Sintering (SLS) technology is the most common additive manufacturing technology for industrial powder base applications. SLS fuses particles together layer by layer through a high energy pulse laser. Similar to SLA, this process starts with a tank full of bulk material but is in a powder form vs. liquid.

Fused filament fabrication (FFF) is an additive manufacturing technology used for building three-dimensional prototypes or models layer by layer with a range of thermoplastics.

FFF technology is the most widely used form of 3D desktop printing at the consumer level, fueled by students, hobbyists and office professionals.

The two most dominant plastics for FFF technology are PLA (Polylactic acid) and ABS (Acrylonitrile-Butadiene-Styrene).

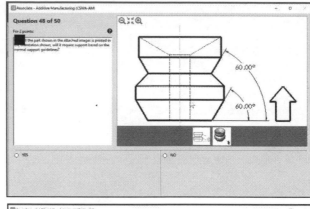

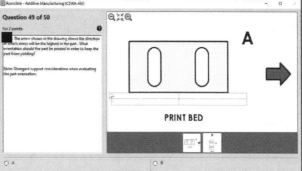

With 1000s of hours using multiple low cost FFF 3D printers, we have found that learning about additive print technology is a great experience. Students face a multitude of obstacles with their first 3D prints. Understanding what went wrong and knowing the capabilities of the 3D printer produces positive results.

Never focus too much on one single issue. These machines are complex, and trouble often arises from multiple reasons. A slipping filament may not only be caused by a bad gear drive, but also by an obstructed nozzle, a wrong feed value, a too low (or too high) temperature or a combination of all these.

Always store filament in an airtight plastic bag (container) with a few silica gel packs. Place the bag in a dry, dark, temperature controlled environment. As an extra precaution, filament manufacturers often recommend using up rolls as soon as possible.

It's best to monitor the first 3 - 5 layers when starting a print. Never assume you will have perfect bed adhesion. If there is a bed adhesion issue, check to see if the bed is level. Check the bed and extruder (hot end) temperatures and then either apply blue painter's tape (non-heated), a glue stick or hair spray. You can also add a raft to the part.

Always clean and level the build plate before every print.

As printed parts cool (PLA, ABS and Nylon), various areas of the object cool at different rates. Depending on the model being printed and the filament material, this effect can lead to warping, curling and or layer delamination.

Acetone is often used in post processing to smooth ABS, also giving the part a glossy finish. ABS can be sanded and is often machined (for example, drilled) after printing. PLA can also be sanded and machined; however, greater care is required.

For high temperature applications, ABS (glass transition temperature of 105°C) is more suitable than PLA (glass transition temperature of 60°C). PLA can rapidly lose its structural integrity as it approaches 55°C - 60°C.

Design the part for your printer. Try to use the 45-degree rule. If your model doesn't have any overhangs greater than 45 degrees, you should not need support. There are exceptions to the 45-degree rule. The most common ones are straight overhangs, and fully suspended islands.

Use various CAD modeling techniques (ribs, fillets, pads, etc.) during the design process to eliminate or to minimize the need for supports and clean up.

From a structural situation, select the correct % infill, infill pattern and orientate the part on the build plate to print the layers perpendicular in relation to the movement of the build platform.

3DXpert is a SOLIDWORKS Add-in. 3DXpert for SOLIDWORKS provides an extensive toolset to analyze, prepare and optimize your design for additive manufacturing. It also provides the ability to print an assembly file as a single part.

Key tools are:

- **Native Data Transfer**
- **Position & Modification**
- **Optimize Structure**
- **Arrange Build Plate**

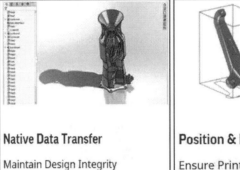

Native Data Transfer

Maintain Design Integrity

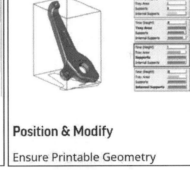

Position & Modify

Ensure Printable Geometry

Design Supports

Ensure Quality Prints with Minimal Supports

Optimize Structure

Minimize Weight & Material Usage and Apply Surface Textures

Arrange Build Plate

Best Utilization of Tray Area and Printer Time

Sample Exam Questions

These questions are examples of what to expect on the certification exam. The questions should serve as a check for your knowledge of the exam material.

1. One of the most common reasons for a Fused Filament Fabrication (FFF) print failure is build plate adhesion. Identify a few common solutions:

A. Blue painter's tape.

B. Hair spray.

C. Glue stick.

D. All of the Above.

2. The acronym FFF stands for:

A. Fused Filament Fabrication.

B. Filament Fabrication Force.

C. Force Filament Fabrication.

D. None of the above.

3. True or False: There are many materials that are being explored for 3D printing; however, the two most dominant plastics for FFF technology are PLA (Polylactic acid) and ABS (Acrylonitrile-Butadiene-Styrene).

A. True.

B. False.

4. True or False: Selective Laser Sintering (SLS) is the preferred rapid-prototyping method of metals and exotic materials.

A. True.

B. False.

5. It is recommended that you store used filament in a:

A. Airtight container.

B. Dry, dark controlled environment.

C. Near an open window.

D. No special requirements are required.

6. True or False: PLA (Polylactic Acid) is a non-biodegradable thermoplastic.

A. True.

B. False.

7. True or False: ABS (Acrylonitrile-Butadiene-Styrene) is an oil-based thermoplastic.

A. True.

B. False.

8. True or False: Thermoplastic filament comes in two standard diameters: 1.75mm and 3mm (true size of 2.85mm).

A. True.

B. False

9. What is the most command nozzle (hot end) diameter size in FFF.

A. .01mm

B. .4mm

C. .04mm

D. .001mm

10. Ultimaker 3 uses Near Field Communication (NFC) technology in their filament roll. The NFC technology provides what information to the printer?

A. Filament color.

B. Filament type.

C. Name of the machine.

D. Name of the model.

11. True or False: PVA (Polyvinyl Alcohol) filament is a water-soluble synthetic polymer.

A. True.

B. False.

12. True or False: PVA (Polyvinyl Alcohol) filament is not hygroscopic.

A. True.

B. False.

13. PVA (Polyvinyl Alcohol) filament is normally used as a water-soluble synthetic polymer. What is the best way to remove PVA from the part?

A. Time. Wait, and the PVA will fall off.

B. Submerge in hot water.

C. Leave in the sun. The sun will melt the PVA.

D. Submerge in a bath of cold circulating water.

14. Selective Laser Sintering (SLS) technology is the most common additive manufacturing technology for industrial _____ applications.

A. Power base.

B. Resin-based.

C. Thermoplastic.

D. None of the above.

15. True or False: SLA technology is a form of 3D printing in which a computer-controlled low-power, highly focused ultraviolet (UV) laser is used to create layers (layer by layer) from a liquid polymer (photopolymerization) that hardens on contact.

A. True.

B. False.

16. True or False: All SLA printed parts require a post-cure.

A. True.

B. False.

17. Formlab Form 2 recommends to wash the part with _____ in the post-cure process.

A. Distilled water.

B. Isopropyl Alcohol.

C. Acetone.

D. Nothing.

18. True or False: PLA gives off a very strong unpleasant odor.

A. True.

B. False.

19. True or False: PLA has a lower printing temperature (20C - 30°C) than ABS.

A. True.

B. False.

20. True or False: Avoid using PLA if the print is exposed to temperatures of 60°C or higher or might be bent, twisted or dropped repeatedly. ABS is a better material for this application.

A. True.

B. False.

21. True or False: ABS in general is more forgiving to large temperature fluctuations and moisture than PLA during a build cycle.

A. True.

B. False.

22. How often should you level your build plate on an FFF printer?

A. After you have a problem.

B. Never.

C. Before every build.

D. Every 10 prints.

23. When printing ABS, what chemical can you use to smooth the surface?

A. Bleach.

B. Machine oil.

C. Acetone.

D. None of the above.

24. Why do you need a heated build plate when printing ABS?

A. You do not need a heated build plate.

B. The print bed should be kept near the glass transition temperature and well below the melting temperature.

C. To keep the build area warm.

D. None of the above.

25. True or False: ABS is better suited than PLA for items that are frequently handled, dropped, or heated. It can be used for mechanical parts, especially if they are subjected to stress or must interlock with other parts.

A. True.

B. False

26. True or False: An STL file describes a raw unstructured triangulated (point cloud) surface by the unit normal and vertices (ordered by the right-hand rule) of the triangles using a three-dimensional Cartesian coordinate system.

A. True.

B. False.

27. True or False: In SOLIDWORKS, the STL Save options provide you with the ability to control the number and size of the triangles by setting the various parameters in the CAD software.

A. True.

B. False.

28. True or False: For a smoother STL file, change the Resolution to Custom. Change the deviation to 0.0005in (0.01mm). Change the angle to 5. Smaller deviations and angles produce a smoother file but increase the file size and print time.

A. True.

B. False.

29. It's best to monitor the first _____ layers when starting a print. Never assume you will have perfect bed adhesion.

A. 3 -5 layers.

B. No monitoring is required.

C. 20 - 30 layers.

D. None of the above.

30. Identify some of the ways to transport a file for the computer to the 3D printer.

A. USB.

B. Wi-Fi.

C. Ethernet.

D. All of the above.

31. True or False: Most slicers provide the ability to address rafts and supports.

A. True.

B. False.

32. True or False: The Formlab Form 2 machine does not have the ability to remotely monitor the unit using a webcam or camera. They use their Dashboard feature.

A. True.

B. False.

33. Name a few benefits of using a dual extruder in an FFF printer.

A. Ability to print more than a single filament color.

B. Ability to print a different support material.

C. Reduce overall printing time.

D. All of the above.

34. True or False: The Formlab Form 2 printer can only print in a single color at a time.

A. True.

B. False.

35. True or False: Most FFF printers and the Formlab Form 2 printer provide the ability to add material during the print cycle if needed.

A. True.

B. False.

36. True or False: 3DExpert from SOLIDWORKS is a free download.

A. True.

B. False.

37. The creation of advance part structures is aided by what type of CAD design tool?

A. 3D Help.

B. Generative Design.

C. PrintXpert.

D. Design Guidance.

38. True or False: For high temperature applications, ABS (glass transition temperature of 105°C) is more suitable than PLA (glass transition temperature of 60°C). PLA can rapidly lose its structural integrity as it approaches 60°C.

A. True.

B. False.

39. True or False: Additive manufacturing, sometimes known as *rapid prototyping*, can be slower than Subtractive manufacturing. Both take skill in creating the G-code and understanding the machine limitations.

A. True.

B. False.

40. In FFF technology, the thermoplastic material is heated in the nozzle (hot end) to form liquid, which solidifies immediately when it's deposited onto the build plate or platform. The nozzle travels at a controlled rate and moves in the X and Y direction. The build plate moves in the _____ direction. This creates mechanical adhesion (not chemical).

A. Z.

B. X, Y Z.

C. X.

D. Y, Z.

41 True or False: A Post-curing process is required for FFF technology.

A. True.

B. False.

42. True or False: In general, post-processing is the removal of supports and any needed polishing or sanding.

A. True.

B. False.

43. True or False: In SLA technology, after the IPA wash and dry, perform a post-cure. At a basic level, exposure to sun light triggers the formation of additional chemical bonds within a printed part, making the material stronger and stiffer.

A. True.

B. False.

44. True or False: To ensure proper post-curing, Formlabs recommends their post-curing unit Form Cure. Form Cure precisely combines temperature and 405 nm light to post-cure parts for peak material performance.

A. True.

B. False.

45. True or False: In SLA technology, parts are watertight and fully dense. No infill is required.

A. True.

B. False.

46. True or False: SLS technology is the preferred rapid-prototyping method of metals and exotic materials.

A. True.

B. False.

47. True or False: In FFF technology, the higher the infill percentage, the more material and longer the print time. It will also increase weight and material cost.

A. True.

B. False.

48. True or False: In general with FFF technology, use a 10% - 15% infill with a maximum infill of 60%. 100% infill is not recommended. Part warping can be a concern.

A. True.

B. False.

49. True or False: Part orientation is important on build strength and the amount of raft and support material required for the build. Incorrect part orientations can lead to poor strength, warping, curling, and delamination.

A. True.

B. False.

50. True or False: If maximizing strength is an issue, select the part orientation on the build plate so that the "grain" of the print is oriented to maximize the strength of the part.

A. True.

B. False.

51. True or False: Parts with a lower percentage of infill should have a lower internal force between layers and can reduce the chance of curling, cracking, and layer delamination along with a lower build cost and time.

A. True.

B. False.

52. True or False: In FFF technology, for a consistent quality build, control the build area and environment temperature. Eliminate all drafts and control air flow that may cause a temperature gradient within the build area.

A. True.

B. False.

53. _____ technology is the most common additive manufacturing technology for industrial powder base applications.

A. SLS.

B. FFF.

C. ABS.

D. None of the above.

54. True or False: In FFF technology, you do not need a heated bed (platform) when using PLA.

A. True.

B. False.

55. True or False: In FFF technology, you do not need a heated bed (platform) when using ABS.

A. True.

B. False.

56. True or False: In FFF technology, a raft is a horizontal latticework of filament located underneath the part.

A. True.

B. False.

57. In FFF technology, it's best to monitor the first 3 - 5 layers when starting a print. Never assume you will have perfect bed adhesion. If there is a bed adhesion issue, check to see if the bed is level. Check the bed and extruder (hot end) temperatures and then either apply.

A. Blue painter's tape (non-heated).

B. Glue stick.

C. Hair spray.

D. None of the above.

58. _____ is often used in post processing when using ABS to provide a smooth glossy finish.

A. Motor oil.

B. Water.

C. Acetone.

D. None of the above.

59. True or False: Scaling a part in your slicer can have a negative consequence for mating parts.

A. True.

B. False.

60. True or False: FFF technology builds a part that have inherently anisotropic properties, meaning they are much stronger in the XY direction than the Z direction.

A. True.

B. False.

61. True or False: For functional parts, it is important to consider the application and the direction of the loads. FFF technology parts are much more likely to delaminate and fracture when placed in tension in the Z direction compared to the XY directions (up to 4-5 times difference tensile strength).

A. True.

B. False.

62. True or False: Scaling the part in a slicer could have a negative effect on the overall wall thickness.

A. True.

B. False.

63. True or False: Most low end 3D FFF printers use an extruder (hot end) with Polyether ether Ketone (PEEK) or Polytetrafluoroethylene (Teflon). Both PEEK and PTFE begin to breakdown above 240°C and will burn and emit noxious fumes. Use an all metal hot end above 240°C.

A. True.

B False.

64. True or False: 3DXpert is a SOLIDWORKS Add-in. 3DXpert for SOLIDWORKS provides an extensive toolset to analyze, prepare and optimize your design for additive manufacturing. It also provides the ability to print an assembly file as a single part.

A. True.

B. False.

65. True or False: The SOLIDWORKS STL Save options allow you to control the number and size of the triangles by setting the various parameters in the CAD software.

A. True.

B. False.

66. True or False: An STL (*.stl) file describes only the surface geometry of a three dimensional object without any representation of color, texture, or other common CAD model attributes. The STL format specifies both ASCII and Binary representations.

A. True.

B. False.

67. True or False: Additive Manufacturing (*.amf) is a file format that includes the materials that have been applied to the parts or bodies in the 3D model.

A. True.

B. False.

68. The file format (*.3mf) stands for:

A. Nothing.

B. 3D Manufacturing Format.

C. 2D Manufacturing Format.

D. 3D Subtractive Format.

69. True or False: The Formlab Form 2 SLA printer has the ability to print flexible material.

A. True.

B. False.

70. True or False: The Formlab Form 2 SLA printer resin only comes in one color.

A. True.

B. False.

71. True or False: Ultimaker Cura - Version: 3.1.0 or newer provides the ability to import SOLIDWORKS part and assembly files directly into the build area.

A. True.

B. False.

72. True or False: When designing your part for 3D printing, try to use the 45 degree rule. If your model has overhangs greater than 45 degrees, you need support material.

A. True.

B. False.

73. True or False: The Formlabs Form 2 uses the PreForm Software. PreForm provides the ability to select the One-Click Print option. The One-Click option automatically optimizes the part orientation and support locations.

A. True.

B. False.

74. True or False: The Formlabs PreForm software only supports STL and OBJ files.

A. True.

B. False.

75. True or False: Using the 45 degree rule in FFF technology. If your model doesn't have any overhangs greater than 45 degrees, you should not need support. There are exceptions to the 45 degree rule. The most common ones are straight overhangs, and fully suspended islands.

A. True.

B. False.

76. The two standard diameters for Thermoplastic filaments are:

A. 1mm.

B. 1.75mm.

C. 3mm (true size 2.85mm).

D. 4mm.

77. True or False: In SLA technology, you have the ability to customize percent infill and infill pattern type for the print.

A. True.

B. False.

78. True or False: In FFF technology, you have the ability to customize percent infill and infill pattern type for the print.

A. True.

B. False.

79. True or False: In FFF technology, darker colors and glow in the dark filament materials, often requires higher extruder temperatures (5C - 10°C).

A. True.

B. False.

80. True or False: Unlike its predecessor STL (*.stl) format, Additive Manufacturing Format (*.amf) has native support for color, materials, lattices, and constellation (groups).

A. True.

B. False.

81. True or False: Unlike its predecessor STL (*.stl) format, 3D Manufacturing Format (*.3mf) has native support for color, materials, lattices, and constellation (groups).

A. True.

B. False.

82. True or False: SOLIDWORKS's Design Guidance functionality produces much more organic and unique geometry than if modeled by the user.

A. True.

B. False.

83. What is the main advantage of using a heated bed for ABS?

A. Print faster.

B. Better print quality.

C. Helps the part from cooling too fast and deforming.

D. None of the above.

84. True or False: The Ultimaker and Formlabs Form 2 provides the ability to add material midway through a print if needed.

A. True.

B. False.

85. True or False: The Ultimaker and Formlabs Form 2 provides the ability to remotely monitor the machine using a built-in webcam\camera.

A. True.

B. False.

86. True or False: It is important to ensure that there is a drain hole when printing a hollow part using an SLA machine. The location of the drain hole is important based on its printing orientation.

A. True.

B. False.

87. True or False: Once the part is completed in the Formlabs Form 2 printer, the remaining resin in the tank should be discarded due to contamination issues.

A. True.

B. False.

88. True or False: Increasing the number of triangles in an STL file will decrease the overall print time.

A. True.

B. False.

89. True or False: The Formlabs Form 2 printer uses a Dashboard to remotely monitor the printer. It also tracks material usage and explore past and future purchases.

A. True.

B. False.

90. True or False: Increasing the number of triangles in an STL file will decrease the overall print time.

A. True.

B. False.

91. True or False: You are using an SLA machine to print a hollow part. It is very important to design a drain hole at the bottom of the part based on it printing orientation.

A. True.

B. False.

92. You require a part to be strong, flexible, and to stand up to heat. What material should you use in a FDM printer?

A. PLA.

B. ABS.

C. Nylon 618.

D. PVA.

93. True or False: In SOLIDWORKS for a smoother STL file, change the Resolution to Custom. Change the deviation to 0.0005in (0.01mm). Change the angle to 5. Smaller deviations and angles produce a smoother file but increase the file size and print time.

A. True.

B. False.

94. True or False: The part displayed is perpendicular to the build plate. The part will not require support material based on standard support guidelines.

A. True.

B. False.

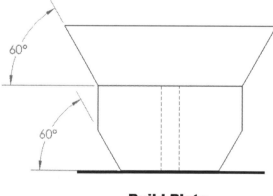

Build Plate

95. The arrow shown in the picture displays the direction in which shear stress will be the highest in the part. What orientation should you print the part in order to keep the part from yielding? Disregard support considerations in this example.

A. A.

B. B.

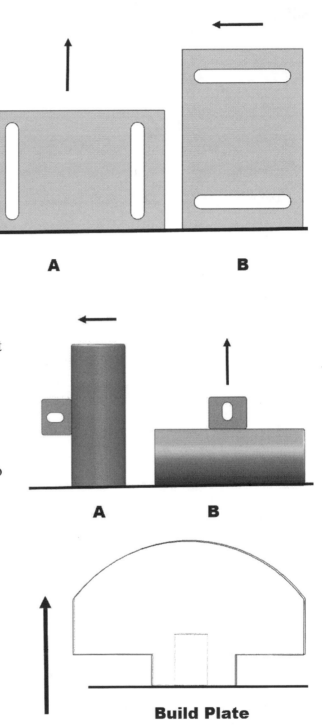

96. The arrow shown in the picture displays the direction in which shear stress will be the highest in the part. What orientation should you print the part in order to keep the part from yielding? Disregard support considerations in this example.

A. A.

B. B.

97. The part displayed is perpendicular to the build plate. The part will **not** require support material based on the standard support guidelines.

A. True.

B. False.

Build Plate

Notes:

Appendix

SOLIDWORKS Keyboard Shortcuts

Below are some of the pre-defined keyboard shortcuts in SOLIDWORKS:

Action:	Key Combination:
Model Views	
Rotate the model horizontally or vertically	**Arrow** keys
Rotate the model horizontally or vertically 90 degrees	**Shift + Arrow** keys
Rotate the model clockwise or counterclockwise	**Alt** + left of right **Arrow** keys
Pan the model	**Ctrl + Arrow** keys
Magnifying glass	**g**
Zoom in	**Shift + z**
Zoom out	**z**
Zoom to fit	**f**
Previous view	**Ctrl + Shift + z**
View Orientation	
View Orientation menu	**Spacebar**
Front view	**Ctrl + 1**
Back view	**Ctrl + 2**
Left view	**Ctrl + 3**
Right view	**Ctrl + 4**
Top view	**Ctrl + 5**
Bottom view	**Ctrl + 6**
Isometric view	**Ctrl + 7**
NormalTo view	**Ctrl + 8**
Selection Filters	
Filter edges	**e**
Filter vertices	**v**
Filter faces	**x**
Toggle Selection Filter toolbar	**F5**
Toggle selection filters on/off	**F6**
File menu items	
New SOLIDWORKS document	**Ctrl + n**
Open document	**Ctrl + o**
Open From Web Folder	**Ctrl + w**
Make Drawing from Part	**Ctrl + d**
Make Assembly from Part	**Ctrl + a**
Save	**Ctrl +s**
Print	**Ctrl + p**
Additional items	
Access online help inside of PropertyManager or dialog box	**F1**
Rename an item in the FeatureManager design tree	**F2**

Action:	Key Combination:
Rebuild the model	**Ctrl + b**
Force rebuild - Rebuild the model and all its features	**Ctrl + q**
Redraw the screen	**Ctrl + r**
Cycle between open SOLIDWORKS document	**Ctrl + Tab**
Line to arc/arc to line in the Sketch	**a**
Undo	**Ctrl + z**
Redo	**Ctrl + y**
Cut	**Ctrl + x**
Copy	**Ctrl + c**
Paste	**Ctrl + v**
Delete	**Delete**
Next window	**Ctrl + F6**
Close window	**Ctrl + F4**
View previous tools	**s**
Selects all text inside an Annotations text box	**Ctrl + a**

In a sketch, the **Esc** key un-selects geometry items currently selected in the Properties box and Add Relations box.

In the model, the **Esc** key closes the PropertyManager and cancels the selections.

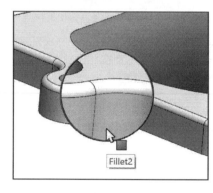

Use the **g** key to activate the Magnifying glass tool. Use the Magnifying glass tool to inspect a model and make selections without changing the overall view.

Use the **s** key to view/access previous command tools in the Graphics window.

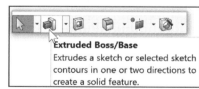

Modeling - Best Practices

Best practices are simply ways of bringing about better results in easier, more reliable ways. The Modeling - Best Practice list is a set of rules helpful for new users and users who are trying to experiment with the limits of the software.

These rules are not inflexible, but conservative starting places; they are concepts that you can default to, but that can be broken if you have good reason. The following is a list of suggested best practices:

- Create a folder structure (parts, drawings, assemblies, simulations, etc.). Organize into project or file folders.

- Construct sound document templates. The document template provides the foundation that all models are built on. This is especially important if working with other SOLIDWORKS users on the same project; it will ensure consistency across the project.

- Generate unique part filenames. SOLIDWORKS assemblies and drawings may pick up incorrect references if you use parts with identical names.

- Apply Custom Properties. Custom Properties is a great way to enter text-based information into the SOLIDWORKS parts. Users can view this information from outside the file by using applications such as Windows Explorer, SOLIDWORKS Explorer, and Product Data Management (PDM) applications.

- Understand part orientation. When you create a new part or assembly, the three default Planes (Front, Right and Top) are aligned with specific views. The plane you select for the Base sketch determines the orientation.

- Learn to sketch using automatic relations.

- Limit your usage of the Fixed constraint.

- Add geometric relations, then dimensions in a 2D sketch. This keeps the part from having too many unnecessary dimensions. This also helps to show the design intent of the model. Dimension what geometry you intend to modify or adjust.

- Fully define all sketches in the model. However, there are times when this is not practical, generally when using the Spline tool to create a freeform shape.

- When possible, make relations to sketches or stable reference geometry, such as the Origin or standard planes, instead of edges or faces. Sketches are far more stable than faces, edges, or model vertices, which change their internal ID at the slightest change and may disappear entirely with fillets, chamfers, split lines, and so on.

- Do not dimension to edges created by fillets or other cosmetic or temporary features.

- Apply names to sketches, features, dimensions, and mates that help to make their function clear.

- When possible, use feature fillets and feature patterns rather than sketch fillets and sketch patterns.

- Apply the Shell feature before the Fillet feature, and the inside corners remain perpendicular.

- Apply cosmetic fillets and chamfers last in the modeling procedure.

- Combine fillets into as few fillet features as possible. This enables you to control fillets that need to be controlled separately, such as fillets to be removed and simplified configurations.

- Create a simplified configuration when building very complex parts or working with large assemblies.

- Use symmetry during the modeling process. Utilize feature patterns and mirroring when possible. Think End Conditions.

- Use global variables and equations to control commonly applied dimensions (design intent).

- Add comments to equations to document your design intent. Place a single quote (') at the end of the equation, then enter the comment. Anything after the single quote is ignored when the equation is evaluated.

- Avoid redundant mates. Although SOLIDWORKS allows some redundant mates (all except distance and angle), these mates take longer to solve and make the mating scheme harder to understand and diagnose if problems occur.

- Fix modeling errors in the part or assembly when they occur. Errors cause rebuild time to increase, and if you wait until additional errors exist, troubleshooting will be more difficult.

- Create a Library of Standardized notes and parts.

- Utilize the Rollback bar. Troubleshoot feature and sketch errors from the top of the design tree.

- Determine the static and dynamic behavior of mates in each sub-assembly before creating the top-level assembly.

- Plan the assembly and sub-assemblies in an assembly layout diagram. Group components together to form smaller sub-assemblies.

- When you create an assembly document, the base component should be fixed, fully defined or mated to an axis about the assembly origin.

- In an assembly, group fasteners into a folder at the bottom of the FeatureManager. Suppress fasteners and their assembly patterns to save rebuild time and file size.

- When comparing mass, volume and other properties with assembly visualization, utilize similar units.

- Use limit mates sparingly because they take longer to solve and whenever possible, mate all components to one or two fixed components or references. Long chains of components take longer to solve and are more prone to mate errors.

Helpful On-line Information

The SOLIDWORKS URL:
http://www.SOLIDWORKS.com
contains information on Local
Resellers, Solution Partners,
Certifications, SOLIDWORKS
user's groups and more.

Access 3D ContentCentral using
the Task Pane to obtain
engineering electronic catalog model and part information.

Use the SOLIDWORKS Resources tab in the Task Pane to
obtain access to Customer Portals, Discussion Forums, User
Groups, Manufacturers, Solution Partners, Labs and more.

Helpful on-line SOLIDWORKS information is available
from the following URLs:

- http://www.swugn.org/

List of all SOLIDWORKS User groups.

- https://www.solidworks.com/sw/educatio
  n/certification-programs-cad-
  students.htm

The SOLIDWORKS Academic Certification
Programs.

- http://www.solidworks.com/sw/in
  dustries/education/engineering-
  education-software.htm

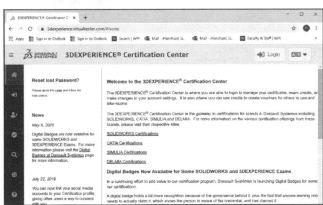

To obtain additional SOLIDWORKS
Certification exam information, visit
https://3dexperience.virtualtester.com
/#home

*On-line tutorials are for educational
purposes only. Tutorials are copyrighted by
their respective owners.

SOLIDWORKS Document Types

SOLIDWORKS has three main document file types: Part, Assembly and Drawing, but there are many additional supporting types that you may want to know. Below is a brief list of these supporting file types:

Design Documents	**Description**
.sldprt	SOLIDWORKS Part document
.slddrw	SOLIDWORKS Drawing document
.sldasm	SOLIDWORKS Assembly document

Templates and Formats	**Description**
.asmdot	Assembly Template
.asmprp	Assembly Template Custom Properties tab
.drwdot	Drawing Template
.drwprp	Drawing Template Custom Properties tab
.prtdot	Part Template
.prtprp	Part Template Custom Properties tab
.sldtbt	General Table Template
.slddrt	Drawing Sheet Template
.sldbombt	Bill of Materials Template (Table-based)
.sldholtbt	Hole Table Template
.sldrevbt	Revision Table Template
.sldwldbt	Weldment Cutlist Template
.xls	Bill of Materials Template (Excel-based)

Library Files	**Description**
.sldlfp	Library Part file
.sldblk	Blocks

Other	**Description**
.sldstd	Drafting standard
.sldmat	Material Database
.sldclr	Color Palette File
.xls	Sheet metal gauge table

Appendix: Check your understanding Answer key

The Appendix contains the answers to the questions at the end of the **CSWA**, **CSWA-SD**, **CSWA-S**, **CSWA-AM** chapters.

CSWA Chapter 2

1. Identify the illustrated Drawing view.

The correct answer is B: Alternative Position View

2. Identify the illustrated Drawing view.

The correct answer is B: Break View.

3. Identify the illustrated Drawing view.

The correct answer is D: Aligned View.

4. Identify the view procedure. To create the following view, you need to insert a:

The correct answer is B: Closed Profile: Spline.

5. Identify the view procedure. To create the following view, you need to insert a:

The correct answer is B: Closed Spline .

6. Identify the illustrated view type.

The correct answer is A: Crop View.

7. To create View B from Drawing View A insert which View Type?

The correct answer is Aligned Section View.

8. To create View B it is necessary to sketch a closed spline on View A and insert which View type?

The correct answer is Broken out Section View.

9. To create View B it is necessary to sketch a closed spline on View A and insert which View type?

The correct answer is Horizontal Break View.

CSWA Chapter 3

1. Calculate the overall mass of the part, volume, and locate the Center of mass with the provided information using the provided Option 1 FeatureManager.

   - Overall mass of the part = 1105.00 grams

   - Volume of the part = 130000.00 cubic millimeters

- Center of Mass Location: X = 43.46 millimeters, Y = 15.00 millimeters, Z = -37.69 millimeters

2. Calculate the overall mass of the part, volume, and locate the Center of mass with the provided information using the provided Option 3 FeatureManager.

   - Overall mass of the part = 269.50 grams

   - Volume of the part = 192500.00 cubic millimeters

   - Center of Mass Location: X = 35.70 millimeters, Y = 27.91 millimeters, Z = -1.46 millimeters

3. Calculate the overall mass of the part, volume, and locate the Center of mass with the provided information.

   - Overall mass of the part = 1.76 pounds

   - Volume of the part = 17.99 cubic inches

   - Center of Mass Location: X = 0.04 inches, Y = 0.72 inches, Z = 0.00 inches

4. Calculate the overall mass of the part, volume, and locate the Center of mass with the provided illustrated information.

   - Overall mass of the part = 1280.91 grams

   - Volume of the part = 474411.54 cubic millimeters

   - Center of Mass Location: X = 0.00 millimeters, Y = -29.17 millimeters, Z = 3.18 millimeters

5. Calculate the overall mass of the part, volume, and locate the Center of mass with the provided information.

   - Overall mass of the part = 248.04 grams

   - Volume of the part = 91868.29 cubic millimeters

   - Center of Mass Location: X = -51.88 millimeters, Y = 24.70 millimeters, Z = 29.47 millimeters

6. Calculate the overall mass of the part with the provided information.

   - Overall mass of the part = 3015.53 grams

7. Calculate the overall mass of the part with the provided information.

   - Overall mass of the part = 319.13 grams

8. Calculate the overall mass of the part, volume, and locate the Center of mass with the provided information.

   - Overall mass of the part = 0.45 pounds. Volume of the part = 4.60 cubic inches. Center of Mass Location: X = 0.17 inches, Y = 0.39 inches, Z = 0.00 inches

9. Calculate the overall mass of the part, volume, and locate the Center of mass with the provided information.

   • Overall mass of the part = 0.28 pounds. Volume of the part = 2.86 cubic inches. Center of Mass Location: X = 0.70 inches, Y = 0.06 inches, Z = 0.00 inches

CSWA Chapter 4

1. Calculate the overall mass of the part, volume, and locate the Center of mass with the provided information.

   • Overall mass of the part = 1.99 pounds

   • Volume of the part = 6.47 cubic inches

   • Center of Mass Location: X = 0.00 inches, Y = 0.00 inches, Z = 1.49 inches

2. Calculate the overall mass of the part, volume, and locate the Center of mass with the provided information.

   • Overall mass of the part = 279.00 grams

   • Volume of the part = 103333.73 cubic millimeters

   • Center of Mass Location: X = 0.00 millimeters, Y = 0.00 millimeters, Z = 21.75 millimeters

3. Calculate the overall mass of the part, volume, and locate the Center of mass with the provided information.

   • Overall mass of the part = 1087.56 grams

   • Volume of the part = 122198.22 cubic millimeters

   • Center of Mass Location: X = 44.81 millimeters, Y = 21.02 millimeters, Z = -41.04 millimeters

4. Calculate the overall mass of the part, volume and locate the Center of mass with the provided information.

   • Overall mass of the part = 2040.57 grams

   • Volume of the part = 755765.04 cubic millimeters

   • Center of Mass Location: X = -0.71 millimeters, Y = 16.66 millimeters, Z = -9.31 millimeters

5. Calculate the overall mass of the part, volume and locate the Center of mass with the provided information. Create Coordinate System1 to locate the Center of mass for the model.

   • Overall mass of the part = 2040.57 grams

   • Volume of the part = 755765.04 cubic millimeters

- Center of Mass Location: X = 49.29 millimeters, Y = 16.66 millimeters, Z = -109.31 millimeters

6. Calculate the overall mass of the part, volume and locate the Center of mass with the provided information.

   - Overall mass of the part = 37021.48 grams

   - Volume of the part = 13711657.53 cubic millimeters

   - Center of Mass Location: X = 0.00 millimeters, Y = 0.11 millimeters, Z = 0.00 millimeters

7. Calculate the overall mass of the part, volume and locate the Center of mass with the provided information.

   - Overall mass of the part = 37021.48 grams

   - Volume of the part = 13711657.53 cubic millimeters

   - Center of Mass Location: X = 225.00 millimeters, Y = 70.11 millimeters, Z = -150.00 millimeters

CSWA Chapter 5

1. Calculate the overall mass and volume of the assembly. Locate the Center of mass using the illustrated coordinate system.

   - Overall mass of the assembly = 843.22 grams

   - Volume of the assembly = 312304.62 cubic millimeters

   - Center of Mass Location: X = 30.00 millimeters, Y = 40.16 millimeters, Z = -53.82 millimeters

2. Calculate the overall mass and volume of the assembly. Locate the Center of mass using the illustrated coordinate system.

   1. Overall mass of the assembly = 19.24 grams

   2. Volume of the assembly = 6574.76 cubic millimeters

   3. Center of Mass Location: X = 40.24, Y = 24.33, Z = 20.75

3. Calculate the overall mass and volume of the assembly. Locate the Center of mass using the illustrated coordinate system.

   4. Overall mass of the assembly = 19.24 grams

   5. Volume of the assembly = 6574.76 cubic millimeters

   6. Center of Mass Location: X = 40.24, Y = -20.75, Z = 24.33

CSWA-SD Chapter 6

1. Environmental Product Declarations, or EPDs, are an increasingly used method for communicating sustainability results with:

A. **Suppliers and Customers**

B. Engineers

C. Managers

D. None of the Above

2. The commonly referenced definition of sustainable development put forth by the Brundtland Commission reads as follows:

A. "Sustainability requires closed material loops and energy independence"

B. **"Sustainable development is development that meets the needs of the present without compromising the ability of future generations to meet their own needs"**

C. "Sustainable development is the use of environmental claims in marketing"

D. None of the above

3. The study of sustainable development broadly covers these three elements:

A. Land, air, and water

B. Natural, man-made, hybrid

C. **Environment, social equity, economics**

D. Animal, vegetable, mineral

4. This answer choice is NOT part of a long-term, working definition of a "sustainable company" ideal:

A. Generates wastes that are useful as inputs by industry or nature

B. Sources recycled waste material and minimal virgin resources

C. **Follows all current environmental regulations**

D. Uses minimal energy that is ultimately from renewable sources

5. "The intelligent application of the principles of sustainability to the realm of engineering and design" is a working definition for the following concept:

A. **Sustainable design**

B. Sustainable business

C. Life cycle assessment

D. SOLIDWORKS Sustainability

6. A focus on product design that ensures the ultimate recyclability of a product you're developing is a sustainable design technique most specifically called:

A. Design for Environment (DfE)

B. **Design for Disassembly (DfD)**

C. Life Cycle Assessment (LCA)

D. Design for Total Life Assessment (TLA)

7. The sustainable design technique that promotes systematically using natural inspiration and technologies found in nature to design products is known as:

A. **Biomimicry**

B. Cradle to Cradle

C. Environmental Management System (EMS)

D. Intelligent Design

8. The sustainable design technique that can most simply be characterized by the concept that the waste from one entity equals the food of another is:

A. **Cradle to Cradle**

B. Design for Disassembly (DfD)

C. Life Cycle Assessment (LCA)

D. Intelligent Design

9. The sustainable design technique that focuses on re-formulating the raw materials we use to design out their toxicity and environmental impacts is known as:

A. **Green chemistry**

B. Design for Environment (DfE)

C. Life cycle assessment (LCA)

D. Cradle to cradle

10. The following is an example of green marketing:

A. A brochure of a product painted green, printed on 100% post-consumer recycled paper

B. An ad touting the cost savings you can get from driving an efficient vehicle

C. **A label that indicates how many trees will be saved by purchasing this product**

D. None of the above

11. A green product is defined as one that:

A. Is made of 100% recycled content, and is itself recyclable

B. Uses no energy or only renewable energy

C. Has been designed using SOLIDWORKS Sustainability

D. **There is no such thing as a green product - the only "green" product is the one that's never made**

12. LCA stands for:

A. Life Cycle Analysis, because LCA is an exact science, similar to Finite Element Analysis (FEA)

B. **Life Cycle Assessment, because LCA is an approximate and pragmatic method, like medicine**

C. Left Cymbal Assassination, because LCA practitioners rove the world destroying half of all percussion equipment

D. None of the above

13. Photochemical oxidation (smog) and ozone layer depletion are examples of environmental impacts that fall into the following domain:

A. **Air impacts**

B. Terrestrial & aquatic impacts

C. Natural resource depletion

D. Climate effects

14. The "global warming potential" (GWP) from greenhouse gases emitted throughout a product's lifecycle, such as carbon dioxide and methane, is a measure of the product's tendency to affect:

A. Human toxicity

B. **Climate change**

C. Ionizing radiation

D. Air acidification

15. The following: "(1) raw material extraction, (2) material processing, (3) part manufacturing, (4) assembly, (5) transportation, (6) product use, and (7) end of life" describes a product's:

A. Environmental indicators

B. Metrics

C. **Lifecycle stages**

D. Good times

CSWA-S Chapter 7

1. What is the Modulus of Elasticity?

- The slope of the Deflection-Stress curve.

- **The slope of the Stress-Strain curve in its linear section**.

- The slope of the Force-Deflection curve in its linear section.

- The first inflection point of a Strain curve.

2. What is Stress?

- A measure of power.

- A measure of strain.

- A measure of material strength.

- **A measure of the average amount of force exerted per unit area**.

3. Which of the following assumptions are true for a static analysis in SOLIDWORKS Simulation with small displacements?

- **Inertia effects are negligible and loads are applied slowly**.

- The model is not fully elastic. If loads are removed, the model will not return to its original position.

- **Results are proportional to loads**.

- **All the displacements are small relative to the model geometry**.

4. What is Yield Stress?

- **The stress level beyond which the material becomes plastic**.

- The stress level beyond which the material breaks.

- The strain level above the stress level which the material breaks.

- The stress level beyond the melting point of the material.

5. A high quality Shell element has _____ nodes.

- 4
- 5
- **6**
- 8

6. Stress σ is proportional to _____ in a Linear Elastic Material.

- **Strain**.
- Stress.
- Force.
- Pressure.

7. The Elastic Modulus (Young's Modulus) is the slope defined as _____ divided by
_____.

- Strain, Stress.
- **Stress, Strain**.
- Stress, Force.
- Force, Area.

8. Linear static analysis assumes that the relationship between loads and the induced response is _____.

- Flat.
- **Linear**.
- Doubles per area.
- Translational.

9. In SOLIDWORKS Simulation, the Factor of Safety (FOS) calculations are based on one of the following failure criteria.

- **Maximum von Mises Stress**.
- **Maximum shear stress (Tresca)**.
- **Mohr-Coulomb stress**.
- **Maximum Normal stress**.

10. The Yield point is the point where the material begins to deform at a faster rate than at the elastic limit. The material behaves _____ in the Plastic Range.

- Flatly.
- Linearly.
- **Non-Linearly**.
- Like a liquid.

11. What are the Degrees of Freedom (DOFs) restrained for a Solid?

- None.
- **3 Translations**.
- 3 Translations and 3 Rotations.
- 3 Rotations.

12. What are the Degrees of Freedom (DOFs) restrained for Truss joints?

- None.
- **3 Translations**.
- 3 Translations and 3 Rotations
- 3 Rotations.

13. What are the Degrees of Freedom (DOFs) restrained for Shells and Beams?

- None.
- 3 Translations.
- **3 Translations and 3 Rotations**.
- 3 Rotations.

14. Which statements are true for Material Properties using SOLIDWORKS Simulation?

- **For solid assemblies, each component can have a different material**.
- For shell models, each shell cannot have a different material and thickness.
- **For shell models, the material of the part is used for all shells**.
- For beam models, each beam cannot have a different material.

15. A Beam element has _____ nodes (one at each end) with _____ degrees of freedom per node plus_____ node to define the orientation of the beam cross section.

- 6, 3, 1

- 3, 3, 1

- **3, 6, 1**

- None of the above.

16. A Truss element has _____ nodes with _____ translational degrees of freedom per node.

- **2, 3**

- 3, 3

- 6, 6

- 2, 2

17. In general, the finer the mesh the better the accuracy of the results.

- **True**.

- False.

18. How does SOLIDWORKS Simulation automatically treat a Sheet metal part with uniform thickness?

- **Shell**.

- Solid.

- Beam.

- Mixed Mesh.

19. Use the mesh and displacement plots to calculate the distance between two _____ using SOLIDWORKS Simulation.

- **Nodes**.

- Elements.

- Bodies.

- Surfaces.

20. Surface models can only be meshed with _____ elements.

- **Shell**.
- Beam.
- Mixed Mesh.
- Solid.

21. The shell mesh is generated on the surface (located at the mid-surface of the shell).

- **True**.
- False.

22. In general, use Thin shells when the thickness-to-span ratio is less than _____.

- **0.05**
- .5
- 1
- 2

23. The model (a rectangular plate) has a length to thickness ratio of less than 5. You extracted its mid-surface to use it in SOLIDWORKS Simulation. You should use a _____.

- Thin Shell element formulation.
- **Thick Shell element formulation**.
- Thick or Thin Shell element formulation, it does not matter.
- Beam Shell element formulation.

24. The model, a rectangular sheet metal part, uses SOLIDWORKS Simulation. You should use a:

- **Thin Shell element formulation**.
- Thick Shell element formulation.
- Thick or Thin Shell element formulation, it does not matter.
- Beam Shell element formulation.

25. The Global element size parameter provides the ability to set the global average element size. SOLIDWORKS Simulation suggests a default value based on the model volume and _____ area. This option is only available for a standard mesh.

- Force.

- Pressure.

- **Surface**.

- None of the above.

26. A remote load applied on a face with a Force component and no Moment can result in: Note: Remember (DOFs restrain).

- **A Force and Moment of the face**.

- A Force on the face only.

- A Moment on the face only.

- A Pressure and Force on the face.

27. There are _____ DOFs restrained for a Solid element.

- **3**

- 1

- 6

- None

28. There are _____ DOFs restrained for a Beam element.

- 3

- 1

- **6**

- None

29. What best describes the difference(s) between a Fixed and Immovable (No translation) boundary condition in SOLIDWORKS Simulation?

- There are no differences.

- There are no difference(s) for Shells but it is different for Solids.

- **There is no difference(s) for Solids but it is different for Shells and Beams**.

- There are only differences(s) for a Static Study.

30. Can a non-uniform pressure or force be applied on a face using SOLIDWORKS Simulation?

- No.

- Yes, but the variation must be along a single direction only.

- **Yes. The variation can be in two directions and is described by a binomial equation**.

- Yes, but the variation must be linear.

31. You are performing an analysis on your model. You select five faces, 3 edges and 2 vertices and apply a force of 20lbf. What is the total force applied to the model using SOLIDWORKS Simulation?

- 100lbf

- 1600lbf

- 180lbf

- **200lbf**

32. Yield strength is typically determined at _____ strain.
- 0.1%

- **0.2%**

- 0.02%

- 0.002%

33. There are four key assumptions made in Linear Static Analysis: 1. Effects of inertia and damping are neglected, 2. The response of the system is directly proportional to the applied loads, 3. Loads are applied slowly, gradually, and_____ .

- **Displacements are very small. The highest stress is in the linear range of the stress-strain curve**.

- There are no loads.

- Material is not elastic.

- Loads are applied quickly.

34. How many degrees of freedom does a physical structure have?

- Zero.

- Three - Rotations only.

- Three - Translations only.

- **Six - Three translations and three rotational**.

35. What criteria are best suited to check the failure of brittle material in SOLIDWORKS Simulation?

- Maximum von Mises Stress and Maximum Shear Stress criteria.

- Maximum von Misses Stress and Maximum Shear Strain criteria.

- Maximum von Misses Stress and Maximum Normal Stress criteria.

- **Mohr-Coulomb Stress and Maximum Normal Stress criteria**.

36. You are performing an analysis on your model. You select three faces and apply a force of 40lb. What is the total force applied to the model using SOLIDWORKS Simulation?

- 40lb

- 20lb

- **120lb**

- Additional information is required.

37. A material is orthotropic if its mechanical or thermal properties are not unique and independent in three mutually perpendicular directions.

- True

- **False**

38. An increase in the number of elements in a mesh for a part will:

- Decrease calculation accuracy and time.

- **Increase calculation accuracy and time**.

- Have no effect on the calculation.

- Change the FOS below 1.

39. SOLIDWORKS Simulation uses the von Mises Yield criteria to calculate the Factor of Safety of many ductile materials. According to the criteria:

- **Material yields when the von Mises stress in the model equals the yield strength of the material.**

- Material yields when the von Mises stress in the model is 5 times greater than the minimum tensile strength of the material.

- Material yields when the von Mises stress in the model is 3 times greater than the FOS of the material.

- None of the above.

40. SOLIDWORKS Simulation calculates structural failure on:

- Buckling.

- Fatigue.

- Creep.

- **Material yield.**

41. Apply a uniform total force of 200lb on two faces of a model. The two faces have different areas. How do you apply the load using SOLIDWORKS Simulation for a Linear Static Study?

- Select the two faces and input a normal to direction force of 200lb on each face .

- Select the two faces and a reference plane. Apply 100lb on each face.

- **Apply equal force to the two faces. The force on each face is the total force divided by the total area of the two faces.**

- None of the above.

42. Maximum and Minimum value indicators are displayed on Stress and Displacement plots in SOLIDWORKS Simulation for a Linear Static Study.

- **True.**

- False.

43. What SOLIDWORKS Simulation tool should you use to determine the result values at specific locations (nodes) in a model using SOLIDWORKS Simulation?

- Section tool.
- **Probe tool**.
- Clipping tool.
- Surface tool.

44. What criteria are best suited to check the failure of ductile materials in SOLIDWORKS Simulation?

- Maximum von Mises Strain and Maximum Shear Strain criteria.
- **Maximum von Misses Stress and Maximum Shear Stress criteria**.
- Maximum Mohr-Coulomb Stress and Maximum Mohr-Coulomb Shear Strain criteria.
- Mohr-Coulomb Stress and Maximum Normal Stress criteria.

45. Set the scale factor for plots_____ to avoid any misinterpretation of the results, after performing a Static analysis with gap/contact elements.

- Equal to 0.
- **Equal to 1**.
- Less than 1.
- To the Maximum displacement value for the model.

46. It is possible to mesh _____ with a combination of Solids, Shells and Beam elements in SOLIDWORKS Simulation.

- **Parts and Assemblies**.
- Only Parts.
- Only Assemblies.
- None of the above.

47. SOLIDWORKS Simulation supports multi-body parts. Which of the following is a true statement?

- You can employ different mesh controls to each Solid body.

- You can classify Contact conditions between multiple Solid bodies.

- You can classify a different material for each Solid body.

- **All of the above are correct**.

48. Which statement best describes a Compatible mesh?

- A mesh where only one type of element is used.

- **A mesh where elements on touching bodies have overlaying nodes**.

- A mesh where only a Shell or Solid element is used.

- A mesh where only a single Solid element is used.

49. The Ratio value in Mesh Control provides the geometric growth ratio from one layer of elements to the next.

- **True**.

- False.

50. The structures displayed in the following illustration are best analyzed using:

- Shell elements.

- Solid elements.

- **Beam elements**.

- A mixture of Beam and Shell elements.

51. The structure displayed in the following illustration is best analyzed using:

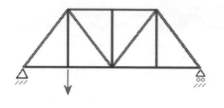

- Shell elements.

- Solid elements.

- **Beam elements**.

- A mixture of Beam and Shell elements.

52. The structure displayed in the following illustration is best analyzed using:

- **Shell elements**.

- Solid elements.

- Beam elements.

- A mixture of Beam and Shell elements.

Sheet metal model

53. The structure displayed in the following illustration is best analyzed using:

- Shell elements.

- **Solid elements**.

- Beam elements.

- A mixture of Beam and Shell elements.

54. Surface models can only be meshed with _____ elements.

- **Shell elements**.

- Solid elements.

- Beam elements.

- A mixture of Beam and Shell elements.

55. Use the _____ and _____ plots to calculate the distance between two nodes using SOLIDWORKS Simulation.

- **Mesh and Displacement**.
- Displacement and FOS.
- Resultant Displacement and FOS.
- None of the above.

56. You can simplify a large assembly in a Static Study by using the _____ or _____ options in your study.

- **Make Rigid**, **Fix**.
- Shell element, Solid element.
- Shell element, Compound element.
- Make Rigid, Load element.

57. A force "F" applied in a static analysis produces a resultant displacement URES. If the force is now 2X and the mesh is not changed, then the resultant displacement URES will:

- Double if there are no contact specified and there are large displacements in the structure.
- Be divided by 2 if contacts are specified.
- The analysis must be run again to find out.
- **Double if there is no source of nonlinearity in the study (like contacts or large displacement options).**

58. To compute thermal stresses on a model with a uniform temperature distribution, what type/types of study/studies are required?

- **Static only**.
- Thermal only.
- Both Static and Thermal.
- None of these answers is correct.

59. In an h-adaptive method, use smaller elements in mesh regions with high errors to improve the accuracy of results.

- **True**.

- False.

60. In a p-adaptive method, use elements with a higher order polynomial in mesh regions with high errors to improve the accuracy of results.

- **True**.

- False.

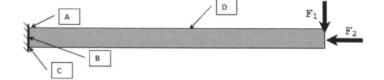

61. Where will the maximum stress be in the illustration?

- A

- B

- **C**

- D

62. Is axisymmetric the same as cyclic symmetry?

- Yes.

- **No**.

63. Can you perform a dynamic simulation using a static SOLIDWORKS Simulation study?

- Yes.

- **No**.

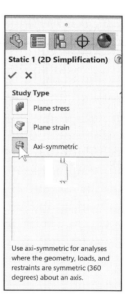

64. The concept of the h-method (available for solid part and assembly documents) is to use smaller elements (increase the number of elements) in regions with high relative errors to improve accuracy of the results.

- **True**.

- False.

65. Various adaptive methods can be applied to improve accuracy of the results. The p-method increases the _____ with each iteration.

- Number of elements.
- Meshing speed.
- **Polynomial order**.
- None of the above.

66. Various adaptive methods can be applied to improve accuracy of the results. The h-adaptive method increases the _____ with each iteration.

- **Number of elements**.
- Meshing speed.
- Polynomial order.
- None of the above.

67. In SOLIDWORKS Simulation, how is a sheet metal part treated in a static study with uniform thickness?

- Solids.
- Beams.
- **Shells**.
- Mixed mesh.

68. Global size in the Mesh PropertyManager refers to the _____.

- Smallest element size of the mesh.
- Largest element size of the mesh.
- **Average element size of the mesh**.
- None of the above.

69. Global size in the Mesh PropertyManager is only available for a Standard Mesh.

- **True**.
- False.

70. Can you use a Static study in SOLIDWORKS Simulation to simulate the deflection of a bridge with a uniform force?

- **Yes.**
- No.

71. Can you calculate the distance between two nodes in a SOLIDWORKS Simulation results plot using the Probe tool?

- **Yes, for all parts**.
- No.
- Yes, only strain plots.
- Yes, only for mesh and displacement plots.

72. True or False: Use the probe tool to display the numerical value of the plotted field at the closest node or element's center to the selected model location. You can graph the results or save them to a file.

A. **True**.

B. False.

73. Which of the following assumptions are correct for a SOLIDWORKS static Simulation analysis? There can be more than a single answer.

A. **All displacements are small relative to the model geometry unless Large Displacements are activated**.

B. **Inertia effects are negligible and loads are applied slowly**.

C. **Results are proportional to loads. With the exception of when the following conditions are present: No penetration, Large Displacements, Shrink fit or Virtual wall contacts**.

D. None of the above.

74. Is this model fixed?

A. **Yes**.

B. No.

75. True or False: Check for interferences between bodies when using a compatible mesh with the curvature-based mesher. If you specify a bonded contact condition between bodies, they should be touching.

A. **True**.

B. False.

76. True or False: If interferences are detected, meshing stops, and you can access the Interference Detection PropertyManager to view the interfering parts. Make sure to resolve all interferences before you mesh again.

A. **True**.

B. False.

77. True or False: For static studies, you should define either **gravity** or **centrifugal loads** in order for the distributed mass to be effective. Simulation calculates the static load based on the acceleration due to gravity or the angular velocity and angular acceleration parameters defined in the Centrifugal PropertyManager

A. **True**.

B. False.

78. A thin sheet metal plate (Shells) is stiffened by structural members (H-beams). What SOLIDWORKS Simulation contact type or connector would you apply between the shell surface and each structural member?

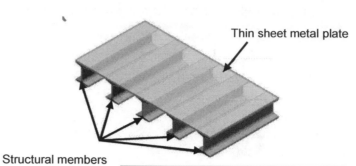

Thin sheet metal plate

Structural members

A. **Bonded Contact (Interactions)**.

B. Rigid Connector.

C. Thermal Resistance.

D. None of the above.

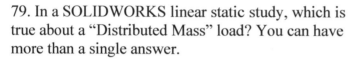

Table 1: User Interface Updates for Contact Terms	
2020	**2021**
Contacts	Interactions
Contact Visualization Plot	Interaction Viewer
Contact set	Local interaction
Component contact	Component interaction
Global contact	Global interaction
Allow penetration	Free
No penetration	Contact
Incompatible mesh	Independent mesh
Compatible mesh	Continuous mesh
Compatible bonding	Enforce continuous mesh between touching faces

79. In a SOLIDWORKS linear static study, which is true about a "Distributed Mass" load? You can have more than a single answer.

A. Moment effects due to mass are considered. Thus, the X, Y, and Z location of the mass needs to be specified.

B. A "Distributed Mass" increases the stiffness of the face to which it is applied.

C. **It has to be used in conjunction with gravity or centrifugal load**.

D. **The mass is assumed to be uniformly distributed on selected faces based on their areas**.

80. Which results are available for a Bolt connector in a SOLIDWORKS linear static study?

A. Stresses.

B. Torque.

C. **Axial forces**.

D. **Shear forces**.

E. **Bending Moments**.

F. **Calculated Factor of Safety**.

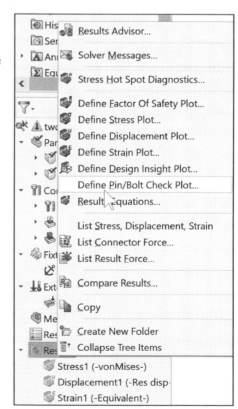

81. If you modify the study (force, material, etc.) you only need to re-run to update the results. You do not need to re-mesh unless you modified contact conditions.

A. **True**.

B. False.

82. What criteria are best suited to check the failure of ductile material in SOLIDWORKS Simulation?

- **Maximum von Mises Stress and Maximum Shear Stress criteria**.

- Maximum von Misses Stress and Maximum Shear Strain criteria.

- Maximum von Misses Stress and Maximum Normal Stress criteria.

- Mohr-Coulomb Stress and Maximum Normal Stress criteria.

83. You have a 10-inch-long beam of cross section A. The beam is fixed at one end. A tensile force of 200 lbf is applied to the other end. The beam material is Alloy Steel. The tensile stress is found to be approximately 1000 psi. You modify the beam material to titanimum Ti-SAI-2.55n. Everything else in the study remains the same. What would the tensile force be?

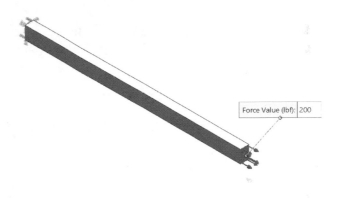

A. **Approximately the same original stress**.

B. Approximately half of the original stress.

C. Approximately 2X of the original stress.

D. Approximately 3X of the original stress.

84. A force "F" applied in a static linear analysis produces a URES: Resultant Displacement. If you double the force and the mesh is not changed, then the URES: Resultant Displacement will:

A. Be divided by 2, if contacts (Interactions) are specified.

B. Double if there is no source of nonlinearity in the study (like contacts or large deployment options).

C. Double if there are no contacts specified and there are large displacements in the structure.

D. **The analysis must be run again to find out**.

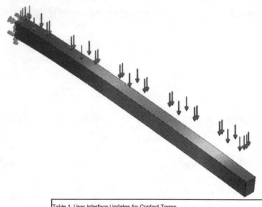

Table 1: User Interface Updates for Contact Terms

2020	2021
Contacts	Interactions
Contact Visualization Plot	Interaction Viewer
Contact set	Local interaction
Component contact	Component interaction
Global contact	Global interaction
Allow penetration	Free
No penetration	Contact
Incompatible mesh	Independent mesh
Compatible mesh	Continuous mesh
Compatible bonding	Enforce continuous mesh between touching faces

85. If you want to view a portion of stress values (range) on your model, which of the following would you do?

A. Change the model and re-run the study.

B. Create a section plot with two section clipping planes. The first plane could be the upper limit range. The second plane would be the lower limit range.

C. **Create an iso plot with two iso values. The first value could be the upper limit. The second value would be the lower limit**.

D. None of the above.

86. Which of the following, best describes the Ratio value in the Mesh Control PropertyManager when using Mesh control? Use Simulation help if needed.

A. The maximum depth to which the mesh control affects the mesh size, expressed a fraction of the overall model size.

B. **The geometric growth ratio from one layer of elements to the next**.

C. The maximum ratio between the largest element and the smallest element in the entire model.

D. The maximum ratio of the longest edge to the shortest edge of any element (i.e. maximum aspect ratio).

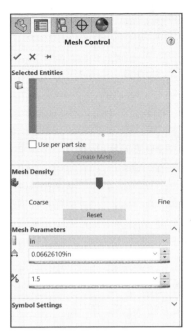

87. Which of the following geometrics can be meshed using beam elements? There can be more than one answer. Note: Beam or Truss elements are suitable for extruded or revolved objects and structural members with constant cross-sections.

A.

B.

C.

D.

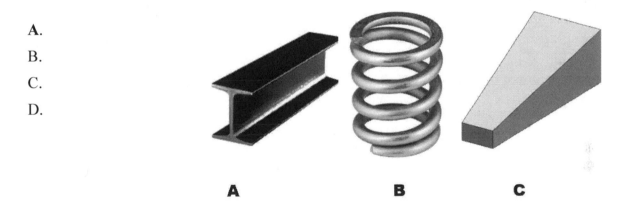

A   B   C

88. The cylindrical faces A are only allowed to rotate about their axis. What Fixture type would you apply to the cylindrical faces of A?

A. Fixed.

B. Roller/Slider.

C. **Fixed Hinge**.

D. On Spherical Face.

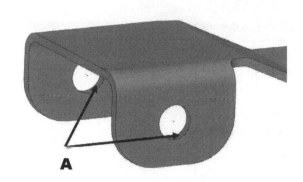

A

In the multiple choice, yes/no and multiple selection section of the exam, it is very important that you apply the SOLIDWORKS Simulation Help tool. This is a great resource.

During the exam, they may ask you to apply a German material standard (DIN) to the model in the Study.

Read the problem carefully. Example: What is the value of the displacement at the location which moves the most? In the example below, it's 0.00952 in

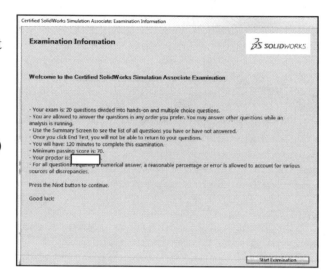

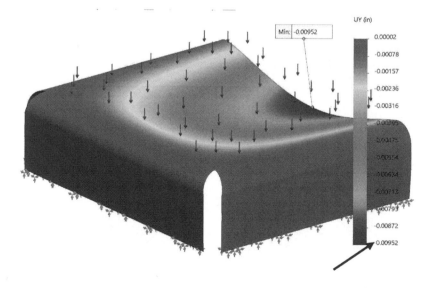

CSWA-AM Chapter 8

1. One of the most common reasons for a Fused Filament Fabrication (FFF) print failure is build plate adhesion. Identify a few common solutions:

A. Blue painter's tape.

B. Hair spray.

C. Glue stick.

D. **All of the Above**.

2. The acronym FFF stands for:

A. **Fused Filament Fabrication**.

B. Filament Fabrication Force.

C. Force Filament Fabrication.

D. None of the above.

3. True or False: There are many materials that are being explored for 3D printing; however, the two most dominant plastics for FFF technology are PLA (Polylactic acid) and ABS (Acrylonitrile-Butadiene-Styrene).

A. **True**.

B. False.

4. True or False: Selective Laser Sintering (SLS) is the preferred rapid-prototyping method of metals and exotic materials.

A. **True**.

B. False.

5. It is recommended that you store used filament in a:

A. **Airtight container**.

B. **Dry, dark controlled environment**.

C. Need an open window.

D. No special requirements are required.

6. True or False: PLA (Polylactic Acid) is a non-biodegradable thermoplastic.

A. True.

B. **False**.

7. True or False: ABS (Acrylonitrile-Butadiene-Styrene) is an oil-based thermoplastic.

A. **True**.

B. False.

8. True or False: Thermoplastic filament comes in two standard diameters: 1.75mm and 3mm (true size of 2.85mm).

A. **True**.

B. False

9. What is the most command nozzle (hot end) diameter size in FFF.

A. .01mm

B. **.4mm**

C. .04mm

D. .001mm

10. Ultimaker 3 uses Near Field Communication (NFC) technology in their filament roll. The NFC technology provides what information to the printer.

A. **Filament Color**.

B. **Filament type**.

C. Name of the machine.

D. Name of the model.

11. True or False: PVA (Polyvinyl Alcohol) filament is a water-soluble synthetic polymer.

A. **True**.

B. False.

12. True or False: PVA (Polyvinyl Alcohol) filament is not hygroscopic.

A. True.

B. **False**.

13. PVA (Polyvinyl Alcohol) filament is normally used as a water-soluble synthetic polymer. What is the best way to remove PVA from the part?

A. Time. Wait, and the PVA will fall off.

B. Submerge in hot water.

C. Leave in the sun. The sun will melt the PVA.

D. **Submerge in a bath of cold circulating water**.

14. Selective Laser Sintering (SLS) technology is the most common additive manufacturing technology for industrial _____ applications.

A. **Power base**.

B. Resin-based.

C. Thermoplastic.

D. None of the above.

15. True or False: SLA technology is a form of 3D printing in which a computer-controlled low-power, highly focused ultraviolet (UV) laser is used to create layers (layer by layer) from a liquid polymer (photopolymerization) that hardens on contact.

A. **True**.

B. False.

16. True or False: All SLA printed parts require a post-cure.

A. **True**.

B. False.

17. Formlab Form 2 recommends to wash the part with _____ in the post-cure process.

A. Distilled water.

B. **Isopropyl Alcohol.**

C. Acetone.

D. Nothing.

18. True or False: PLA gives off a very strong unpleasant odor.

A. True.

B. **False**.

19. True or False: PLA has a lower printing temperature (20C - 30°C) than ABS.

A. **True**.

B. False.

20. True or False: Avoid using PLA if the print is exposed to temperatures of 60°C or higher or might be bent, twisted or dropped repeatedly. ABS is a better material for this application.

A. **True**.

B. False.

21. True or False: ABS in general is more forgiving to large temperature fluctuations and moisture than PLA during a build cycle.

A. True.

B. **False**.

22. How often should you level your build plate on an FFF printer?

A. After you have a problem.

B. Never.

C. **Before every build**.

D. Every 10 prints.

23. When printing ABS, what chemical can you use to smooth the surface?

A. Bleach.

B. Machine oil.

C. **Acetone**.

D. None of the above.

24. Why do you need a heated build plate when printing ABS?

A. You do not need a heated build plate.

B. **The print bed should be kept near the glass transition temperature and well below the melting temperature. It also helps to keep the part from cooling too fast and warping**.

C. To keep the build area warm.

D. None of the above.

25. True or False: ABS is better suited than PLA for items that are frequently handled, dropped, or heated. It can be used for mechanical parts, especially if they are subjected to stress or must interlock with other parts.

A. **True**.

B. False

26. True or False: An STL file describes a raw unstructured triangulated (point cloud) surface by the unit normal and vertices (ordered by the right-hand rule) of the triangles using a three-dimensional Cartesian coordinate system.

A. **True**.

B. False.

27. True or False: In SOLIDWORKS, the STL Save options provide you with the ability to control the number and size of the triangles by setting the various parameters in the CAD software.

A. **True**.

B. False.

28. True or False: For a smoother STL file, change the Resolution to Custom. Change the deviation to 0.0005in (0.01mm). Change the angle to 5. Smaller deviations and angles produce a smoother file but increase the file size and print time.

A. **True**.

B. False.

29. It's best to monitor the first _____ layers when starting a print. Never assume you will have perfect bed adhesion.

A. **3 -5 layers**.

B. No monitoring is required.

C. 20 - 30 layers.

D. None of the above.

30. Identify some of the ways to transport a file for the computer to the 3D printer.

A. USB.

B. Wi-Fi.

C. Ethernet.

D. **All of the above**.

31. True or False: Most slicers provide the ability to address rafts and supports.

A. **True**

B. False.

32. True or False: The Formlab Form 2 machine **does not** have the ability to remotely monitor the unit using a webcam or camera. They use their Dashboard feature.

A. **True**.

B. False.

33. Name a few benefits of using a dual extruder in an FFF printer.

A. **Ability to print more than a single filament color**.

B. **Ability to print a different support material**.

C. Reduce overall printing time.

D. All of the above.

34. True or False: The Formlab Form 2 printer can only print in a single color at a time.

A. **True**.

B. False.

35. True or False: Most FFF printers and the Formlab Form 2 printer provide the ability to add material during the print cycle if needed.

A. **True**.

B. False.

36. True or False: 3DExpert from SOLIDWORKS is a free download.

A. **True**.

B. False.

37. The creation of advance part structures is aided by what type of CAD design tool?

A. 3D Help

B. **Generative Design**

C. PrintXpert

D. Design Guidance

38. True or False: For high temperature applications, ABS (glass transition temperature of 105°C) is more suitable than PLA (glass transition temperature of 60°C). PLA can rapidly lose its structural integrity as it approaches 60°C.

A. **True**.

B. False.

39. True or False: Additive manufacturing, sometimes known as ***rapid prototyping***, can be slower than Subtractive manufacturing. Both take skill in creating the G-code and understanding the machine limitations.

A. **True**.

B. False.

40. In FFF technology, the thermoplastic material is heated in the nozzle (hot end) to form liquid, which solidifies immediately when it's deposited onto the build plate or platform. The nozzle travels at a controlled rate and moves in the X and Y direction. The build plate moves in the _____ direction. This creates mechanical adhesion (not chemical).

A. **Z**.

B. X, Y Z.

C. X.

D. Y, Z.

41 True or False: A Post-curing process is required for FFF technology.

A. True.

B. **False**.

42. True or False: In general, post-processing is the removal of supports and any needed polishing or sanding.

A. **True**.

B. False.

43. True or False: In SLA technology, after the IPA wash and dry, perform a post-cure. At a basic level, exposure to sun light triggers the formation of additional chemical bonds within a printed part, making the material stronger and stiffer.

A. **True**.

B. False.

44. True or False: To ensure proper post-curing, Formlabs recommends their post-curing unit, Form Cure. Form Cure precisely combines temperature and 405 nm light to post-cure parts for peak material performance.

A. **True**.

B. False.

45. True or False: In SLA technology, parts are watertight and fully dense. No infill is required.

- **True**.

- False.

46. True or False: SLS technology is the preferred rapid-prototyping method of metals and exotic materials.

A. **True**.

B. False.

47. True or False: In FFF technology, the higher the infill percentage, the more material and longer the print time. It will also increase weight and material cost.

A. **True**.

B. False.

48. True or False: In general, with FFF technology, use a 10% - 15% infill with a maximum infill of 60%. 100% infill is not recommended. Part warping can be a concern.

A. **True**.

B. False.

49. True or False: Part orientation is important on build strength and the amount of raft and support material required for the build. Incorrect part orientations can lead to poor strength, warping, curling, and delamination.

A. **True.**

B. False.

50. True or False: If maximizing strength is an issue, select the part orientation on the build plate so that the "grain" of the print is oriented to maximize the strength of the part.

A. **True.**

B. False.

51. True or False: Parts with a lower percentage of infill should have a lower internal force between layers and can reduce the chance of curling, cracking, and layer delamination along with a lower build cost and time.

A. **True.**

B. False.

52. True or False: In FFF technology, for a consistent quality build, control the build area and environment temperature. Eliminate all drafts and control air flow that may cause a temperature gradient within the build area.

A. **True.**

B. False.

53. _____ technology is the most common additive manufacturing technology for industrial powder base applications.

A. **SLS.**

B. FFF.

C. ABS.

D. None of the above.

54. True or False: In FFF technology, you do not need a heated bed (platform) when using PLA.

A. True.

B. False.

55. True or False: In FFF technology, you do not need a heated bed (platform) when using ABS.

A. True.

B. **False**.

56. True or False: In FFF technology, a raft is a horizontal latticework of filament located underneath the part.

A. **True**.

B. False.

57. In FFF technology, it's best to monitor the first 3 - 5 layers when starting a print. Never assume you will have perfect bed adhesion. If there is a bed adhesion issue, check to see if the bed is level. Check the bed and extruder (hot end) temperatures and then either apply.

A. **Blue painter's tape (non-heated)**.

B. **Glue stick**.

C. **Hair spray**.

D. None of the above.

58. _____ is often used in post processing when using ABS to provide a smooth glossy finish.

A. Motor oil.

B. Water.

C. **Acetone**.

D. None of the above.

59. True or False: Scaling a part in your slicer can have a negative consequence for mating parts.

A. **True**.

B. False.

60. True or False: FFF technology builds parts that have inherently anisotropic properties, meaning they are much stronger in the XY direction than the Z direction.

A. **True**.

B. False.

61. True or False: For functional parts, it is important to consider the application and the direction of the loads. FFF technology parts are much more likely to delaminate and fracture when placed in tension in the Z direction compared to the XY directions (up to 4-5 times difference tensile strength).

A. **True**.

B. False.

62. True or False: Scaling the part in a slicer could have a negative effect on the overall wall thickness.

A. **True**.

B. False.

63. True or False: Most low-end 3D FFF printers use an extruder (hot end) with Polyether ether Ketone (PEEK) or Polytetrafluoroethylene (Teflon). Both PEEK and PTFE begin to break down above 240°C and will burn and emit noxious fumes. Use an all metal hot end above 240°C.

A. **True**.

B False.

64. True or False: 3DXpert is a SOLIDWORKS Add-in. 3DXpert for SOLIDWORKS provides an extensive toolset to analyze, prepare and optimize your design for additive manufacturing. It also provides the ability to print an assembly file as a single part.

A. **True**.

B. False.

65. True or False: The SOLIDWORKS STL Save options allow you to control the number and size of the triangles by setting the various parameters in the CAD software.

A. **True**.

B. False.

66. True or False: An STL (*.stl) file describes only the surface geometry of a three dimensional object without any representation of color, texture, or other common CAD model attributes. The STL format specifies both ASCII and Binary representations.

A. **True**.

B. False.

67. True or False: Additive Manufacturing (*.amf) is a file format that includes the materials that have been applied to the parts or bodies in the 3D model.

A. **True**.

B. False.

68. The file format (*.3mf) stands for:

A. Nothing.

B. **3D Manufacturing Format**.

C. 2D Manufacturing Format.

D. 3D Subtractive Format.

69. True or False: The Formlab Form 2 SLA printer has the ability to print flexible material.

A. **True**.

B. False.

70. True or False: The Formlab Form 2 SLA printer resin only comes in one color.

A. True.

B. **False**.

71. True or False: Ultimaker Cura - Version: 3.1.0 or newer provides the ability to import SOLIDWORKS part and assembly files directly into the build area.

A. **True**.

B. False.

72. True or False: When designing your part for 3D printing, try to use the 45 degree rule. If your model has overhangs greater than 45 degrees, you need support material.

A. **True**.

B. False.

73. True or False: The Formlabs Form 2 uses the PreForm Software. PreForm provides the ability to select the One-Click Print option. The One-Click option automatically optimizes the part orientation and support locations.

A. **True**.

B. False.

74. True or False: The Formlabs PreForm software only supports STL and OBJ files.

A. **True**.

B. False.

75. True or False: Using the 45 degree rule in FFF technology. If your model doesn't have any overhangs greater than 45 degrees, you should not need support. There are exceptions to the 45 degree rule. The most common ones are straight overhangs, and fully suspended islands.

A. **True**.

B. False.

76. The two standard diameters for Thermoplastic filaments are:

A. 1mm.

B. **1.75mm**.

C. **3mm (true size 2.85mm)**.

D. 4mm.

77. True or False: In SLA technology, you have the ability to customize percent infill and infill pattern type for the print.

A. True.

B. **False**.

78. True or False: In FFF technology, you have the ability to customize percent infill and infill pattern type for the print.

A. **True**.

B. False.

79. True or False: In FFF technology, darker colors and glow in the dark filament materials, often requires higher extruder temperatures (5C - 10°C).

A. **True**.

B. False.

80. True or False: Unlike its predecessor STL (*.stl) format, Additive Manufacturing Format (*.amf) has native support for color, materials, lattices, and constellation (groups).

A. **True**.

B. False.

81. True or False: Unlike its predecessor STL (*.stl) format, 3D Manufacturing Format (*.3mf) has native support for color, materials, lattices, and constellation (groups).

A. **True**.

B. False.

82. True or False: SOLIDWORKS's Design Guidance functionality produces much more organic and unique geometry than if modeled by the user.

A. **True**.

B. False.

83. What is the main advantage of using a heated bed for ABS.

A. Print faster.

B. Better print quality.

C. **Helps the part from cooling too fast and deforming**.

D. None of the above.

84. True or False: The Ultimaker and Formlabs Form 2 provides the ability to add material midway through a print if needed.

A. **True**.

B. False.

85. True or False: The Ultimaker and Formlabs Form 2 provides the ability to remotely monitor the machine using a built-in webcam\camera.

A. True.

B. **False**.

86. True or False: It is important to ensure that there is a drain hole when printing a hollow part using an SLA machine. The location of the drain hole is important based on its printing orientation.

A. **True**.

B. False.

87. True or False: Once the part is completed in the Formlabs Form 2 printer, the remaining resin in the tank should be discarded due to contamination issues.

A. True.

B. **False**.

88. True or False: Increasing the number of triangles in an STL file, will decrease the overall print time.

A. True.

B. **False**.

89. True or False: The Formlabs Form 2 printer uses a Dashboard to remotely monitor the printer. It also tracks material usage and explores past and future purchases.

A. **True**.

B. False.

90. True or False: Increasing the number of triangles in an STL file, will decrease the overall print time.

A. True.

B. **False**.

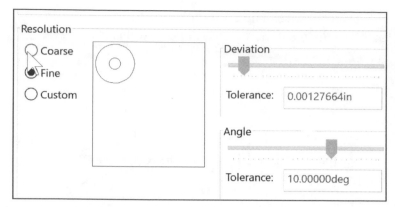

91. True or False: You are using an SLA machine to print a hollow part. It is very important to design a drain hole at the bottom of the part based on it printing orientation.

A. **True**.

B. False.

92. You require a part to be strong, flexible, and to stand up to heat. What material should you use in a FDM printer?

A. PLA.

B. ABS.

C. **Nylon 618**.

D. PVA.

93. True or False: In SOLIDWORKS for a smoother STL file, change the Resolution to Custom. Change the deviation to 0.0005in (0.01mm). Change the angle to 5. Smaller deviations and angles produce a smoother file but increase the file size and print time.

A. **True**.

B. False.

94. True or False: The part displayed is perpendicular to the build plate. The part will not require support material based on standard support guidelines.

A. **True**.

B. False.

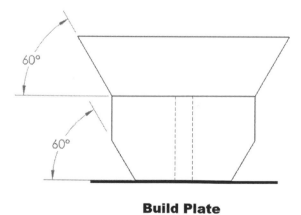

Build Plate

95. The arrow shown in the picture shows the direction in which shear stress will be the highest in the part. What orientation should you print the part in order to keep the part from yielding? Disregard support considerations in this example.

A. A.

B. **B**.

96. The arrow shown in the picture shows the direction in which shear stress will be the highest in the part. What orientation should you print the part in order to keep the part from yielding? Disregard support considerations in this example.

A. **A**.

B. B.

97. The part displayed is perpendicular to the build plate. The part will **not** require support material based on the standard support guidelines.

A. True.

B. **False**.

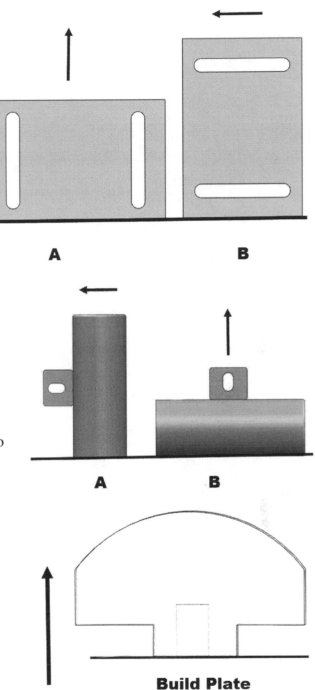

Appendix

Notes:

Index

Index

Index